Meteorology in America, 1800-1870

METEOROLOGY
in America, 1800–1870

James Rodger Fleming

THE JOHNS HOPKINS UNIVERSITY PRESS
BALTIMORE AND LONDON

The Johns Hopkins University Press
701 West 40th Street, Baltimore, Maryland 21211
The Johns Hopkins Press Ltd., London

∞ The paper used in this book meets the minimum requirements of American National Standard for Information Sciences—Permanence of Paper for Printed Library Materials, ANSI Z39.48-1984.

Title page illustration: Triple waterspout, from James P. Espy, *The Philosophy of Storms* (Boston, 1841), xv.

Library of Congress Cataloging-in-Publication Data
Fleming, James Rodger.
Meteorology in America, 1800–1870 / James Rodger Fleming.
p. cm.
Includes bibliographical references.
ISBN 0–8018–3958–0 (alk. paper)
1. Meteorology—United States—History. I. Title.
QC857.U6F54 1990
551.5'0973—dc20 90–30816 CIP

paperback edition, 1999
ISBN 0-8018-6359-7 (pbk)

For Miyoko, Jamitto, and Jason

The astronomer is, in some measure, independent of his fellow astronomer; he can wait in his observatory till the star he wishes to observe comes to his meridian; but the meteorologist has his observations bounded by a very limited horizon, and can do little without the aid of numerous observers furnishing him contemporaneous observations over a wide-extended area.

—James P. Espy, *Second Report on Meteorology to the Secretary of the Navy*, 1851

CONTENTS

List of Illustrations *xi*

List of Tables *xiii*

Acknowledgments *xv*

Introduction *xvii*

1. Early Issues and Systems of Observation *1*
2. The American Storm Controversy, 1834–1843 *23*
3. Observational Horizons in the 1830s and 1840s *55*
4. The Structure of the Smithsonian Meteorological Project *75*
5. Stormy Relations among Theorists and Administrators *95*
6. Cooperative Observations and Contributions to Knowledge *117*
7. Weather Telegraphy *141*
8. The Worldwide Horizon *163*

Appendix. Collective Biography of a Sample of Smithsonian Observers Active in 1851, 1860, and 1870 *175*

Abbreviations *185*

Notes *189*

Bibliography *229*

Index *253*

ILLUSTRATIONS

1.1. First report of observers, January 1806, published in the *Medical and Agricultural Register* *11*
1.2. Army post observations, 1820–1874 *16*
1.3. Page from the weather journal of Josiah Meigs, 1821 *18*
1.4. New York academy observations, 1826–1863 *20*
2.1. Espy's tornado theory *32*
2.2. Outhouse and grounds of Mr. D. Polhemus *33*
2.3. Grêle electrique *34*
2.4. Redfield's second "test" of Espy's theory *41*
2.5. Marginalia in Henry's copy of *Philosophy of Storms* *42*
3.1. Weather map of the Joint Committee, 1837 *58*
3.2. Forry's climate map, 1842 *69*
3.3. Map from Espy's *First Report,* 1843 *71*
4.1. Smithsonian meteorology budgets, 1849–1878 *77*
4.2. Locations of Smithsonian meteorological observers, ca. 1850 *80*
4.3. Smithsonian meteorological observers, 1849–1874 *82*
4.4. Completed Smithsonian "Register of Meteorological Observations," 1857 *84*
4.5. Locations of Smithsonian observers having barometers and other instruments in 1860 *87*
4.6. Locations of Smithsonian observers having barometers and other instruments in 1870 *87*
4.7. "Monarchs of the forest felled by the meteor" *89*
4.8. Number of years individuals participated as observers in the Smithsonian system, 1851, 1860, and 1870 *91*
5.1. Espy's nephelescope (1830s) and double nephelescope (1850s) *99*
5.2. Two illustrations from Robert Hare's unpublished manuscript "On the Suppositious Travelling Whirlwinds Called Cyclones" *103*
6.1. Joseph Henry's map of arable lands *128*
6.2. Rain chart of the United States, 1868 *130*
6.3. Temperature chart of the United States, 1874 *131*
6.4. Curves of secular change in the mean annual temperature *132*
6.5. James Cooper's map of forest regions of North America *134*
6.6. Comparison of Coffin's charts and Ferrel's chart of the general circulation of the atmosphere *137*
7.1. Telegraph map of the United States, December 1846 *144*
7.2. Smithsonian telegraph network, 1860 *145*

7.3. Meteorological map by Increase Lapham using Smithsonian data *154*
7.4. Initial stations of the Signal Service, 1870 *158*
7.5. Loomis's "Storm Paths of the United States, N.E." *159*
7.6. Signal Office map, "Mean Tracks of Low Barometer for Dec. 1871, '72 and '73" *160*

TABLES

I.1. Periodization of the history of meteorology in America, from colonial times to the present *xviii*
2.1. Forshey's questions and the responses of Espy, Redfield, and Hare *30*
2.2. Espy v. Redfield, 1834–1841 *36*
2.3. Stops on Espy's speaking tour, 1837–1841 *43*
2.4. Robert Hare enters the controversy *51*
4.1. Statistical summary for sampled observers, 1851, 1860, and 1870 *91*
4.2. Occupational categories and status indicators of sampled observers *92*
5.1. The American storm controversy revisited, 1849–1857 *96*
6.1. Observers and stations in the Survey of Northern and Northwestern Lakes, 1860 *124*
6.2. Coffin's expenses for data reduction, 1857–1860 *126*
8.1. Dates of founding of weather services and directors, according to Kutzbach, *Thermal Theory*, pp. 12–13 *165*
8.2. Comparisons of observational systems in Europe, Russia, and the United States in the nineteenth century *166*

ACKNOWLEDGMENTS

The key institutions supporting the research and writing of this book were, in chronological order, Princeton University, the Smithsonian Institution, the American Meteorological Society, and Colby College. I would like to thank my mentors, friends, and colleagues at these institutions for their support, advice, and encouragement. Charles Gillispie, Gerald Geison, Nathan Reingold, Marc Rothenberg, Martin Rudwick, John Servos, Paul Theerman, and Sean Wilentz read all of part of the manuscript and provided valuable advice. David Ludlum unselfishly shared his research materials with me. A note of appreciation is also due Marc Rothenberg, Paul Theerman, Kathleen Dorman, John Rumm, Beverly Lepley, and Steven Kottsy at the Joseph Henry Papers; William Moss, William Deiss, Libbey Glenn, William Massa, and their colleagues at the Smithsonian Institution Archives; Marjorie Ciarlante and her colleagues at the National Archives; Richard Sheldon and the staff of the National Historical and Public Records Commission; and the librarians and archivists at Colby College's Miller Library, Princeton University's Firestone Library, Yale University's Beinecke and Sterling libraries, the Smithsonian Libraries, the Library of the American Philosophical Society, the Library of the Franklin Institute, the Library of Congress, and the Library of the National Oceanic and Atmospheric Administration. The staff members at these institutions were, without exception, cheerful, professional, and responsive to my enthusiasm. A special note of thanks is due the staff of the Library of Congress for responding to innumerable inquiries, approving seemingly endless renewals of my stack pass, and providing me with a study desk. Kenneth Spengler, Richard Hallgren, Joseph Smagorinsky, and the executive board of the American Meteorological Society provided crucial support during an important stage of this work. Steven Suddaby checked my statistical analysis.

Aspects of this research were presented at meetings and colloquia sponsored by the American Geophysical Union, the American Meteorological Society, Colby College, the History of Science Society, the International Congress of the History of Science, the Johns Hopkins University, the National Center for Atmospheric Research, Pennsylvania State University, Princeton University, the Smithsonian Institution, the Society for the History of Natural History, the University of Maryland, and the University of Pennsylvania.

The final version of the manuscript was prepared at Colby under the auspices of the Program in Science and Technology Studies in the Department of Physics and Astronomy. The assistance of Sam Atmore, Susan Cole, Theresa

Fox, Michael Gerard, Linda Heaney, Chuck Lakin, Jane Moss, Francis Parker, Sunny Pomerleau, Sam Sharnik, Rurik Spence, Jeff Winkler, and Elizabeth Wright is gratefully acknowledged. I would also like to thank my editors, Henry Y. K. Tom, George Thompson, Trudie Calvert, and Kimberly Johnson at the Johns Hopkins University Press for their courtesy and professionalism.

INTRODUCTION

Meteorology in the nineteenth century experienced a rapid and dramatic expansion of its scientific horizons. On many levels—theoretical, empirical, institutional, technological—it encouraged inquiry, demanded discipline, and attracted controversy. Meteorologists were driven by fundamental questions about climatic change, the nature of storms, and the geography of health and disease. They were sponsored by scientific societies, including the American Philosophical Society, the Franklin Institute, and the new Smithsonian Institution; patronized by state governments and federal agencies expecting practical results; and aided by new techniques such as the weather map and new technologies such as the telegraph. Scientific organizations in Europe had already made halting attempts to link individual observers, theorists, and administrators into groups of common practice. So too in America during the first seven decades of the nineteenth century, systems of observation sponsored by diverse institutions gradually covered the American continent, delineated its climate, and encircled its weather systems. The theoretical, empirical, institutional, and technological dimensions of meteorology were gradually connected in an interrelated web that grew in extent, complexity, and utility throughout the century.

The history of meteorology in the United States (see Table I.1) may be divided into four major periods from colonial times to the present: (1) the era of individual, isolated diarists before 1800; (2) the period of emerging systems and expanding horizons between 1800 and 1870—the focus of this study; (3) the era of government service, 1870 to 1920 (and beyond); and (4) the current disciplinary and professional period, which began in the 1920s.

In the seventeenth and eighteenth centuries, isolated American diarists without reliable instruments, sponsoring institutions, or proper instruction, contributed to meteorological science by keeping records of the local weather and climate. Early in the nineteenth century, while Thomas Jefferson presided over both the nation and its leading scientific society, college professors in New England and officers of the federal government began systematically to collect climatic and phenological statistics over large areas of the country. As the geographic coverage and reliability of meteorological observations increased, and as results appeared in print, meteorologists constructed synoptic maps and developed theories of atmospheric dynamics from the small scale of thunderstorms to the grand scale of the general circulation.

Beginning in 1834, the bitter disputes among American meteorologists

TABLE I.1. Periodization of the history of meteorology in America, from colonial times to the present

Period	Culture	Patrons	Problems	Instruments	Institutions/Journals
Before 1800	Individuals	None	Astro-meteorology, lightning, climate change	Nonstandard	None/private weather diaries
1800–1870	Emerging systems, expanding horizons, volunteers	Army Medical Dept., Amer. Phil. Soc., Frankl. Inst., Navy, N.Y., Smithsonian, Patent Office	American storms, climate, medical geography	Standardized by 1850s, surface observations only, telegraph experiments	Ad hoc/general: *Amer. Journ. Sci.*, *Amer. Phil. Soc. Proc.*, *Frankl. Inst. Journ.*, *Smithsonian Contributions and Reports*
1870–1920	Service and forecasting, military, government	Army Signal Office, U.S. Department of Agriculture	Forecasts, vertical structure, agriculture	Telegraph, balloons, mountain stations, self-recording instruments	Federal and state/specialized: U.S. Weather Bureau, New England Metl. Soc., *Monthly Weather Review*, *Amer. Journ. of Meteo.*
1920–	Disciplinary, professional, subdisciplinary	Government, universities, foundations, industry	Polar front, computer modeling, remote sensing, air pollution, weather modification, climatic change	Aircraft, radio, radar, rockets, computers, satellites	Numerous/very specialized: American Meteorological Society, American Geophysical Union, National Center for Atmospheric Research, NOAA, NASA, *Amer. Metl. Soc. Bulletin*, *J. Atmospheric Sciences*, *J. Climate and Applied Meteo.*, *J. Geophysical Research*, etc.

involved in the "storm controversy" attracted the attention, if not always the admiration, of European savants. Hotly debated issues included the cause of storms, their phenomenology, and the proper methodology for investigating them. Although it came to no clear intellectual resolution, the controversy stimulated an observational "meteorological crusade" conducted by local scientific societies, the Army Medical Department, the Navy Department, the Smithsonian Institution, and others, which transformed meteorological theory and practice.

This period of emerging systems and expanding horizons, from 1800 to 1870, represented a new stage in the meteorological enterprise. No longer were a few isolated enthusiasts calling for weather observations; hundreds, eventually thousands of organized observers were actually collecting them according to a more or less regular plan. Because these people typically reported by mail at the end of each month, the earliest systems did not provide current weather information, although retrospective studies of storm events could be attempted with some of the data. Telegraphic experiments began in 1849 and, although the Civil War interrupted the growth of the systems and caused a delay of about a decade, the United States established a national system of weather "telegrams and reports for the benefit of commerce" by 1870. Over seven decades physicians, educators, naturalists, military officers, settlers, agriculturalists, and others all contributed their mite and called for increased levels of government patronage, while scientists such as James Espy, Elias Loomis, Joseph Henry, James Coffin, and William Ferrel—students of nature and men of action who were not at all timid, imitative, or tame—represented at least in science the bold aspirations and modest contributions of the "American scholar" at midcentury.[1]

In general the mid-nineteenth century has been a neglected period in American science, known to some as the "trough" between the twin "peaks" of legendary colonial glory and acknowledged world leadership.[2] Throughout the century, American scientists in most fields were typically regarded at best as junior partners of their European betters. If American contributions were recognized at all, it was not until the closing decades. By the 1870s, however, as Nathan Reingold has argued, America had become a "mature but small scientific nation with many of its basic institutions and attributes in existence or in embryo"; its scientific infrastructure, as Robert Bruce has documented, had been formed and "launched" in the three decades preceding.[3] Meteorology's role in these developments, although certainly not unnoticed by historians, has heretofore remained largely unexplored. Much of the literature on meteorology in the mid-nineteenth century (with notable exceptions) consists of brief chronologies, biographical sketches, and institutional histories.[4] Until now, little has been done with the available documentary sources.[5]

Two later periods of American meteorological history listed in Table I.1—government service (1870–1920) and the disciplinary and professional era (after 1920)—are beyond the scope of this study and will receive only passing mention here. Although the Smithsonian's annual meteorological budget was never greater than $5,000, the U.S. Army Signal Office spent over $400,000 on meteorological reports in 1874, its fourth year of operation. The budget of the U.S. Weather Bureau in the Department of Agriculture averaged $800,000 in the 1890s and topped $1 million by 1900. On the basis of scale alone, the period from 1870 to 1920 marked a new era. The "disciplinary" period in the atmospheric sciences began in the 1920s. Recent decades have seen the growth of professional societies and specialized journals, well-defined career paths, and graduate education in the atmospheric sciences. These developments are mentioned here as a reminder of the dangers of imposing twentieth-century expectations and organizational models on the science of an earlier time, valuing the past only for its contributions to current scientific practice, or forcing it to fit preconceived notions.

Meteorology has no tradition of crucial experiments and few instances of isolated genius. Meteorologists need to rely on others for information. In meteorology, as in astronomy and the geophysical sciences, research is complicated because the objects under investigation are typically not amenable to laboratory analysis: the atmosphere is too large and too complex. Usually the science is done outdoors, where factors such as viscosity, divergence, turbulence, humidity, and thermal forcing are difficult if not impossible to isolate. In meteorological problems, not only are the force laws not "given," the bodies in motion are not solid. Thus meteorologists pursue a task more difficult than the "inverse problem" of Sir Isaac Newton, reiterated by James Clerk Maxwell on the Cambridge tripos in the 1860s: given the motions of a body, derive the forces and laws governing its behavior. No wonder meteorologists have collected so many measurements!

Because meteorologists require large amounts of information to frame and test their hypotheses, the legitimate scientific need to collect and analyze data is often confused with the questionable practice of mere fact gathering. According to one rather cursory analysis:

> In the history of science there has been one subject that traditionally developed in a true Baconian manner: meteorology. For a long time, at an enormous number of stations all over the world, meteorologists have been collecting data concerning temperature, humidity, rainfall, and wind conditions in a systematic fashion that would have delighted Francis Bacon. But it is a matter of record that this branch of science has not (inductively or in any other way) developed a useful theoretical structure as have physics, chemistry, biology, and geology. We can talk about the weather, but we can neither make very accurate predictions of it nor do anything to change it.[6]

It is not surprising that investigations of the earth's atmosphere and its changes over long periods of time require enormous amounts of data, but it is surprising to see an eminent scholar brand generations of meteorologists as Baconians and dismiss the entire theoretical structure of the atmospheric sciences as useless.

Perhaps the popular perception of meteorology, centered on a television or radio broadcaster attempting to predict tomorrow's weather, has further confused the issue. Weather reporters are communications specialists, not atmospheric scientists. They deal with a public interested in whether a coat or an umbrella will be needed, not in the principles governing the general circulation of the earth's atmosphere, the radiative properties of clouds, or the behavior of a tropical storm. Meteorologists are able to compute the energy budget of a winter storm with precision, but they tread on thin ice if they try to predict the details of the storm's motion. As August Schmauss so aptly stated, "A certain tragedy for meteorology will always lie in the fact that what it knows precisely is of interest only to a small part of humanity, while on the other hand humanity's interest is mainly in those processes which we can approach only with an uncertain degree of probability."[7]

Recall James Espy's comparison of the solitary astronomer and the gregarious meteorologist in the epigraph to this book. Herein lies the key to understanding the development of meteorology. Like the astronomer's telescope, but infinitely more unwieldy, *systems* of observation, requiring the cooperative efforts of hundreds of individuals, were the "instruments" used by meteorologists to gather data over large areas of the country. Eastern cities such as Albany (1825–50), Philadelphia (1834–49), and Washington (1849–present) served as the theoretical and administrative foci—the imperial centers—of systems such as the academies in the state of New York, the joint committee on meteorology of the American Philosophical Society and the Franklin Institute, and the Smithsonian meteorological project. Hundreds of observers on the periphery sent in their results voluntarily. Working at the administrative center of the system, scientists then created "images" of the weather by displaying the collected data on charts and maps. To use the terminology of Bruno Latour, meteorologists formed a "cascade" of mobile inscriptions, beginning with the paper forms filled out by observers to represent their local weather and ending with weather maps and charts showing storm tracks, isotherms, and other constructs. These "representative inscriptions" of increasing complexity and abstraction were used by scientists to support their theoretical, polemical, and practical objectives.[8]

The overall quality of these systems, especially in the first half of the nineteenth century, was not great. A typical observer, equipped with a thermometer, wind vane, and rain gauge (calibrated barometers were rare), recorded only the surface weather conditions, observed on an irregular schedule, and communicated the results by mail. After 1850, however, the Smithsonian supplied calibrated instruments to some observers, supported observations of the upper air at mountain stations and by balloonists, and instituted standardized reporting forms and schedules. As the telegraph networks expanded across the nation, weather inscriptions could also be mobilized quickly. Meteorologists could now claim new utility for their science by warning the public of the approach of storms. Even the storms cooperated, typically traveling from the West or South toward the more heavily populated mid-Atlantic coast, where the warnings were posted. Thus, the term *system,* used by Joseph Henry and others to describe a group of meteorological observers governed by standards established at a central office, provides a natural unit of analysis for meteorology in the nineteenth century.[9]

In some systems there was competition for "observing time" among theorists with divergent interests, each making different and often conflicting demands on the observers. The Army Medical Department, whose real purpose was to evaluate "airs, waters, and places" for the health of the troops, also accommodated Espy's storm studies for several years. The Smithsonian meteorological project, formed to resolve the controversy over storms, was in reality better suited to study the climate and natural history of the North American continent. The U.S. Army Signal Office, which organized a national storm-warning service, became involved in investigating and reporting numerous other threats to domestic tranquillity such as Indian uprisings, plagues of locusts, and striking workers.[10] Government funding also placed organizational, budgetary, and social expectations and constraints on meteorology. Beginning in 1856, for example, funds from the agricultural branch of the Patent Office were given to the Smithsonian project and were particularly important in shaping its research agenda in favor of agricultural climatology. From then on, more farmers and fewer natural philosophers served as meteorological observers, and studies were geared toward practical information and climatological reports for the benefit of agriculture.

Whereas a system of observation had one administrative center and increasingly formal exchanges of information, *networks* of meteorological theorists were more informal, without a uniform center of authority and with more communication links between individuals. For example, the storm controversy protagonists, William Redfield, James Espy, and Robert Hare, and their critics and supporters formed the core of three nonexclusive scientific networks that included the "Lazzaroni," members of domestic and foreign scientific societies (acting more often as a group than as individuals), and other elites and near-elites, both civilian and military. These networks are similar to the "Cambridge network" of Walter F. Cannon, the "invisible colleges" of Derek J. de Solla Price, and the "highly complex web of social relationships" which Michael Mulkay and others associate with the creation of scientific knowledge.[11]

Employing observational systems and theoretical networks as interrelated units of analysis, and using archival and manuscript sources wherever possible, this history explores a previously neglected specialty during a significant period of its development. If it raises new questions about science in America or provides a measure of insight into old ones; if it stimulates in the reader a sense of the "otherness" of a bygone era in science or a sense of empathy and continuity with the past; if it conveys in any measure the curiosity, energy, frustration, and excitement of the meteorologist, the observer, or the system builder; if it accomplishes any of these, I will deem it a success.

Meteorology in America, 1800-1870

CHAPTER 1 Early Issues and Systems of Observation

I am sorry it is not in my power to begin immediately the course of observations you proposed in your last letter. I have not a thermometer even, at present, but shall provide myself directly with one, and as soon as possible with a Barometer. The addition of the Meteorological phænomena, observations with respect to the migration of birds, and the changes in plants ought to render it a pleasing task to the Philosopher, the Sportsman and the Farmer, equally.
—Thomas Mann Randolph, Jr., to Thomas Jefferson, May 3, 1790, *Jefferson Papers.*

From earliest times, explorers and colonists in the New World had speculated on the causes of climatic change, the influence of weather on disease, and the peculiar nature of American storms. Without institutions to coordinate and support their research, however, individual observers and theorists were voices crying in the wilderness, their opinions based on personal impressions, memory, and the general folklore.[1]

The American story begins in Europe. Although meteorology had little or no status in the physical sciences, descriptive records of climatic phenomena, such as the closing of rivers by ice and the time of blossoming, harvesting, and first frost, existed from early times. In the middle of the seventeenth century, savants began to turn away from the practice of preparing new commentaries on Aristotle's *Meteorologica* and instead focused on new techniques for measuring and weighing the atmosphere.[2] Also from the mid-seventeenth century new experimental systems of meteorological observation developed, typically centered in the new scientific societies of Europe. Early compilations included those by Luigi Antinori (1654) at the Accademia del Cimento;[3] by Edme Mariotte (ca. 1670) through personal correspondence;[4] by Johann Kanold, editor of the *Breslauer Sammlung* (1718–30);[5] by James Jurin (1732) and Roger Pickering (1744) at the Royal Society;[6] by Daniel Bernoulli (1734–49) through the "Great Northern Expedition" of Vitus Behring;[7] by Père Louis Cotte and Antoine-Laurent Lavoisier (1776–86) through the Société Royale de Médicine;[8] and by Father Johann Jakob Hemmer (1781–92) through the Societas Meteorologica Palatinae.[9]

These projects in Europe and the early example of Thomas Jefferson and James Madison in Virginia stimulated groups of college professors in New England, the Army Medical Department, the General Land Office, and academies in the state of New York systematically to collect climatic and phenological statistics over large areas of the country in the first quarter of the nineteenth century. These early projects were an attempt to broaden the geographic coverage of observations, standardize their collection, and publish the results. Advocates, administrators, and observers concerned themselves with such basic themes as the utility of the observations (for settlement, agriculture, and health), the quality of the science (in comparison to that in Europe), and the commitment and dedication of the sponsoring institutions. These themes, sounded early in the century, would reverberate throughout the decades.

Also of note is Joseph Henry's early involvement with the New York system, the beginning of his lifetime interest in meteorology. Throughout his long career, his friends, co-workers, and acquaintances included all of the prominent figures in meteorology. As head of the Smithsonian Institution, he became America's leading organizer of meteorological systems.

Climatic Change

Climatic change, especially as it related to human activity such as clearing the forests for settlement, was a major issue in colonial America and remained so until the middle of the nineteenth century. The classical heritage related the climate of an area uniquely to its latitude. Parmenides of Elea (ca. 500 B.C.) and Aristotle (384–22 B.C.) were the first to attempt to describe the climatic differences of the known parts of the world. They believed that *climate*—from *klima,* meaning inclination—was determined only by the height of the sun above the horizon. Because of its seemingly favorable location in latitudes more southerly than in most European nations, the New World was expected to have a warm, exotic climate. Initially, colonial promoters envisioned a rich harvest of wine, silk, olive oil, sugar, and spices from their investment. The Virginia Company's *Declaration of the State of Virginia* contained the following statement: "Wee rest in great assurance, that this Countrey, as it is seated neere the midst of the world, betweene the extreamities of heate and cold; So it also participateth of the benefits of bothe, and is capable (being assisted with skill and industry) of the richest commodities of most parts of the Earth."[10]

Early settlers, however, found the climate harsher and the meteorological phenomena more violent than in the Old World. The Jamestown colonists suffered deprivation, disease, and death during the "extremely cold" winter of 1607–8. Of the 105 persons in the original company, all but 32 perished before the first supply ship arrived the following summer.[11] In 1644–45 the Reverend John Campanius of Swedes' Fort (Delaware) described *Mochijric Schackhan,* or mighty winds, unknown in Europe, which "came suddenly with a dark-blue cloud and tore up oaks that had a girt of three fathoms." Another colonist in New Sweden noted that when it rains "the whole sky seems to be on fire, and nothing can be seen but smoke and flames."[12] James MacSparran, a missionary to Rhode Island for thirty-six years until his death in 1757, spent considerable energy warning colonists against emigrating to America. He found the American climate "intemperate," with "excessive heat and cold, sudden violent changes of weather, terrible and mischievous thunder and lightning, and unwholesome air"—all "destructive to human bodies."[13] As the early colonists soon discovered, the climate of the New World, including the frequency and intensity of storms, was indeed harsher than that of Europe.

According to folk wisdom, it was also possible that rainfall patterns were changing as the forests were cleared. For example, Christopher Columbus, according to his son Ferdinand, knew "from experience" that the afternoon rains in the West Indies were produced by the luxuriant forests of the islands and that removal of these forests would reduce their mist and rain, as had already happened elsewhere.[14] In a similar vein, William Wood, a colonist in Massachusetts from 1629 to 1633, observed: "In former times the rain came seldom but very violently, continuing his drops (which were great and many) sometimes four and twenty hours together, sometimes eight and forty, which watered the ground for a long time after. But of late the seasons be much altered, the rain coming oftener but more moderately, with lesser thunders and lightnings and sudden gusts of wind."[15]

In addition to perceived changes in rainfall, American colonists thought temperature patterns were changing, again caused by the clearing of the forests. There was no general agreement, however, about the direction or magnitude of the change. Cotton Mather believed it was getting warmer: "Our cold is much moderated since the opening and clearing of our woods, and the winds do not blow roughly as in the days of our fathers, when water, cast up into the air, would commonly be turned into ice before it came to the ground."[16] Benjamin Franklin agreed, writing to Ezra Stiles that "cleared land absorbs more heat and melts snow quicker." He thought than many years of observations, however, would be necessary to settle the issue of climatic change.[17] Hugh Williamson of Harvard College also found the winters becoming less severe and the summers warmer because, as the forests were cut down, open fields were better able to absorb and retain heat, thus ameliorating the northwest winds.[18]

Thomas Jefferson speculated that, as the settlements of Virginia progressed inland from the seacoast, an increasing area of cleared and cultivated land was being heated strongly by the sun. Cool air from the mountains to the west and from the ocean to the east rushed in to replace hot rising air over the cultivated land. The western flow, however, was slowed by the greater roughness of the terrain, allowing sea breezes to penetrate farther inland than ever before. Jefferson continued: "We are led naturally to ask where the progress of our sea breezes will ultimately be stopped? No confidence can be placed in any answer to this question. If they should ever pass the Mountainous country which separates the waters of the Ocean from those of the Missisipi [sic], there may be circumstances which might aid their further progress as far as the Missisipi. . . . When this country shall become cultivated, it will, for the reasons before explained, draw to it the winds from the East and West." The result, which Jefferson illustrated with two diagrams, would be zones of rising air over the cultivated areas east and west of the mountains.[19]

Both inadvertent and purposeful changes in the climate were subjects for speculation. Jefferson proposed that the Spanish should open a narrow passage through the isthmus of Panama ("a work much less difficult than some even of the inferior canals of France") and let the force of the tropical current widen the passage. Then Spain could circumnavigate the globe without encountering the Dutch at the Cape of Good Hope or the stormy seas of Cape Horn:

> These consequences would follow. 1. Vessels from Europe, or the Western coast of Africa, by entering the tropics, would have a steady wind and tide to carry them thro' the Atlantic, thro America and the Pacific ocean to every part of the Asiatic coast, and of the Eastern coast of Africa: thus performing with speed and safety the tour of the whole globe, to within about 24° of longitude. . . . 2. The gulph of Mexico, now the most dangerous navigation in the world on account of its currents and moveable sands, would become stagnant and safe. 3. The gulph stream on the coast of the United States would cease, and with that those derangements of course and reckoning which now impede and endanger the intercourse with those states. 4. The fogs on the banks of Newfoundland, supposed to be the vapours of the gulph stream rendered turbid by the cold air, would disappear. 5. Those banks, ceasing to receive supplies of sand, weeds, and warm water by the gulph stream, it might become problematical what effect changes of pasture and temperature might have on the fisheries.[20]

Jefferson, in his *Notes on Virginia,* summarized the positions of those who believed the climate was less harsh than before: "A change in our climate, however, is taking place very sensibly. Both heats and colds are become much more moderate within the memory even of the middle-aged. Snows are less frequent and less deep. . . . The elderly inform me, the earth used to be covered with snow about three months in every year. The rivers, which then seldom failed to freeze over in the course of the winter, scarcely ever do so now."[21] Jefferson's correspondent William Dunbar, however, saw the reverse trend and supported his assertions with observations published in the *Transactions of the American Philosophical Society:*

> It was with us a general remark, that of late years the summers have become hotter and the winters colder than formerly. . . . Doctor Williamson and others have endeavored to show that clearing, draining and cultivation, extended over the face of a continent, must produce the double effect of a relaxation of

> the rigours of winter, and an abatement of the heats of summer; the former is probably more evident than the latter, but admitting the demonstration to be conclusive, I would enquire whether a partial clearing extending 30 or 40 miles square, may not be expected to produce a contrary effect by admitting with full liberty, the sunbeams upon the discovered surface of the earth in summer, and promoting during winter a free circulation of cold northern air.[22]

David Ramsay, a legislator and historian in South Carolina, concurred. He found it "remarkable" that oranges, which were plentiful forty or fifty years earlier, were "now raised with difficulty."[23] The weather itself added impressionistic evidence to the debate. On September 23, 1815, the "great September gale," a storm thought to be the most destructive of the century, devastated the East Coast. In the following year up to twenty inches of snow fell in New England in June causing a general crop failure in the "year without a summer."[24]

Climatic changes were of particular interest to physicians, who believed that the correlation between weather, climate, and the health of their patients was significant. In 1822 Dr. Job Wilson of Salisbury, Massachusetts, offered his opinion, based on sixteen years of meteorological records, that the "extensive clearing and cultivation" of the country had increased the extremes of heat and cold—causing added stress to his patients.[25]

Weather and Health: Medical Topography

The legendary Asclepiad, Hippocrates of Cos (460–ca. 370 B.C.), advocated a holistic view of health and disease, waged war on magico-religious medical practice, and sought to establish a scientific and rational basis for medicine. Hippocrates viewed health as an expression of the balanced interplay between an organism as a whole and its environment. He emphasized the importance in health and disease of water, place, topography, and orientation to sun and wind: "Whoever wishes to investigate medicine properly, should proceed thus: in the first place to consider the seasons of the year and what effects each of them produces for they are not at all alike, but differ much from themselves in regard to their changes. Then the winds, . . . the qualities of the waters . . . [and the] situation [of a city], how it lies as to the winds and the rising of the sun."[26] Although Hippocratic medicine lay dormant for hundreds of years, a sixteenth century revival popularized the notion that conditions in the atmosphere—especially seasonal changes and rapid changes of temperature—were related to the recurrence and spread of disease.[27] The revival coincided with the beginning of regular collection of vital data in England. Government decrees required each parish to keep a day book recording christenings, marriages, and burials. In the seventeenth century, bills of mortality were begun in London and other cities along with "natural and political" commentaries on the data.[28]

Because of the premise that weather and climate were causes of illness and death, it is not surprising that physicians and natural philosophers proposed that more regular and systematic measurements of weather, climate, and mortality be undertaken. The prevailing sentiment was voiced by John Fothergill (1712–80), London physician, natural historian, and confidant of Benjamin Franklin: "I know of nothing that would more effectually conduce to state the different degrees of healthiness or unhealthiness in different parts of this nation so clearly, as a proper bill of mortality . . . The records of the seasons in respect to heat and cold, dryness and moisture, made by ingenious men in different parts of the kingdom, compared with such annual bills, would afford many useful reflections to the faculty, [and] much benefit the community in general."[29]

In America the first series of instrumental meteorological observations by Dr. John Lining were related to his medical concerns. Lining, who came to Charleston about 1730, found the climatic conditions radically different from those of his native Scotland. Beginning in 1740 he decided to observe both the weather and the intake and outgo of his own body for a period of one year in an effort to understand their relation to epidemic disease: "I began these experiments . . . [to] discover the influence of our different seasons upon the human body by which I might arrive at some certain knowledge of the cause of our epidemic diseases which regularly return at their stated seasons as a good clock strikes twelve when the sun is on the meridian."[30] Lining used a portable barometer, a Fahrenheit thermometer, and a Heath thermometer divided into ninety equal parts with the freezing point at sixty-five and the "temperate point" at forty-nine. His hygroscope was a whip cord, divided into one hundred equal parts, which expanded and contracted five inches with variations in the humidity. Lining took observations upon arising, at 3:00 P.M., and at bedtime, maintaining his record for eight years.

Lining's associate Dr. Lionel Chalmers continued the observations from 1750 to 1759 and published the findings in two volumes, *An Account of the Weather and Diseases of South Carolina* (London, 1776). In a related effort, Dr. Edward Holyoke, a promoter of medical education in Massachusetts, provided a monthly tabulation of meteorological data, observations on the prevailing diseases, vital data on the number of persons dying, and the "causes" of the 139 deaths in Salem in 1786.[31] Leading medical institutions such as the College of Physicians in Philadelphia also joined the cause. The college charter, written in 1787, forged an explicit link between recording meteorological observations, investigating American diseases and epidemics, and advising the government on medical matters.[32] As the old century passed, speculation on medical geography was epitomized by Noah Webster's two-volume review of the relationships between "pestilential epidemics and sundry other phenomena of the physical world." Webster thought that weather conditions, comets, volcanic eruptions, earthquakes, and meteors were all epidemiological factors.[33]

Theories of Storms: Franklin's Legacy

Benjamin Franklin's investigations of the electrical nature of lightning, although well known, were only one aspect of his broad interests in geophysical phenomena. Franklin formulated hypotheses on the role of electricity in causing precipitation and auroral phenomena and on the role of moist, heated air in causing waterspouts, whirlwinds, thunderstorms, and the equator-to-pole atmospheric circulation. Franklin also provided a conduit to America for European scientific opinion and philosophical instruments.[34] His speculations on the role of electricity, heat, and the whirling motion of the winds provide the backdrop for the theories of succeeding generations of American meteorologists.

One of Franklin's many interests was comparing meteorological observations from different locales. Writing to Jared Eliot in Connecticut in 1747, Franklin thought it both advantageous and interesting to know the weather from "the several parts of the Country."[35] He noted that storms along the coast sometimes persisted for three or four days and moved from the southwest toward the northeast even though the wind blew from the northeast. Franklin's deduction came from his attempt, in 1743, to observe an eclipse of the moon in Philadelphia. Storm clouds blowing in from the northeast obscured the view, but the eclipse was observed in Boston as scheduled; the storm arrived there four hours after it was noticed in Philadelphia. By collecting reports of the weather from travelers and compiling newspaper accounts from New England to Georgia, he concluded that storms with winds from the northeast began in the southwest, in Georgia or the Carolinas, and traveled toward the northeast at a rate of one hundred miles per hour.[36]

Franklin gave two analogies to explain the cause of these storms. The first was the motion of water in a long canal when a gate is suddenly opened. All the water moves toward the gate, but there is a successive delay in the time when motion begins. The water nearest the gate begins to move first, causing progressive motion of the water in the length of the canal. The water at the head of the canal is the last to move. A second example supposed the air in a room to be at rest until a fire is set in the chimney. The heated air in the chimney rises immediately, and the air flows toward the chimney, and so on to the back of the room. Thus Franklin's hypothesis to explain northeast storms supposed "some great heat and rarefaction of the air in or about the Gulph of Mexico," which caused the air to rise and be replaced by a successive current of cooler, denser air from the north. The strike of the coast and the ridge of the Appalachian Mountains guided the northeast flow of air.

Franklin did not suppose that all storms were generated in the same manner, and he provided a counterexample of "thundergusts" arriving from the northwest.[37] Franklin speculated that whirlwinds and waterspouts were similar phenomena consisting of progressive and circular motions caused by the rising of rarefied columns of "violently heated" and moisture-laden air. The distinction between the two was that whirlwinds occurred on land and waterspouts over water.

Natural-Theological and Astrological Meteorology

Eighteenth-century sermons frequently dwelt on divine displeasure or Providence manifest in specific events in nature such as stormy winds, thunderclaps, drought, and rain.[38] During inclement weather churches often rang their bells to ward off the demons of the air and parishioners gathered to offer prayers. Andrew Dickson White credited Franklin's kite experiments with dealing the "death-blow" to theological meteorology: "The 'Prince of the Power of the Air' tumbled from his seat; the great doctrine which had so long afflicted the earth was prostrated forever."[39]

In contrast to the theological meteorology of earlier centuries, early nineteenth-century meteorology treatises followed a broader tradition in natural philosophy and were typically used for moral instruction or to provide a glimpse of the divine in the workings of the laws of nature. William Prout's Bridgewater treatise, *Chemistry, Meteorology, and the Function of Digestion, Considered with Reference to Natural Theology*, argued that weather and climate provided evidence of design; Reverend John Lauris Blake's *Wonders of the Earth* contained accounts of earthquakes, volcanos, storms, and whirlwinds primarily for the moral guidance of the young. In general, professional scientists shared the religious sentiments of the day and employed natural theological arguments in their explanation and justification of science. For example, Joseph Henry found a teleological purpose for science in the search for unity in nature which would reveal the "operation of a single and simple law of the *Divine Will*." If, as Marc Rothenberg reported, American astronomers found a "linkage" between "their science and their God," there is no reason to suspect that meteorologists saw things any differently.[40]

Astrology also played a role in the development of observational science. Astrologers of the fourteenth century, searching for connections between the heavens and the earth, were among the first to keep daily weather diaries. Their practice received a boost when printed ephemerides became available in the late fifteenth century. In the early seventeenth century, Johannes Kepler and others conducted astrologically motivated weather research which was widely quoted, if not imitated.[41] Almanacs, second in popularity only to the Bible, typically published astrological tables for amusement and seasonal and yearly weather forecasts for better planting and harvests. The German *Wetterbüchlein* (1505, seventeen editions), and the *Bauern-Praktik* (1508, sixty editions), two successful publications rooted in popular lore, based their predictions for the year on the weather for Christmas eve and Christmas night. Some almanacs included blank pages each month that could be used for notes on the weather. The *Farmer's Almanac,* now the *Old Farmer's Almanack* (1792–present), is a descendant of this genre.[42]

The Climatological Observations of Thomas Jefferson and His Correspondents

On the western side of the Atlantic, Thomas Jefferson and the Reverend James Madison, president of the College of William and Mary, are credited with making the first simultaneous meteorological measurements in 1778. During a period of approximately six weeks, the two men measured the barometric pressure, temperature, and winds at Monticello and Williamsburg and noted an "almost unvaried difference" in the average height of the mercury in the barometers—a difference Jefferson attributed to the greater elevation of his home. Jefferson concluded that these observations proved the variations in the weight of air to be "simultaneous and corresponding in these two places." He also noted that observations taken over a nine-month period showed that the prevailing winds were from the northwest in the mountains and from the northeast at Williamsburg and that killing frosts were less common at Monticello than at lower elevations.[43]

Jefferson, who firmly believed that understanding the weather and its changes required a comparison of observations made in different locations, also tried to recruit Madison's more famous namesake in 1784: "I wish you had a thermometer. Mr. Madison of the college and myself are keeping observations for a comparison of climate. We observe at Sunrise and at 4. o'clock P.M. which are the coldest and warmest points of the day. If you could observe at the same time it would show the difference between going North and Northwest on this continent. I suspect it to be colder in Orange or Albemarle than here." He asked Madison to keep a diary with columns for the date; observations at sunrise and 4 P.M. of thermometer, barometer, wind and weather; the appearance of shoots, flowers, or falling of leaves of plants; the appearance or disappearance of birds, their migrations, and so forth; and miscellaneous observations of the aurora or other rarities.[44] Madison's weather diaries, which span almost two decades, have been preserved.[45] Jefferson also exchanged observations with Isaac Zane, Jr., in Philadelphia and Benjamin Vaughan in London. His northernmost correspondent was Hugh Williamson in Quebec, his southernmost William Dunbar in Natchez, Mississippi. Among other meteorological journals available to him were those of David Rittenhouse, Peter

Legaux, William Adair, William Bartram, George Hunter, John Gottlieb Ernestus Heckewelder, and John Breck Treat.[46] Jefferson had a vision of a national meterological system. Beginning with his home state of Virginia, he hoped to supply observers in each county of each state with accurate instruments. The entire system would be supervised by the American Philosophical Society and funded by the federal government. However, the revolutionary war and other responsibilities prevented him from beginning "this long-winded project."[47] As president of the American Philosophical Society from 1797 to 1815, Jefferson continued to press for a national system. As President of the United States, he provided the Lewis and Clark expedition with scientific instruments and instructions. He also encouraged the explorations of Dunbar, Thomas Freeman and Peter Custis, and Zebulon Montgomery Pike. Despite these contributions, prominent historians have had little to say about Jefferson's meteorology.[48]

New England Collegiate Systems

During Jefferson's presidency, the first regular collections and compilations of meteorological observations from several different locations in the United States were initiated by editors of medical and literary journals in New England. In 1806 Daniel Adams, editor of the *Medical and Agricultural Register* (Boston, 1806–7), began such a series to cast "some new light on the climate and the diseases of the country."[49] Correspondents extending in an arc fifty to one hundred miles from Boston reported temperature, winds, and weather along with marriages, births, and deaths. Some observers also supplied a narrative account of the month's weather and diseases.[50] The greatest number of reports—seven—was received for January and March 1807. Figure 1.1 reproduces the first published report for January 1806.

Fredrick Hall, editor and publisher of the *Literary and Philosophical Repertory* and professor of mathematics and natural philosophy at Middlebury College in Vermont, also collected meteorological journals from his associates Chester Dewey at Williams College and Jeremiah Day at Yale College. The three professors agreed to take observations at 7:00 A.M., 2:00 P.M. and 9:00 P.M.—the last observation to be compared with an additional temperature measurement at 9:00 A.M. to calculate the daily mean. Rain and snow were to be caught and measured, and the "day of the moon's age" was to be noted.[51] Hall published the data from 1810 to 1816 in his *Repertory*.[52] The last issue of the *Repertory* has data on the cold summer of 1816 which are not available elsewhere.

Hall's journal, however, did not circulate beyond New England. In December 1814, for example, the Literary and Philosophical Society of New York was "not aware of there being any complete regular series of observations kept in this country . . . We know, at least, of no journal in which such a series is

Result of Meteorological and other Observations, for January, 1806; *made at* MASON, *(N. H.)* 50 *Miles northwest of Boston, by the Rev.* EBENEZER HILL; *at* LEOMINSTER, *(Mass.)* 45 *Miles westwardly from Boston, by* ABIJAH BIGELOW, *Esq.; at* CONCORD, *(Mass.)* 18 *Miles northwest of Boston, by Dr.* ISAAC HURD:—*For the Medical and Agricultural Register.*

January, 1806.	*Mean degs. at sun-rise.*	*Mean degs. at 2 P. M.*	*Mean deg. of the mo.*	*Greatest heat in the month.*	*Least heat in the month.*	*Prevailing winds.*	*Marriages.*	*Births.*	*Deaths.*
Mason	25 4/7	33 3/7	29 21/31	30th day 55°	18th day 4°	W.N.W.&N.			1
Leominster	25 2/3	34 1/16	29 16/31	30 51	18 4				
Concord	20 1/2	30 2/3	25 12/31	30 48	16 0/12	N.E. & N.W.	3	2	2

WEATHER.

Mason.	Inc.	*Leominster.*	*Concord.*	Inc.
3d day, snow, N. fair 10 A. M.	1½	3d day, snow, N. fair 10 A. M.	3d day, snow, N.	1½
		4th, rain in the evening.		
7th, snow, N. E.	4	7th, snow, N.E. 4 P. M. rain.	7th, snow, N. E. sprink. of rain.	2½
9th, snow, N. W. 10th, a little snow at evening.	3	9th, snow, wind northwd. 10th, a little snow at evening.	9th, snow, N. E. 10th, snow evening, S. E.	6
13th, sprink. of rain and misty, snow at night, S. W. 14th, snow day and night, N.	18	13th, rainy morning, snow at night. 14th, snow and hail day and night.	13th, rain, snow at 4 P. M. wind W. by N. 14th, snow, N. E. 15th, snow in morn.	9
15th, cloudy, fair.		15th, snow in morning.		
19th, snow began at evening, N. 20th, snow, W. of N.	7	19th, snow 4 P. M. N. 20th, snow, even. mist with rain.	19th, snow 5 P. M. N. N. E. 20th, severe storm.	8½
21st, cloudy, freezing mist, still.		21st, cloudy, trees covered with ice.	21st, cloudy, N.N.E.	
22d, snow 7 P. M. N. 23d, snow day and night, W. of N. 24th, snow till 9 A. M. fair.	22	22d, snow at night. 23d, snow, wind northwardly. 24th, cloudy.	22d, cloudy, snow, N. N. E. 23d, violent storm, 9 P. M. rain thro' the night, snow settled 8 inches. 24th, snow, wind N. N. E.	22
Total of snow *ft.* 4	7½		Total of snow *ft.* 4	1½

Mason.—The depth of snow was taken either from actual measurement or from the judgment of men who had been where it did not drift; in which case, the mean difference of their opinions has been taken.

Fig. 1.1. First report of observers, January 1806, published in the *Medical and Agricultural Register* 1, (1806): 31.

registered. It appears to us very desirable that this deficiency should be supplied: and we should willingly hope that an example, fairly held up by this society, would be followed by similar associations in other parts of the United States."[53] In part to remedy this perceived deficiency, the first volume of the *North American Review,* published in 1815, contained the meteorological journals of John Farrar, Hollis Professor at Harvard College; Parker Cleaveland, professor of mathematics and natural philosophy at Bowdoin College, Brunswick, Maine; and Dr. Jonathan Eights, an Albany physician.[54] The three observers had thermometers and barometers, but the instruments were not of common construction and the times of observation were not standardized. The published observations of Farrar and Cleaveland extend from February 1815 to March 1818. Eights observed only from February to July 1815. "Professor" Bishop of the university in Lexington, Kentucky, contributed observations from June to August 1817.[55] Although these early efforts were short-lived and covered only a limited area, their impact should not be underestimated. For the first time in America, observations from geographically distinct locales were being collected and published in journals and a generation of students at the colleges was being exposed to the subject for the first time.

Although interest in systematic observation was expanding, Chester Dewey noted in an article in the *North American Review* that observations were being made at very different times in different locations. Harvard, Williams, Bowdoin, and Middlebury colleges used 7:00 A.M. and 2:00 and 9:00 P.M.; Yale used "the supposed coldest and warmest times of the day" as indicated by Six's self-registering thermometer; the Literary and Philosophical Society of New York suggested 8:00 A.M. and 1:00 and 6:00 P.M.; and the Royal Society made observations at 8:00 A.M. and 3:00 P.M. from November through March and at 7:00 A.M. and 3:00 P.M. from April through October. Others used 6:00 A.M. and 6:00 P.M. or sunrise and sunset. Dewey argued that these published thermometrical observations, although great in number, were of little use without standard hours of observation. The underlying problem was how to approximate the true mean temperature using only three daily observations. On six different occasions Dewey took hourly observations for periods of 120 hours and compared the mean result with the mean calculated from several selected hours. He concluded that temperatures recorded at the hours of 7:00 A.M. and 2:00 P.M. and 9:00 P.M.—as he did at Williams College—supplemented by a maximum and minimum thermometer would most closely approximate the results of a full twenty-four-hour series.[56]

With such great variation in the standards of observation across the country, the first issue of Benjamin Silliman's *American Journal of Science and Arts* (1818) advocated that investigators use the time of blossoming of fruit trees and other phenological phenomena in their comparative studies. In 1817 Jacob Bigelow, a professor of botany at Harvard, corresponded with observers across the country and published a table of the dates of blooming of peach trees for eleven locations from Alabama Territory to Montreal, Canada, and Brunswick, Maine, covering fourteen degrees of latitude and seventeen degrees of longitude. Bigelow found a difference of seasons between the northern and southern extremes of about two and a half months but concluded that longitudinal differences had little effect on the floral calendar. In his report on Bigelow's investigation, Silliman added the dates of blossoming for Valencia, London, and Geneva for comparison and concluded that Bigelow's method afforded "an excellent criterion of the actual temperature, on a scale more extensive than it is practicable to obtain from thermometrical registers" and should be used to investigate climatic change in the United States.[57]

The Army Medical Department, 1814–1836

Dr. James Tilton, a graduate of the College of Philadelphia, had distinguished himself as a hospital surgeon during the revolutionary war through his innovative practice of housing only a few patients at a time in small, well-ventilated log huts. After the war, he became a member of the Continental Congress from his native state of Delaware. He came out of retirement to serve as chief of the Army Medical Department during the War of 1812.[58] Soon after hostilities began, one of the hospital surgeons, Joseph Lovell, filed a detailed report on the adverse effects of the climate on the health of the troops stationed in western New York State:

> The division of the army stationed at Fort George . . . was encamped on the bank of the Niagara . . . The surrounding country is flat, and the camp was deprived of the lake breezes . . . During the month of June it rained almost incessantly; while the latter part of July, and the whole of August were extremely hot; the whole of September was however remarkably mild and pleasant. Thus after having been wet for nearly a month, the troops were exposed for six or seven weeks to intense heat during the day, and at night to a cold and chilly atmosphere, in consequence of the fog arising from the lake and river . . . The diseases consequent to this alternate exposure to a dry hot, and cold damp atmosphere, were such as might have been expected; typhus and intermittent fevers, diarrhea and dysentery.[59]

Environmental theories of disease and its treatment were widespread among army physicians. Mumps were thought to be caused by "exposure to severe cold weather, & storms of snow and rain."[60] Even the ill effects of compounds such as sugar of lead as a treatment for chronic diarrhea (it paralyzed and killed patients) were blamed not on the medicine itself but on its unsupervised use in a fickle climate. The opinion of a surgeon quoted in James Mann's *Medical Sketches of the Campaign of 1812, 13, 14,* was the dominant one: "A knowledge of geography in general, and topography [and climate], are particularly important to the physician and surgeon."[61]

Reports such as these motivated Surgeon General Tilton to issue a general order, dated May 2, 1814, directing all hospital surgeons, mates, and post surgeons under his command to "keep a diary of the weather" and report quarterly as part of their official duties. Medical officers were instructed to "keep a book in which shall be registered all the reports transmitted by him; and to make from time to time such remarks on meteorological phenomena, and the appearance of epidemicks, as may be deemed useful in promoting medical science."[62] With the war in progress, efficient and widespread compliance with this order was impracticable. Tilton apparently demanded quarterly reports in an effort to exert a semblance of control over his "independently minded" surgeons. Regular reporting would help the surgeon general keep track of the locations of his staff members, the number of patients hospitalized, their diseases, how many died, and the siting and cost of the hospitals.[63] The army surgeons under Tilton can hardly be credited with forming a functioning medical-climatological system. One of the few surgeons to comply with the new orders was Benjamin Waterhouse in Cambridge, Massachusetts, whose diary of the weather dated March 1816 is the earliest meteorological journal preserved in the Army Medical Library.[64]

One author asserted that by 1817 studies of environmental influences on disease seemed to have "somewhat lessened the prevalent enthusiasm for the idea that the origin of fevers might be in 'those changes in the air that are pointed out by the thermometer, barometer, or hygrometer.' "[65] Without proper documentation, however, this is difficult to believe, for Lovell's war report for 1817 discussed 838 cases of "fevers and other important complaints." He classified 266 cases as inflammatory fevers, "which are the almost inevitable consequence of a cold and changeable climate," and 340 cases as diarrhea and dysentery, whose probable cause "will be found to arise from an *undue exposure to cold and moisture.*" In discussing the high mortality at Sackett's Harbor during the war (where the provisions and water were not particularly bad), Lovell concluded that "it must have been the climate—the weather—that produced the mischief." Lovell maintained that to save lives, medical doctors needed a tacit knowledge "of minutiae which depend neither on General Regulations, nor specific orders." He concluded that medical staff should be required to keep detailed journals recording the weather, prevalent diseases, and medical practice: "To effect this purpose it should be made the duty of every surgeon and mate having the charge of a hospital . . . to transmit an account of the local situation of his station, of the climate, the diseases most prevalent in the vicinity, and their probable causes, the state of the weather during the time reported with respect to temperature; winds, rain, etc.; to state at large the general symptoms of the complaints among the troops, as well as every peculiarity of disease; to investigate and as far as possible report their causes; the means employed to obviate them, with the success; as well as the practice adopted and the result."[66]

On April 14, 1818, a general reorganization of the U.S. Army placed Lovell in the position of surgeon general. During his administration (1818–36), the meteorological work of the Army Medical Department increased dramatically as he put his medical theories into practice. Lovell's revised orders for the medical department, issued in September 1818, involved his entire staff, from the post surgeons to himself, in the collection and compilation of meteorological statistics and reports of the sick and their treatment. The assistant surgeon general was to sort through these statistics to determine the causes and means of preventing diseases: "From an examination of the book containing the diary of the weather, medical topography of the station or hospital, account of the climate, complaints prevalent in the vicinity, &c., and from suitable inquiries concerning the clothing, subsistence, quarters, &c., of the soldiers, he [the assistant surgeon general] will discover as far as practicable the probable causes of disease, and recommend the best means of preventing them."[67]

Early results of these meteorological observations were published in the *Medical Repository* from 1821 to 1824.[68] Initially, reports arrived from seventeen army posts—from Plattsburgh in the North to Baton Rouge in the South and from Portsmouth in the East to Council Bluffs in the West—an area embracing sixteen degrees of latitude and twenty-six degrees of longitude. Within a year the number of stations had increased to twenty-seven. Never before had observations been made over so large an area. The compilation was presented to the medical community without much commentary as interesting statistics "from which to deduce the history of epidemic diseases, in so far as they depend on atmospheric causes."[69] The record contained many gaps. Officers in the Army Medical Department were, of course, first of all physicians with numerous and often pressing duties and interests other than the collection of meteorological data. Common excuses included sickness or absence of the observer and breakage of instruments. Moreover, military posts were often abandoned or relocated as the frontier advanced.

The first official publication on meteorology by the Army Medical Department—a compilation of meteorological registers for the years 1822–25, was issued by the surgeon general's office in 1826. Again the question of climatic change was posed: "whether in a series of years there be any material change in the climate of a given district of country; and if so, how far it depends upon cultivation of the soil, density of population, &c. . . . The United States . . . appear to offer an opportunity of bringing the question to the test of experiment and observation. . . . For here within the memory of many now living the face of whole districts of country has been entirely changed; and in several of the States two centuries have effected as much as two thousand years in many parts of

Europe. In this respect, the 'Landing of the Pilgrims' in 1620, is as remote a period as that of the invasion of Gaul or of Great Britain by Julius Caesar."[70] The chief object was to record and preserve the climatic facts in the face of a rapidly receding wilderness and aboriginal population.

The American Journal of Science for 1827 published Lovell's table of temperature, winds, and "weather" for eighteen stations covering nearly twenty degrees of latitude and thirty degrees of longitude. Another table displayed the comparative temperatures of seventeen places in Europe, Russia, Africa, and China in nearly the same latitude which showed that the temperature was higher in Europe for a given latitude. The editors praised the collection as "the best base for general conclusions respecting the climate of the U.S. hitherto published."[71] The *Meteorological Register,* distributed free to the scientific institutions of Europe, immediately received a favorable notice by Alexander von Humboldt, who cited the American system as a model for the Russians to emulate: "If only, following this fine example, there could be similar calibrated thermometer observations at the behest and expense of a mighty monarch in the eastern part of our old continent—in the widespread space, equal to half the lunar surface, between the Vistula and the Lena . . . ; then all of climatology would gain a new and improved stature in a few years." Humboldt reiterated this opinion two years later at the Imperial Academy of Science in St. Petersburg.[72] Even with this high praise, fourteen years would elapse until the next publication of meteorological statistics by the Army Medical Department. Figure 1.2 shows the number of observers involved between the years 1820 and 1874.

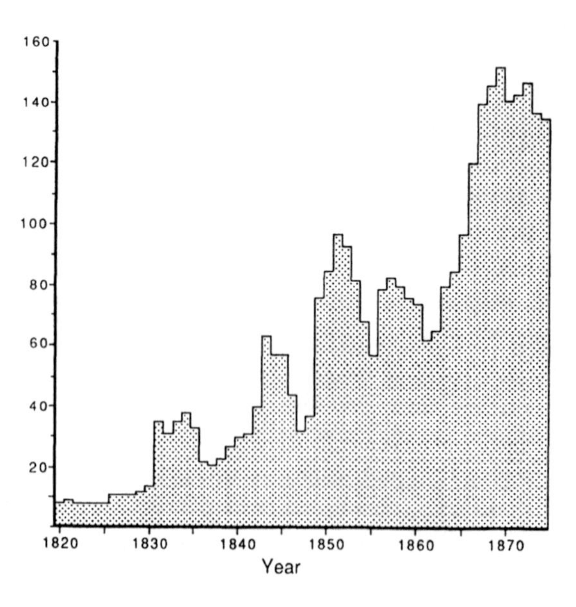

Fig. 1.2. Army post observations, 1820–1874. From data in the U.S. Army Medical Department *Meteorological Registers* (1826, 1840, 1851, 1855) and from Lewis J. Darter, Jr., comp., *List of Climatological Records in the National Archives* (Washington, D.C., 1942).

Josiah Meigs and the General Land Office, 1817

On April 29, 1817, Josiah Meigs, commissioner of the General Land Office in the Treasury Department of the United States, issued a circular requesting that the registers of the twenty regional land offices under his charge keep a regular meteorological record.[73] Meigs was trained in mathematics, astronomy, and natural philosophy at Yale and taught there as a tutor and professor. He had been the official meteorological observer for the Connecticut Academy of Arts and Sciences and had helped promote a scheme to collect regular observations along the East Coast before becoming the acting president of the University of Georgia. He had also served as first president of the School of Literature and the Arts in Cincinnati. Meigs wrote of his plans to his old friend Daniel Drake: "You will see at a *glance* of your bright mind that something very clever might be made of this. If my plan be adopted, and the *Registers* be furnished with the requisite Instruments for *Temperature, Pressure, Rain, Wind, &c.,* the expense of which would be a mere trifle compared to what YOU and I, at least, think the *value* of the object, we may in a course of years know more than we shall be able to know on any other plan."[74]

Meigs's circular requested three daily observations (morning, 2:00 P.M., and evening) of temperature, winds, and weather. Meigs sent blank forms with the circular and asked that particular attention be given to the column entitled *miscellaneous observations,* which requested, among other things, phenological observations on the foliation and flowering of plants, the migration of birds and fishes, the hibernation of animals, and unusual numbers of insects. In addition to accounts of unusual or remarkable rains, storms, and lightning, Meigs asked for a wide variety of additional but possibly related observations: "10. All facts concerning earthquakes and subterranean changes; 11. Concerning epidemic and epizootic distempers; 12. The fall of stones, or other bodies, from the atmosphere. *Meteors,* their direction, apparent velocity, &c. and, *particularly,* the interval between their apparent explosion and the hearing the report; 13. Discoveries relative to the antiquities of the country; 14. Memorable facts as to the *topography* of the country."[75] All these desiderata were to be transmitted with the monthly official returns of the registers to the General Land Office in Washington.

Land offices were located in Michigan (Detroit), Ohio (Wooster, Steubenville, Marietta, Zanesville, Chillicothe, and Cincinnati), Indiana (Jeffersonville and Vincennes), Illinois (Kaskaskia, Shawneetown, and Edwardsville), Missouri (St. Louis), Louisiana (New Orleans, Opelousas, and "north of Red river"), and Mississippi Territory (Huntsville, Washington, Saint Stephens, and "in the territory lately acquired from the Creeks"). Other volunteer observers also contributed their journals. These observations were dispersed over thirteen degrees of latitude and ten degrees of longitude. Meigs tried, without success, to

METEOROLOGICAL REGISTER for January 1821. at Washington City.

D.	THERMOMETER. M.	2 P.M.	E.	WINDS. M.	2 P.M.	E.	WEATHER. M.	2 P.M.	E.	RAIN—Inches.	MISCELLANEOUS REMARKS.
1	-20	38	36	0	W	W	Clr	Clr	Clr	29.95	
2	32	34	30	0	NW	N	Clr	Clr	Clr	29.84	
3	18	22	21	NW	NW	NW	Clr	Clr	Clr	29.98	Potomac frozen over
4	12	30	20	0	NW	NW	Clr	Hazy	Clr	29.75	
5	18	30	28	NW	NW	NW	Clr	Clr	[illegible]	[illegible] 30 [illegible]	
6	25	27	26	NE	NE	NE	Cldy	Snow	Snow	29.95 29.88	Snow began at 9.15 A.M.
(7)	22	27	27	W	N	W	Snow	Clr	Clr	29.72 29.75	[illegible] about 12 Inches
8	4	25	25	0	NW	NW	Foggy	Clr	Clr	29.74 29.70	[illegible]
9	20	38	30	[illegible]	0	0	Cldy	Cldy	[illegible]	29.72	
10	20	37	30	NW	NW	NW	Clr	[illegible]	Clr	29.95 29.99	
11	24	36	35	NW	0	0	Snow	Cldy	Fog	29.96 [illegible]	
12	16	36	30	0	NW	0	Clr	Clr	Clr	29.90 29.86	
13	18	36	34	0	0	0	Snow	Rain	Rain	29.61	Snow began at [illegible]
(14)	32	38	34	0	0	0	Cldy	Cldy Misty	Cldy	29.33 29.37	
15	24	32	28	NE	NW	0	Cldy	Clr	Clr	29.82 29.93	Light Snow 1/2 inch
16	21	[illegible]	22	NE	NE	NW	Cldy	Cldy	Cldy	30.18 30.15	[illegible] Snow all day [illegible]
17	16	26	21	NE	NW	NW	Snow	Cldy	Cldy Snow	29.92 29.80	Snow began at 9 [illegible] about 2 Inches
18	15	23	18	NW	NW	NW	Clr	Clr	Clr	30.12 30.16	[illegible] 12 at 9 PM
19	(6-)	23	14	NW	NW	NW	Clr	Hazy	Hazy	30.24 30.40	
20	20	30	30	0	0	0	Clr	Hazy	Hazy	30.60 [illegible]	30.04
(21)	28	38	34	0	0	0	Cldy	Clr	Clr	29.98 29.70	Hail & Sleet last [illegible] Beautiful weather [illegible]
22	30	36	34	W	0	0	Cldy	Cldy	Rain	29.82 29.77	Slight Rain 29.70.
23	32	40	36	W	W	W	Clr	Clr	Clr	29.62 29.75	Snow last Night 29.82
24	14	15	12	NW	NW	NW	Cldy	Cldy	Clr	30.12 30.14	High Wind last Night
25	-5	8	8	NW	NW	NW	Clr	Clr	Clr	30.50 30.28	[illegible] 20 [illegible]!!
26	-6	28	25	0	SW	SW	Clr	Clr	Cldy	30.27 30.46	
27	29	40	38	S	S	0	Cldy	Cldy	partly Clr	30.10 29.97	Thawing rapidly
(28)	24	39	40	0	W	0	Fog	[illegible]	Clr	30.03 30.05	White Frost this Morning [illegible]
29	22	40	38	0	E	0	Clr	Clr	Clr	30.26 30.30	[illegible]
30	28	50	48	0	SW	0	Clr	Clr	Clr	30.30 30.27	
31	[illegible]	55	50	0	S	S	Fog	Cldy	Cldy	30.14 30.04 29.70	[illegible]
	315 307 622	486 519 1005	444 468 912								

30.18 29.31 [illegible]

30.50 29.33 1.17

30.50 29.72 .78

324 17 307

Fig. 1.3. Page from the weather journal of Josiah Meigs, 1821.

get a resolution passed by Congress to fund the system and provide money for instruments.

Results were published in *Niles' Weekly Register* for 1818 and 1819.[76] Notable among them is Meigs's notice of an "almost universal prevalence of cold" between Detroit and Augusta, Georgia (a distance of 650 miles), on December 21, 1818.[77] He was planning to write a paper on the results of these observations for the American Philosophical Society but died before doing so.[78] According to Meigs's biographer, the original observations of the General Land Office system were presented to the American Institute of New York in 1858, but the records are missing.[79] Josiah Meigs's personal weather record for 1821, taken in Washington, was given to the Smithsonian Institution and, with the exception of one other, is the only original report of this system known to be extant.[80] Part of that journal is reproduced in Figure 1.3.

The Academies in the State of New York, 1825–1850

In 1801—long before Humboldt's paper on isotherms—Simeon DeWitt, a graduate of Queens College (1776, A.M. 1778), who had served as a geographer in the revolutionary army, advocated the construction of meteorological charts by which to compare the climates of North America and Europe: "It is a fact well known, that the western parts of this country do not correspond in climate, with those lying in the same latitude along the sea coast. If, therefore . . . a chart be constructed from observations made in places near the Atlantic, correspondent observations made in the interior parts of the country will enable us to ascertain to what latitude they are to be referred. If a distinct chart were constructed, on the same plan, for the principal places in Europe, one might with equal facility, by the help of both charts, compare their climates with those of America."[81] In 1825 DeWitt, now surveyor general of the state of New York and vice-chancellor of the educational system, introduced a resolution to the regents of the university requiring each academy in the state to submit annual meteorological reports.[82] He proposed that each of the incorporated academies in New York be furnished with a thermometer (by Kendall of New Lebanon) and rain gauge at the expense of the state. The regents designated a committee consisting of DeWitt, Gerrit Y. Lansing, and John Greig to provide observational guidelines and inaugurate the system.[83] At a subsequent meeting, the Board of Regents decided that academies must submit meteorological observations before they could receive their dividends from the public fund.[84] The *American Journal of Science* commended the regents and published their plans for a statewide meteorological magazine with uniform monthly reports from each county and a speculative proposal for establishing a national meteorological journal.[85] From 1825 to 1841, Gideon Hawley, the secretary of the regents, was in charge of the system. He was succeeded by Dr. T. Romeyn Beck,

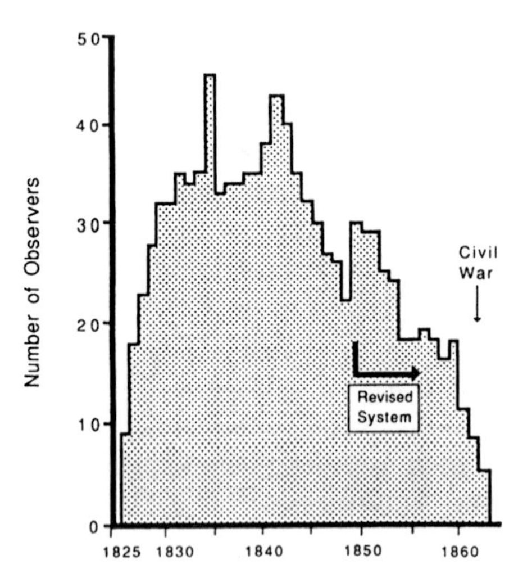

Fig. 1.4. New York academy observations, 1826–1863. Details on specific locations are given in Franklin B. Hough, comp., *Results of a Series of Meteorological Observations . . . in the State of New York,* first series, *1826 to 1850* (Albany, 1855); second series, *1850 to 1863* (Albany, 1872).

an Albany physician, principal of the Albany Academy, and advocate of research in meteorology, which he called "the natural history of the atmosphere."[86] Sixty-two academies cooperated in full or in part between 1825 and 1850. A revised system in cooperation with the Smithsonian Institution began in 1850 and continued until 1863 (see Chapter 6). Figure 1.4 shows the number of academies reporting observations to the New York Board of Regents between 1826 and 1863.

The Meteorological Interests of Joseph Henry

After graduating from the Albany Academy in 1822, Joseph Henry served as T. R. Beck's "Chemical Assistant" and worked as a curator at the Albany Lyceum of Natural History and the Albany Institute.[87] During the summer of 1825 Henry participated in the survey of the proposed "Great State Road" to connect the Hudson Valley and Lake Erie, gaining firsthand experience of the influence of topography on local weather and climate: "The configuration of the surface of a country you know has a great influence on its mean temperature as well as on the direction of its prevailing winds."[88] In 1826 he was appointed professor of mathematics and natural philosophy at the Albany Academy and from 1829 to 1833 assisted Beck in compiling the meteorological reports for the state of New York and publishing abstracts of the results in the *Annual Report* of the regents.[89] Henry demonstrated his interest in meteorology in articles such as "Topographical Sketch of the State of New-York," in which he called attention to the meteorological observations of the New York academies. Stephen Alexander, Henry's cousin and brother-in-law, considered the paper "especially useful in perfecting the science of meteorology."[90] In other meteorological investigations, Henry studied the relationship between auroral displays and the magnetism of the earth and introduced Daniell's hygrometer to the institute.[91]

Although Henry's growing reputation in science was primarily attributable to his work as a brilliant experimentalist in electrical and magnetic phenomena, his lifelong interest in meteorological phenomena was also an important aspect of his scientific work. His letter of application in 1832 for a teaching position at Princeton cited his research interests, responsibilities, and accomplishments, notably, "In connection with Dr. T. R. Beck I have the principal direction of the Meteorological observations made by the different academies of the State of N.Y. to the Regents of the University. In this work I am considerably interested and have hoped at some future time to deduce many facts from it of importance to the science of meteorology."[92] In Henry's candid writings to his friend Alexander Dallas Bache, however, we find a self-deprecating evaluation of his early meteorological work: "You will probably duly appreciate the important service which I then rendered to science when I inform you that my labours principally consisted in reading the proof sheets of the work."[93]

At Princeton Henry personally investigated thunderstorms, lightning, aurora, and other atmospheric phenomena related to electricity and magnetism. Throughout his life he advocated the construction of a physical atlas of America which would delineate the topography, geology, magnetic lines, climate, and distribution of plants and animals. Later he used the resources of the Smithsonian Institution to advance this goal.[94] Nathan Reingold, first editor of the Joseph Henry Papers, thinks it is more accurate to describe Henry as a "typical geophysicist" of this era, albeit with an uncommon talent and interest in experimental physics, than as an "experimental physicist with some minor activities in meteorology, surveying and the like."[95] Reingold has also suggested that Henry's early experiences in Albany were formative for his research program in Washington.[96]

Climatic change, medical geography, and speculation on the nature of storms were fundamental scientific issues in the colonial and early national periods. Little could be learned of these complex phenomena, however, until widespread systems of observation were developed. Although storms were not studied systematically, cooperative observers—college professors, journal editors, army post surgeons, registers of the General Land Office, students, and teachers at academies—collected and compiled climatic and medical statistics covering large areas of the country in the first three decades of the century. In the 1830s a bitter and prolonged controversy over American storms led to further expansion, refinement, and refocusing of the observational programs.

The loud debate and the expanding observational systems soon attracted the attention of scientists in Europe. If, as one historian wrote, tornadoes, hurricanes, and waterspouts were "simple topics" for geophysical speculation in the colonial period, this was manifestly not the case after 1830.[97]

CHAPTER 2 The American Storm Controversy, 1834–1843

Meteorology has ever been an apple of contention, as if the violent commotions of the atmosphere induced a sympathetic effect on the minds of those who have attempted to study them.
—Joseph Henry, U.S. Patent Office, *Annual Report, Agricultural,* 1858

In the 1830s an international controversy developed among meteorologists over the nature and causes of storms. The American component of the controversy centered on competing theoretical positions advanced by three prominent scientists: William C. Redfield, James Pollard Espy, and Robert Hare.[1] The dispute between Espy and Redfield held center stage between 1834 and 1841, then Hare entered the controversy with a vengeance and attacked both Espy and Redfield. American theorists argued over the *primum mobile* of storms: was it gravity, caloric, or electricity? They argued over methodology: were they searching for the *quo modo* or the *causa verum?* And they argued over basic definitions of the phenomena under investigation: were they looking at hurricanes, thunderstorms, tornadoes, winter storms, or some other "meteor?"

The American storm controversy—like its counterpart in Britain, the Devonian controversy—was an extremely well-documented scientific debate that was particularly heated in the period 1834–43.[2] The Devonian controversy was "resolved" by consensus by 1847. The American storm controversy came to no clear intellectual conclusion but led instead to new observational attempts to resolve it. The Joint Committee on Meteorology of the American Philosophical Society and the Franklin Institute (hereafter Joint Committee) was an institutional leader in this effort (see Chapter 3). The storm controversy was also important because of the pattern of support, both domestic and international, that emerged for the meteorological theorists. The Philadelphia circle—Espy, Hare, Bache, and Sears C. Walker—and the New England circle—Redfield, Denison Olmsted, Benjamin Peirce, and Charles H. Davis—represented early, "pre-Lazzaroni" alignments in American science. Internationally, Redfield found approval within a circle of the British Association for the Advancement of Science, which included John Herschel and William Reid. Espy received mixed reviews from this group but was supported at the French Académie des Sciences in the persons of François Arago, Jacques Babinet, and C-S-M Pouillet.[3] The European tours of Bache, Henry, Espy, and Hare brought the storm controversy fully to the attention of the European scientific community.

Some historians have stereotyped all meteorologists as "Baconian," suggesting their general lack of theoretical ability. This use of Bacon is misplaced, overly general, and misleading, but unfortunately it is not an isolated example. For example, George Daniels, writing on science in the Jacksonian period, labeled the storm studies of Elias Loomis as typically "baconian."[4] Loomis compiled observations over a widespread area, but he also advanced hypotheses concerning overlapping warm and cold air masses. Moreover, he tested the theories of Espy and Redfield by studying a single storm in exhaustive detail, rather than seeking to collect facts indiscriminately about all storms. Another author argued, unconvincingly, that Redfield's Baconianism, coming from a "leader" of the scientific community, and "Espy's incorrect interpretation of the motion of storm winds delay[ed] the acceptance of the method of hypothesis in American science." Sir David Brewster, the "leader of the British Baconians," was said to have supported Redfield and Reid because of their "successful application" of the "true" scientific method.[5] Not only is this interpretation ahistorical, it is wrong. It is little more than a caricature of a complex controversy. In reality, Brewster was an "aggressive debunker" of Bacon, who argued "energetically and often" that Bacon had no influence on the practice of science or its development.[6] Joseph Henry concurred.[7] The philosophy of Bacon did *not* set the agenda for the storm controversy. Although Redfield employed the rhetoric of Baconianism in defense of his amateur status, he did not use it as a working philosophy. Nor was Redfield a real leader of American science—he ran a transportation business in New York. The author's corollary on Espy is also far wide of the mark. Espy did not block hypothesis testing in America. Indeed, he stimulated the thought of at least two generations of meteorologists.[8] Observation, experiment, induction, and deduction were all part of the repertoire of nineteenth-century meteorologists. Hypothesis and theory, although perhaps discredited in the popular mind, were considered by Henry and others "of essential importance" to the advance and application of science.[9]

Theoretical Positions: Redfield, Espy, and Hare

William Redfield, James Espy, and Robert Hare, the principals in the American storm controversy, presented radically different interpretations of the nature and causes of storms. Redfield, a transportation engineer, studied storms as an avocation. He focused his attention almost exclusively on the whirlwind storms of the Atlantic Ocean. Redfield established a personal data-gathering system and corresponded extensively with other scientists, especially William Reid, governor of Bermuda and author of several books on storms.[10] Redfield gathered information from ships' logs and coastal stations which he used to reconstruct the path and wind patterns of hurricanes. Although Redfield believed that gravitational forces were the source of all atmospheric disturbances, in his investigations he claimed to be merely describing the phenomena of whirlwind storms, and he employed Baconianism in defense of his facts. Indeed, he attacked the other theorists for their speculation and made his lack of theory into a virtue.

Beginning in 1831, Redfield advanced a kinematic theory of storms in which the global wind pattern, produced solely by gravitational forces on the rotating earth, was taken to be a given. Barometric pressure, temperature, and moisture were all functions of the wind currents. Even without the differential heating of the earth's surface by the sun, the winds would continue to whirl because of gravity. According to Redfield, the tendency of all fluid matter to run in whirls, or circuits, when subjected to unequal or opposing forces sometimes leads to violent rotative movements "and is the only *known cause*" of tempests, tornadoes, or hurricanes.[11] Redfield's studies focused on the rotary gales of the North Atlantic, which he believed were caused by the northeast trade winds pressing against the islands of the Caribbean. Like whirls produced by water flowing in a rocky streambed, the trade winds are deflected into a circular pattern. The winds in these atmospheric whirls revolve about a central axis of motion: from right to left in the Northern Hemisphere and the reverse in the Southern Hemisphere. The violence of such a storm generally increases toward its center, but at the very center of the whirlwind is found a region of clear, calm air, the harbinger of a sudden and violent shift in wind direction and speed. These "aerial vortices" or whirlwinds also demonstrate an overall progressive motion which depends on the currents of the upper air. Thus, Redfield maintained, the common wind vane could not be trusted to indicate a storm's true motion. Once generated, these storms follow a parabolic course: northwesterly until reaching the Atlantic coast of the United States, then along the eastern seaboard, turning to the northeast between twenty-five and thirty-five degrees of latitude, and proceeding along the coast toward Nova Scotia. Waterspouts, tornadoes, and even winter storms were thought to be smaller whirlwinds following the same principles.

James Espy was the only disputant actively involved in institutionally sponsored observational systems. He directed the Joint Committee in Philadelphia in the 1830s and worked in Washington, D.C., with the Army Medical Department, the navy, and the Smithsonian Institution in the 1840s and 1850s. He was also a popular lecturer on the lyceum circuit. Although Espy's basic physical insights were considered sound, especially by his contemporaries Henry and Bache, his presentation of them was not. He consistently offended other investigators by his claims and persistently linked his theoretical ideas to impractical schemes for artificial rainmaking.[12]

Espy's theory, in contrast to Redfield's, emphasized thermally induced vertical convection and the "steam power" of the cooling and condensing of moist air. As heater columns of air rise, winds from all directions rush inward, producing centripetal wind patterns associated with cumulus clouds, tornadoes, hailstorms, and other meteors:

> When the air near the surface of the earth becomes more heated or more highly charged with aqueous vapor, which is only five eighths of the specific gravity of atmospheric air, its equilibrium is unstable, and up-moving columns or streams will be formed. As these columns rise, their upper parts will come under less pressure, and the air will therefore expand; as it expands it will grow colder, about one degree and a quarter for every hundred yards of its ascent . . . . The ascending columns will carry up with them the aqueous vapor which they contain, and, if they rise high enough, the cold produced by expansion from diminished pressure, will condense some of this vapor into cloud.[13]

Thus the motive power of heat was primary in Espy's system. The wind field was derived from the pressure distribution and ultimately from the temperature, density, and moisture distributions of the atmosphere. Espy clearly acknowledged the earlier work of John Dalton and J. L. Gay-Lussac on the laws governing the compression and expansion of gases, but he did not agree that water vapor diffused freely in the atmosphere or otherwise behaved as an ideal gas in a laboratory retort. His theory of centripetal wind patterns surrounding low-pressure areas followed work done in 1816 by H. W. Brandes, professor of mathematics at Breslau, with data from the late eighteenth century.[14]

Among the three protagonists, Robert Hare, professor of chemistry at the University of Pennsylvania, was both the most polemical and the least committed to gathering new observational data. His meteorological theories were largely developed by analogy to his electrical experiments and laboratory demonstrations. Although most scientists considered electricity an important but largely unknown factor in atmospheric processes, they were not convinced by Hare's arguments. His main contributions to science were in chemical apparatus, not in meteorological theory, but his perennial opposition to the theories of Redfield and Espy kept the storm controversy in the public realm and served to sharpen the responses of those he attacked.

Hare's electrical theory of storms postulated that the atmosphere behaved like a charged Leyden jar. There were two electrical oceans of opposite charge: the celestial and the terrestrial. Clouds, the mediators between the two oceans, were electrified by the celestial ocean and caused an accumulation of opposite charge on the terrestrial surface beneath the cloud. This electric accumulation counteracted gravity, caused a local diminution of atmospheric pressure, and, when combined with the effects of vaporization and condensation of water, caused inward- and upward-rushing currents of air. These currents produced rain, hail, thunder, lightning, and, in extreme cases, tornadoes. One of Hare's controversial claims was that he had discovered a new electrical "discharge by convection" in the atmosphere, which formed the motive power of the tornado. This discharge was to be the complement of the famous electrical discharge by conduction discovered by Franklin in lightning strokes.

Forshey's Comparisons

In 1840, at the height of the storm controversy, Caleb G. Forshey, a professor of engineering at Jefferson College, Mississippi, attempted to compare the competing meteorological theories of Espy, Redfield, and Hare. Forshey's questions, and the answers he received, provide a basis for comparison which is both convenient and historically based.[15] Forshey's five questions, addressed to the three theorists were as follows:

1. What is the principal cause in the production of rain; and what other causes chiefly modify the action of this principal cause?
2. What effect has electricity in the production or modification of *rains and winds?*
3. What is the *primum mobile* of tornadoes?
4. What generates and maintains a gyration in whirlwinds, spouts and tornadoes?
5. How are the trade winds produced and sustained?[16]

It is reasonable to assume that Forshey had Espy in mind when he wrote these questions. He was both familiar with and extravagantly supportive of Espy's storm theories. For example, Forshey wrote in his eyewitness account of a tornado at Natchez, Mississippi, published in Espy's *Philosophy of Storms:* "Had the heavens obeyed Mr. Espy's summons, and every wind rushed to the point he assigned it, and had the Omnipotent clothed him for the moment with his own dread powers, the demonstrations of his 'Philosophy of Storms' could not have been sublimer or more triumphant."[17] Although Espy was touring Europe in the summer of 1840 and no direct response to Forshey has been found, answers to all five questions figure prominently in Espy's book *The Philosophy of Storms,* published in 1841.

Espy's response to question 1 (on the principal cause of rain) brought out the main themes of his *Philosophy of Storms:* "Any cause which produces an upmoving column of air, whether that cause be natural or artificial, will produce rain, when the complement of the dew point is small, the air calm below and above, and the upper part of the atmosphere of its ordinary temperature. . . . All storms are produced by steam power . . . *in all very great and widely extended rains or snows, the wind will blow towards the centre of the storm* [causing up-moving columns of air]" (xxiv, 8). He considered electrical effects (question 2) to be secondary phenomena requiring more study: "How far electricity is concerned as a *cause* in the production of [meteorological phenomena] I am not prepared to say. As an *effect* of the sudden condensation of large masses of aqueous vapor in the air, it is pretty well understood. . . . But as all effects in nature become themselves causes, and as the utility of atmospheric electricity has not yet been discovered, we must be careful not to attribute on the one hand, to the action of electricity effects which are plainly accounted for by the dynamical agency necessarily resulting from the diminished weight of a suddenly formed cloud, nor on the other to deny that any effects whatever are produced by the immense quantities of electricity developed by the condensation of the vapor" (369–70). Question 3 (on the *primum mobile* of tornadoes) had a clear answer in Espy's system: "The tornado cloud form[s] only when the dew point is very high, that is when the steam power in the air is very great. . . . When the air near the surface of the earth becomes very much heated or very highly charged with aqueous vapor, . . . [an] ascending column [of air is produced] . . . and [is] kept up for a long time" (141, 307). Question 4 (on gyration) also generated an unequivocal response: "The effects of these phenomena indicate a moving column of rarefied air, without any whirling motion at or near the surface of the earth" (319). Even global phenomena were explained by Espy's theory; thus the answer to question 5 (on the trade winds): "The air in the torrid zone is about 80° in temperature at a mean, and as air in the frigid zone is about zero. . . . This will cause the air at the equator to stand more than seven miles higher from the surface of the earth to the top of the atmosphere than at the north pole. The air, therefore, will roll off from the torrid zone both ways towards the poles causing the barometer to fall in low latitudes. . . . This will cause the air to run in below towards the equator [from the northeast] and of course rise there" (xxii).

Redfield's responses to the same five questions, which fortunately exist in their original manuscript form, were given in reverse order. By beginning his answers with the role of the trade winds, Redfield revealed his emphasis on gravitational and kinematic forces that cause whirling motion. Redfield prefaced his remarks with the claim that "the inquiry for the '*quo modo*,' [is] the only true path which is to lead us to the '*causa verum*'." Concerning the trade winds (question 5), Redfield answered: "*By terrestrial gravitation, in connexion with the diurnal and orbital movements of the earth's surface and mainly, as apart from the real or supposed influences of equatorial heat and calorific rarefaction.*" On gyrations (question 4), Redfield advanced his own speculative theories in the quest for the *causa verum:*

> Was the question proposed in sole relation to great whirlwind storms or gales, I would again answer in nearly the above terms; omitting the allusion only to equatorial heat. To the question as here presented and confined to the class of narrower whirlwinds, I answer: *Terrestrial gravitation or atmospheric pressure:* aided subordinately and more or less by (1) the elasticity of the atmosphere; (2) the differences of density and pressure at the surface of the earth and at greater elevations; (3) *undue excess of calorific elasticity in the lowest stratum,* owing, in many cases, to a sudden diminution of temperature, by geographical transfer, in an overlying stratum; (4) calorific influences of condensation: the appar-

> ent effects of which last appear to have been by some [i.e., Espy], extravagantly misconstrued or overrated. . . .
>
> Of the more incipient origin or causes of gyration it may be remarked, that the moving masses of all aerial fluids and fluviatile bodies abound with the incipient rudiments or *nuclei* of gyration, in various stages of development, and that these rudimentary gyrations seem to be the necessary results of variable and unequal corpuscular motion.

On the *primum mobile* of tornadoes (question 3) Redfield answered: "This I deem to have been answered above"; on the role of electrical effects (question 2): "So far as I know or apprehend, electrical effects are mainly the incidental *results* of 'rains and winds,' or of changes of temperature and hygrometric condition in different portions of the air, which amounts to nearly the same." Finally, Redfield suggested that there were numerous causes of rain (question 1): "Solar evaporation and aerial absorption or vaporization, followed by a reduction of temperature; the latter depending, mainly, upon the various phenomena of geographical transfer, as connected with aerial stratifications, changes of elevation, absence of the sun's rays, and stormy or vorticular gyrations."[18]

Forshey visited Robert Hare in Philadelphia on September 3, 1840. Forshey noted in his journal, "The Doctor gives me an outline of his theory of Tornados and supports it by many plausible arguments."[19] But he did not record the details of the interview. Hare's responses are reconstructed from his published papers and begin with the answer to question 2 on electricity, which Hare had emphasized the most: "The atmosphere is an electric in a hollow globular form, . . . [located between] two oceans of electricity, of which one may be called the celestial, the other the terrestrial electric ocean. For an adequate cause of diversity in the states of the electric ocean, it must be sufficient to refer to the vaporization and condensation of water. . . . The proximity of a stratum of clouds electrified by the celestial ocean, must cause an accumulation of electricity in any portion of the terrestrial surface immediately subjacent; and by counteracting gravitation, cause a local diminution of atmospheric pressure which is, it is well known, a precursor and demonstrably a cause of wind and rain."[20] Then, in logical succession, Hare responded to question 3, on the *primum mobile* of tornadoes: "A tornado is the effect of an electrified current of air, superseding the more usual means of discharge between the earth and clouds in those sparks and flashes which are called lightning. I conceive that the inevitable effect of such a current would be to counteract within its sphere the pressure of the atmosphere, and thus enable this fluid, in obedience to its elasticity, to rush into the rarer medium above."[21] Hare supported Espy in his response to question 4, on gyration: "It would be inconsistent with the facts to suppose such a [gyratory] motion [in tornadoes], unless as a *contingent* result, and . . . it could only be a casual effect of the currents rushing toward the axis

TABLE 2.1. Forshey's questions and the responses of Espy, Redfield, and Hare

Forshey's questions	Espy's answers	Redfield's answers	Hare's answers
1. What is the principal cause of rain?	Any cause, natural or artificial, that produces an upmoving column of moist air unleashes the steam power of the atmosphere—especially excess heating and winds blowing toward the center of a storm.	Numerous causes: solar evaporation, aerial absorption or vaporization, cooling caused by geographical transfer, absence of the sun's rays, and vorticular gyrations.	Following Hutton, rain is caused by the overrunning and mixing currents of air of different temperatures.
2. What is the role of electricity in producing or modifying rains and winds?	Electricity is a secondary phenomenon caused by the sudden condensation of large masses of aqueous vapor in the air.	Electrical effects are the incidental result of rains or winds.	Electricity is the primary cause of wind and rain; electrified clouds counteract gravity and cause a local diminution of atmospheric pressure.
3. What is the *primum mobile* of tornadoes?	Tornadoes form only when the air has excess heat and moisture and the steam power of the atmosphere is very great.	Terrestrial gravitation or atmospheric pressure aided by a number of factors; *nuclei* of gyration inherent in all aerial fluids.	A tornado is the effect of an electrified current of air.
4. What generates and maintains a gyration in whirlwinds, spouts, and tornadoes?	The effects of these phenomena indicate *no* whirling motion at or near the surface of the earth.	Same as 3 above.	Gyratory motion is not an essential feature of tornadoes.
5. How are the trade winds produced and sustained?	The trade winds are caused by heating in low latitudes resulting in low pressure, rising air, and air running in below toward the equator.	The trade winds are caused by terrestrial gravitation and the diurnal and orbital movements of the earth's surface.	The trade winds are caused by the ascent of warm, moist air from the oceans near the equator and the afflux of air from neighboring regions.

of the tornado . . . gyration [is not] an essential feature of tornadoes."[22] Hare's response to question 1, on rain, explicitly follows James Hutton's theory (1788) of overrunning currents of air. This theory supposes that two currents of air of different temperatures, both nearly saturated with vapor, mingle together and that rain occurs because, at their mean temperature, all the vapor cannot be retained, and therefore the surplus is precipitated.[23]

> Consistently with the hypothesis which I suggested in my essay on the gales of the United States, the enduring rains which accompany those gales are attributed to the contact of an upper warm and moist current of air, with a lower current of the same fluid at an inferior temperature, and moving in an opposite direction. The air of the adjoining country first precipitates itself upon the surface of the Gulf [of Mexico], then that from more distant parts [the whole north-eastern portion of the North American continent]. Thus a current from the north-eastward is produced below. In the interim the air displaced by this current rises, and being confined to the high land of Spanish America and in part possibly by the trade winds, from passing off in any southerly course, it is of necessity forced to proceed over our part of the continent, forming a south-western current above us. At the same time its capacity for heat being increased by the rarefaction arising from its altitude, much of its moisture will be precipitated and the lower stratum of the south-western current, mixing with the upper stratum of the cold north-eastern current below, there must be a prodigious condensation of aqueous vapor.[24]

Finally, in reverse order to Redfield's emphasis, Hare advocated the caloric theory of the trade winds: "In the vicinity of the equator, where in consequence of the never ceasing ascent of warm moist air from the ocean, that afflux of this fluid from neighboring regions takes place, to which trade winds are attributed."[25]

Forshey's questions and the responses of Espy, Redfield, and Hare are summarized in Table 2.1 Given the radically different theoretical positions of the three controversialists, their diverse backgrounds and interests, and the lack of disciplinary norms in meteorology, it is not surprising that controversy was so widespread. Indeed, theoretical disagreements may be considered the norm.

The Case of the Brunswick Spout

> Meteorologists still do not agree on the precise mechanism by which tornadoes are initiated.
> —Louis J. Battan, *Weather,* 1985

> Do not forget we are to have a Whirlwind hunt and a magnetic intensity feast in the open air.
> —Henry to Bache, October 16, 1839, *JHP*

A violent "meteor"—was it a whirlwind? a hurricane? a spout? a tornado?—passed near New Brunswick, New Jersey, in 1835 and left a trail of fallen trees and damaged property in its path. Henry, Bache, Espy, Hare, Redfield, and Walter Johnson all visited the site of the storm, expressed opinions about the whirlwind nature of the phenomenon, and all but Henry wrote articles deducing the nature of the storm from its traces.[26] The "Brunswick Spout" became the center of controversy among competing storm theories and established the pattern for the investigation of tornadoes for the next forty years.

In the *Minutes* of the American Philosophical Society for 1835, Espy described "the late destructive storm at & near New Brunswick," from observations made by himself and Bache, "showing first that on the course of the storm, there was an upward motion in the centre and outward at the top, and inward below; . . . The facts proving this point were stated; & diagrams, made from angles taken with a compass were exhibited."[27] Espy found the most "perfect regularity" in the pattern of toppled trees lying across each other, which "to a careless observer"—or to one not familiar with Espy's theory—"might have indicated a whirlwind of confusion."[28] Espy's tornado theory is illustrated in Figure 2.1.

Bache reported the results of his survey of the storm's track and cautiously limited his conclusions to noting the lack of evidence for whirling motion at or near the surface of the ground, but agreed with Espy that there was "a motion *in all directions towards the storm.*" Although his measurements supported Espy's theory, his strongest claim was "to have first *plotted* a tornado by compass measurement; independent necessarily of theoretical bias in the results."[29] Figures 2.2a and 2.2b are Bache's diagrams illustrating damage to a farm and its buildings.[30]

Hare called the tornado a "vertical hurricane," which, he agreed with the accounts of "his friends" Bache and Espy, exhibited a central vertical current at its axis and a horizontal conflux of air toward that axis from the surrounding space. He noted the effects of gyratory forces but thought them incidental. In

Fig. 2.1. Espy's tornado theory. From James P. Espy, *The Philosophy of Storms* (Boston, 1841), 304.

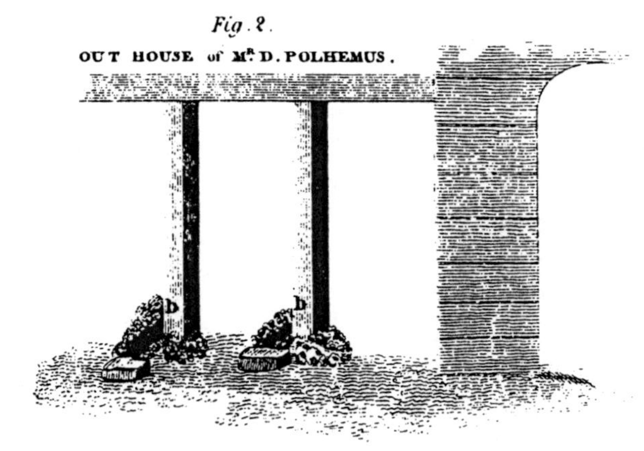

Fig. 2.2a. Outhouse of Mr. D. Polhemus.—Ground Plan, showing the effects of the Brunswick tornado on the outhouse of Mr. D. Polhemus. From A. D. Bache, "Notes and Diagrams, Illustrative of the Directions of the Forces Acting at or Near the Surface of the Earth, in Different Parts of the Brunswick Tornado of June 19, 1835," *Amer. Phil. Soc., Transactions* n.s. 5 (1837): Plate XXIII, following p. 417.

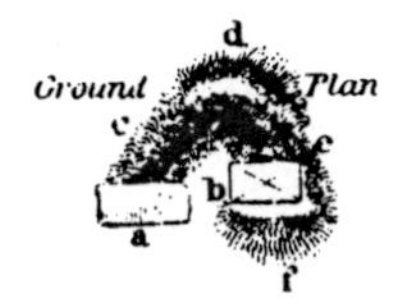

Bache's explanation: (a) A flat stone, on which the post b originally stood. (b) Present position of the foot of the post. (c) Groove made in earth to northward of a by the post. (d) A mound of manure heaped up at the end of the groove c. (e) A second groove north of east in direction. (f) A mound heaped up by the post b two feet high.

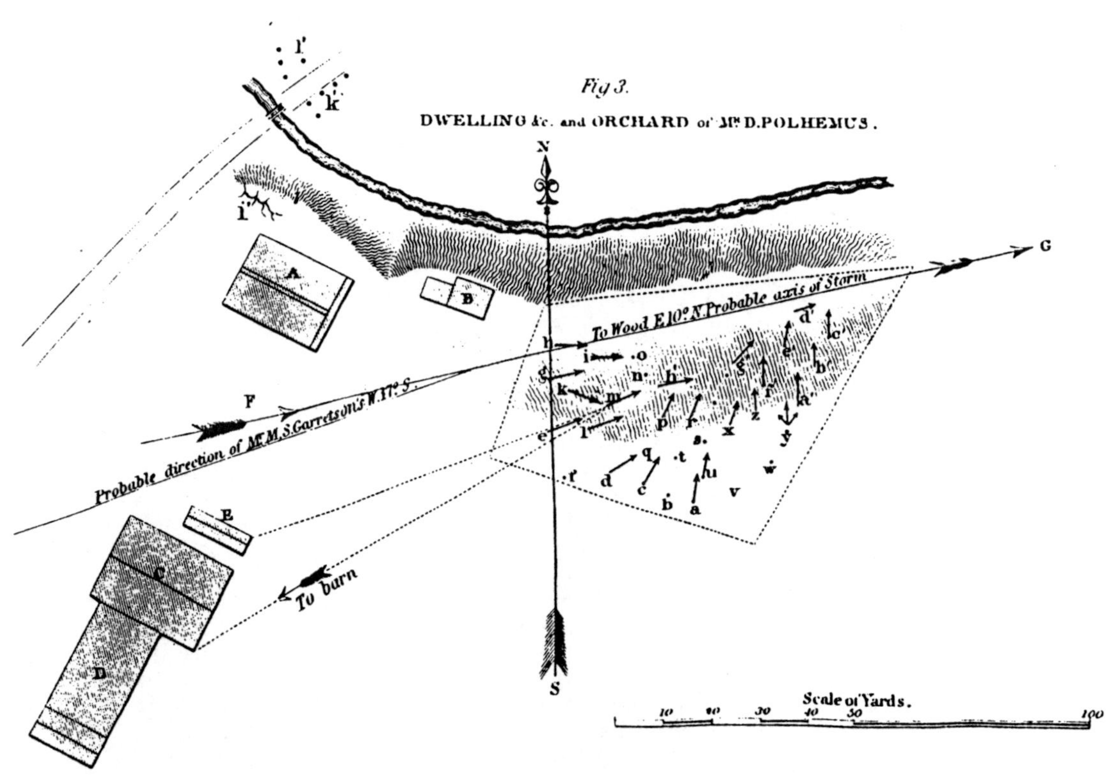

Fig. 2.2b. Grounds of Mr. D. Polhemus, illustrating the effects of the Brunswick tornado on the buildings and trees of Mr. D. Polhemus. From ibid.

(*continued on next page*)

support of his theory of electricity, Hare drew an analogy between the forces at work in the tornado and the phenomenon of "grêle electrique" in which pith balls are lifted by static electricity in a Leyden jar: so in the tornado, "water, trees, houses, haystacks and barns may be powerfully affected" (see Figure 2.3).[31]

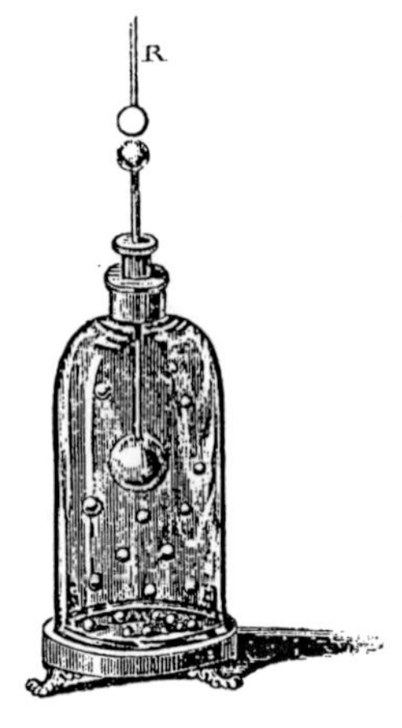

Fig. 2.3. Grêle electrique. From Robert Hare, "On the Causes of the Tornado, or Water Spout," *Amer. Phil. Soc., Transactions* n.s. 5 (1837): 377.

Redfield claimed that he had examined the track of the Brunswick storm twice: "I have observed on each occasion, numerous facts which appear to demonstrate the *whirling* character of this tornado, as well as the *inward* tendency of the vortex at the surface of the ground."[32] But Redfield presented no supporting evidence and was chided by Henry for this omission:

> I saw Mr. Redfield in New York and had some conversation on the subject of storms &c. I mentioned the fact of his remarks on the Brunswick tornado as

Fig. 2.2b (*continued*)

Bache's explanation: (A) Dwellinghouse of Mr. Polhemus, slightly injured. (B) Outhouses not injured. (C) Barn, shingles torn off, not many in number. (D) Shed shown in Figure 2. (E) Open work corn crib, not injured. (a) N. 12° E. Uprooted (the directions of the arrows indicate those of the trunks of the trees). (b) Tree uprooted, too crooked to determine its direction. (c) N. 6° E. (d) N. 27-1/2° E. (e) E. 20° N. (f) Tree standing near the fence. (g) E. 6° N. (h) E. 3° N. (i) E. 3-1/2° N. (k) E. 18-1/2° S. (l) E. 25° N. (m) E. 18° N. Tolerably straight. Shingles from barn found at the foot of m. Southeast angle of the barn bears W. 30° S. (n) Tree standing. (o) Low tree standing: small. (p) N. 32° E. (q) Broken, not uprooted. (r) N. 27° E. Dead: bushy. (s) Standing. High and stout. (u) N. 24° E. (v) Plum tree near, standing. (x) N. 5° E. (z) N. 9° W. Small, firmly rooted in north side. (y) Thick and bushy: broken off into three parts, the smallest of which points west of north, the next north, and the largest east of north; the bark is stripped off below the fracture. (a′) N. 4° E. (b′) Same general direction as a′. (c′) N. 2-1/2° E. Small roots. (d′) E. 22° N. Very large roots. (e′) N. 10° E. (f′) N. 10° W. A small tree near this, in the same row is untouched. (g′) E. 35° N. (h′) E. 3° N. Three trees at the south end of the row f′ g′ are standing. (i′) A very large black cherry tree, uprooted, and lying nearly parallel to the house. (k′ l′) groups of willows, the limbs and branches of which are torn off, and thrown to southward and eastward.

not being according to the proper philosophical spirit. He said that at the time he wrote the article he had some misgivings in reference to it but that his time was so much occupied that he could not go into the facts in detail. He appears to have somewhat modified his views and now says (I think for the first time to me) that there is an inward and a spiral upward motion of the air or in other words his views now appear to be identical with those of Franklin as given in his letters particularly the one for Feby 4th 1753.[33]

Plucked Chickens and Projectiles

Some investigators noticed that the violent winds accompanying tornadoes killed barnyard fowl and stripped them of their feathers. Elias Loomis, astronomer and meteorologist at Western Reserve College, examined the results of a "hurricane" (tornado) in Stow, Ohio, in 1837. His rather gruesome account of general mayhem and destruction stated: "There were . . . geese, hens and turkeys, in considerable numbers [lying dead among the ruins], and several of the fowls were picked almost clean of their feathers, as if it had been done carefully by hand."[34] Espy confirmed this phenomenon in a letter to Loomis: "A gentleman told me that he saw turkies walking about naked after the passage of a tornado which occurred many years ago; but he added that they soon died. This was the first intimation I ever had of the fact, and he told it to me as a strange phenomenon, which came under his own observation. I know not how to explain it. There is no doubt also of persons having been stripped entirely naked in the same way without being even hurt *seriously.*"[35]

From this evidence Loomis attempted to calibrate the speed of tornado winds. He used a six-pound cannon charged with five ounces of powder and added a freshly killed chicken for a ball. Loomis reported:

> As the gun was small, it was necessary to press down the chicken with considerable force, by which means it was probably somewhat bruised. The gun was pointed vertically upwards and fired; the feathers rose twenty or thirty feet, and were scattered by the wind. On examination they were found to be pulled out clean, the skin seldom adhering to them. The body was torn into small fragments, only a part of which could be found. The velocity is computed at five hundred feet per second, or three hundred and forty one miles per hour. A fowl, then, forced through the air with this velocity, is torn entirely to pieces; with a less velocity, it is probable most of the feathers might be pulled out without mutilating the body.[36]

Espy v. Redfield, 1834–1841

The protagonists in the storm controversy invested an inordinate amount of time attacking each other in print. The first flurry of polemical activity consisted of nine publications by Espy and Redfield in which each author criticized both the theoretical position and the personal character of the other. These publications, listed in Table 2.2, appeared between 1834 and 1841 and constitute the core of the first phase of the controversy. Among Espy's supporters were Bache, Henry, and, during Espy's European tour, a committee of scientists from the French Academy. Redfield's theory was embraced by Denison Olmsted in the United States and by Sir William Reid and Sir John Herschel in Britain.

Espy launched an initial attack against Redfield early in 1834, but it was in connection with atmospheric tides, not storms.[37] In particular, Espy questioned Redfield's assertion that in large portions of the Pacific Ocean, the oceanic tides are "exempt from lunar influence" and that atmospheric tides resulted from the "rotation of the earth, and its connexion with the solar system." Espy dismissed the first assertion and corrected the latter "mistaken notion" by employing the diurnal heating of air by the sun to explain the effects.[38] In reply, Redfield gave ground to Espy on tides but launched an offensive on whirlwind storms. He acknowledged that his statements on tides contained "a partial inaccuracy" and perhaps were not "sufficiently guarded." Redfield shifted attention to his whirlwind storm theory and attacked the "received theories" founded on vague generalizations or unproven and untenable

TABLE 2.2. Espy v. Redfield, 1834–1841, the first group of polemic exchanges in the American storm controversy (see bibliography for full citations)

Espy v. Redfield (1834–1841)
1. "Notes of an Observer—Meteorology," *Journ. Frankl. Inst.*
2. "Essays on Meteorology": Nos. I and II, "Theory of Hail," No. III, "Examination of Hutton's, Redfield's and Olmsted's Theories," No. IV, "North East Storms, Volcanoes, and Columnar Clouds," *Journ. Frankl. Inst.*
3. "Examination of Col. Reid's Work on the Law of Storms," *Journ. Frankl. Inst.*
4. *The Philosophy of Storms.*

Redfield v. Espy (1835–1839)
1. "On the Evidence of Certain Phenomena," *Journ. Frankl. Inst.*
2. "In Reply to Mr. Espy, on the Whirlwind Character of Certain Storms," *Journ. Frankl. Inst.*
3. "On the Courses of Hurricanes; with Notices of the Tyfoons of the China Sea, and Other Storms," *Amer. Journ. Sci.*
4. "Remarks on Mr. Espy's Theory of Centripetal Storms, Including a Refutation of His Positions Relative to the Storm of September 3rd, 1821: with Some Notice of the Fallacies Which Appear in His Examination of Other Storms," *Journ. Frankl. Inst.*
5. "Further Notice of the Fallacies of Mr. Espy's Examination of Storms," *Journ. Frankl. Inst.*

hypotheses "from the time of Halley downwards." Confronting the caloric theory and theorizing that mechanical gravitation explained the wind currents of the globe, Redfield lamented the "grand error into which the whole school of meteorologists appear to have fallen in ascribing to heat and rarefaction the origin and support of the great atmospheric currents which are found to prevail over a great portion of the globe. . . . Sir John [Herschel] . . . has erred, like his predecessors, in ascribing mainly, if not primarily, to heat and rarefaction, those results which should have been ascribed solely to *mechanical gravitation,* as connected with the rotative and orbital motion of the earth's surface."[39]

With the initial sparring concluded, Espy began to produce his "Essays on Meteorology," which expounded, in grand Newtonian fashion, "a law in meteorology" that was to bring unity to the science by explaining "at once, with a simplicity which nothing can equal, all the *seven* [meteorological] phenomena" (rain, hail, snow, waterspouts, landspouts, winds, and barometric fluctuations).[40] Espy's third essay in this series was an attack against the traditional theory of rain propounded by Hutton, Olmsted's Huttonian theory of hailstorms, and Redfield's whirlwind theory of storms.[41] Espy reviewed Redfield's observations on whirlwinds during the Connecticut storm of 1821, which showed winds blowing in opposite directions forty miles apart. He then calculated the centrifugal forces that would be generated in a whirlwind and concluded that "it would soon be destroyed by its outward motion, unless some mighty cause exists, of which we have no knowledge, to generate new motion in the air."[42] Moreover, frictional forcing of the gyration by the West India Islands and the American coast could not account to Espy's satisfaction for the whirlwind's persistence in time, rainfall patterns, or extremely low pressure.

In reply, Redfield noted the courtesy of Espy's earlier remarks and decided "to waive the objections which I entertain to any controversial discussion." Redfield then went on to accuse Espy of misquoting him at critical places, confounding the "FACTS" established by observers, and appropriating his meteorological generalizations, presenting them as the original results of the Joint Committee.[43] Clearly the storm controversy was heating up.

Redfield gained an important ally and a lifelong friend in William Reid, a lieutenant-colonel in the British engineer service. In 1831, Reid witnessed firsthand the destructive power of a great hurricane in Barbados. When he learned that Redfield had published on the theory of rotatory storms, he wrote Redfield a letter of encouragement: "I have written to Sir John Herschel on the subject . . . as well as to others; and I have little doubt but that you will now soon have the great satisfaction of seeing your labours successful in proving the true nature of storms." He also offered to confirm and extend Redfield's studies by gathering observations from lighthouses along the coast of Europe. Before Redfield could respond, Reid wrote again to the effect that he had found copies of Espy's papers in London and would read them attentively.[44] Redfield was delighted that the inquiry into the true nature of storms was beginning to excite interest on Reid's side of the Atlantic.[45]

Redfield maintained a strong rhetorical commitment to Baconianism and used it defensively throughout the storm controversy. In his first letter to Reid, Redfield attempted to consolidate his position and dismiss the opposition: "I suppose that our method of investigations, like that which Mr. Whewell has adopted in relation to tides, is in accordance with the strict principles of the Baconian or inductive philosophy. It is safe therefore, to disregard all loose theories, and to adopt a system of meteorological physics which is grounded on direct observations." He soon revealed, however, the "idols" in his own approach: "I am not sure that Prof. Loomis comprehends our national system of storms, but he is a very thorough and persevering young man, and by the aid of your book I think he will be fairly *indoctrinated.*" He later discussed his aversion to theory and hypothesis: "A majority of readers are content with nothing short of a theory, or some hypothetical pretensions to one. While all join in praise of the Baconian system of investigation, there are but few that adhere to its rules, and still fewer that encourage this adherence in others. I am alluding however to the great public, and not to the more accurate class of scientific inquirers." Redfield subsequently informed Reid that in regard to Espy, "I have contented myself thus far with an attitude towards him which is strictly defensive," but he characterized Espy as "an ardent and at the same time an amiable man," whose interpretation of the facts could not be sustained and whose conclusions were at times "exceedingly vulnerable if not absurd." When Redfield indicated that he was ready to do battle over the nature of the Brunswick tornado, Reid advised him to spend his time more productively: "I do not understand Mr. Espy's papers. I hope you will not lose your time by taking any portion of it up in Controversy but employ what [time] you can spare in working out what you have begun."[46] Redfield did not heed this advice.

On August 20, 1838, Reid dashed off a letter in haste and great excitement telling Redfield of their victory before the British Association:

> Knowing how much it will gratify you to learn that Sir John Herschel's section of the British Association was opened today by a paper which I read on the subject [of] the Law of Storms. I make no delay in acquainting you with this. The reception of this paper was very gratifying to me as I am sure it will be to you. I laid before the meeting an [outline] of Storms and also the work I have printed, in which the Chapter II is headed *Redfields Storms.* The work consists of twelve chapters & extends to above 400 pages. In about a fortnight or 3 weeks it will be ready for publication and I shall have the pleasure of sending you a Copy. Professor Bache was present and very handsomely thanked me for the way I had introduced the subject and his Countryman. . . . I have felt, ever since I read your first pamphlet that the work was deeply indebted to you and I endeavored to convey that impression at the meeting of the British Association.[47]

Bache's remarks at the meeting also included a request that Reid examine Espy's rival theory of storms, which he found "entirely in accordance" with observations he had made on the track of the Brunswick spout in June 1835. In that storm, Bache "had found no evidence of a whirling motion at the surface of the ground."[48] Sir J. F. W. Herschel, the president of the section, "hailed the communication of Col. Reid," publicly thanked Redfield, and criticized Espy. He thought that Reid, though judicious as an observer, should advance some theoretical speculations: "In the present assembly, a theory, if it served no better purpose, helped memory, suggested views, and was even useful by affording matter for controversy, which might produce brilliant results, by the very collision of intellect." As if to demonstrate his point, Herschel, in a flight of fancy, proposed that circumstances connected with spots on the sun forcibly impressed his mind with the idea of "*tornadoes in the solar atmosphere,* which, by scattering and opening out the luminous superficial matters, laid bare the opake and dark mass beneath." He further speculated that the atmosphere of the sun might well contain features analogous to the "trade winds" of earth.[49] After such a coup, Redfield found Bache's position (in support of Espy) to be one of "peculiar delicacy," and he prepared to "[assail] the meteorologists of Philadelphia in their own entrenchments."[50]

Espy, now on the defensive, responded to the criticism of Herschel and reworked the data in Reid's *Law of Storms* to sustain his centripetal theory.[51] Espy thought that Herschel's comments before the British Association stemmed from a lack of proper information: "I am sure that as soon as Sir John shall see my papers on the subject, he will see and confess that the objection has already been fully answered. If I am right in this matter, Sir John owes it not merely to me, but to the cause of science, on a point which he acknowledges to be of immense importance, to come out and correct his mistake; for such is the weight of his name, that many will not think it worth while to examine a system which has been condemned by Sir John Herschel."[52] Espy's primary strategy, however, lay in undercutting the empirical foundations of the rotatory theorists. In particular, Espy wanted to rework the Connecticut storm of 1821, which had led Redfield to see storms as "great whirlwinds" and had inspired Reid's book.[53] He prepared revised analyses and maps of Reid's data which demonstrated that the observations supporting the rotatory theory of storms also could be used to prove his centripetal theory. By this strategy Espy hoped to discredit Reid's book, which Redfield was distributing in the United States.[54] A widely read article by Charles Davis, however, called into question the reliability of Espy's charts.[55]

During the early years of the storm controversy, relations between Espy and Redfield were strained but remained formally polite. In 1836 Espy invited Redfield and other scientists to Philadelphia to observe the activities of the Franklin Kite Club of which he was chairman and to "compare notes" on storms.[56] Redfield did not attend. Espy asked Redfield to insert a circular letter calling for volunteer observers in one of the New York newspapers and invited

Redfield to meet him at the site of the Newark tornado to "go over the ground together." Redfield declined, citing pressing business engagements.[57] In May 1837 Redfield stopped in Philadelphia, ostensibly to close the "narrow . . . ground of difference" between his position and that of Espy. He met with Hare and others and called on Espy but did not find him at home.[58]

After Espy's attack on the whirlwind theory, however, Redfield's initial politeness turned to rage, and he voiced his grievances to Henry: "It is a matter of sincere regret with me that our friend Espy is not content with his own position, but thinks it necessary to renew his attacks upon my statements and conclusions relating to the storm of 1821, as you have seen in the March No of the Journal of the Franklin Institute. To this I have felt it necessary to reply and to speak with some freedom as well as candor of his general course of proceedings in this matter, all which I was desirous to avoid."[59] Hoping for a final resolution of the whirlwind controversy in his favor, Redfield published an article in British and American journals which contained a key section entitled "Test of Mr. Espy's Theory."[60] The test was an attempt to distinguish between radial and rotatory storm theories by observing the pattern of wind shifts in an Atlantic hurricane moving along the coast of the United States. Espy responded by suggesting that radial storms may be highly eccentric ellipses—not circular.[61]

Then in rapid succession Redfield published two articles against Espy. In the first, he lamented Espy's "erroneous inductions" regarding the New Brunswick tornado and argued that Espy's tornadoes seemed to blow inward only when the facts were conflated and not properly arranged in temporal sequence. Redfield referred to his rival's "injudicious support" from highly respectable sources—a reference to Bache, Henry, Hare, and Walter Johnson—and the "favor and guardianship of the Philadelphia press."[62] In the second article, Redfield defended his analysis of the storm of August 17, 1830, and accused Espy of manipulating the data to confuse the issue: "I had long been curious to know, how Mr. Espy would deal with the facts which pertain to this storm . . . but must now acknowledge that I did not anticipate so complete an evasion of all the distinguishing points at issue, and so barren an effort at confusing and mystifying the most distinct phenomena of this storm, as is manifested in his present examination." Then Redfield examined the Joint Committee's reports on the storm of March 17 and 18, 1838, and found that their conclusions did not necessarily support Espy's theory because the observations covered too large an area with insufficient detail to resolve the features of the storm: "[Should the facts of this storm] be adduced in favor of Mr. Espy's theory, I would say, that in this, as in some former cases, the field of action of the whirlwind storm will have been in part mistaken. I would also remark, that the points at issue, do not relate to the common and often irregular winds, which, in different localities, accompany a general fall of rain and snow; or which sometimes attend the progress of a

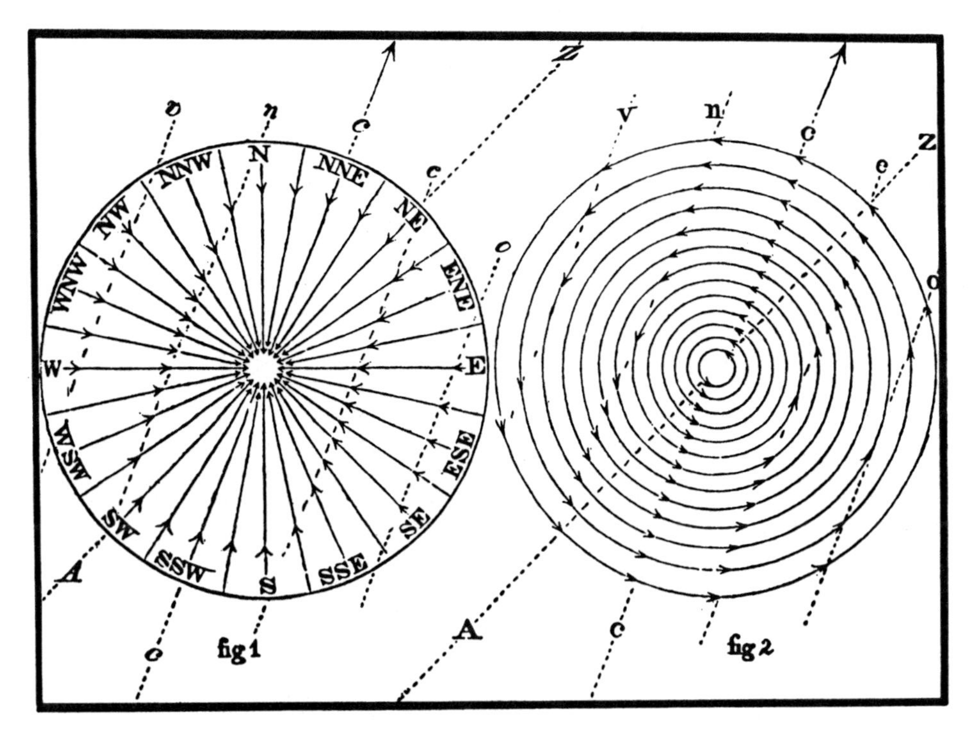

Fig. 2.4. Redfield's second "test" of Espy's theory. From *Journ. Frankl. Inst.* n.s. 23 (1839): 326. The circle on the left represents Espy's centripetal theory and the circle on the right, Redfield's centrifugal theory. Possible paths through the storm are indicated as v, n, c, e, and o. The test refers to observed and predicted wind shifts as the storm passes a stationary observer. A path through the center of the storm, A–Z, gives a characteristic wind shift of NE–SW for Espy and SE–NW for Redfield.

whirlwind storm, exterior to its limits."[63] Figure 2.4 is a diagram published by Redfield which renews the challenge to "test" Espy's theory of hurricane winds. Espy did not publish a new answer to this attack but republished his original response in *Philosophy of Storms.*[64]

Espy accepted Redfield's challenge, again arguing that the test applied only to round storms. Moreover, marginalia in Henry's library copy of *Philosophy of Storms,* undoubtedly written by Espy, show that a critical "Error," a typographical one, had crept into the text: S.E. should read N.E., and N.W. should read S.W. (see Figure 2.5).[65] In the sketches at the bottom of the page, Espy showed that a rotary storm could indeed exhibit wind shifts from N.E. to N.W. and that an elongated centripetal storm passing an observer along the line c–d would produce wind shifts from N.E. to S.W. Rather than conceding the point to Redfield and risking defeat on other theoretical points, Espy modified his theory to include large storms in the form of elongated ellipses or long lines of minimum pressure, maintaining that in-rushing winds still best explained

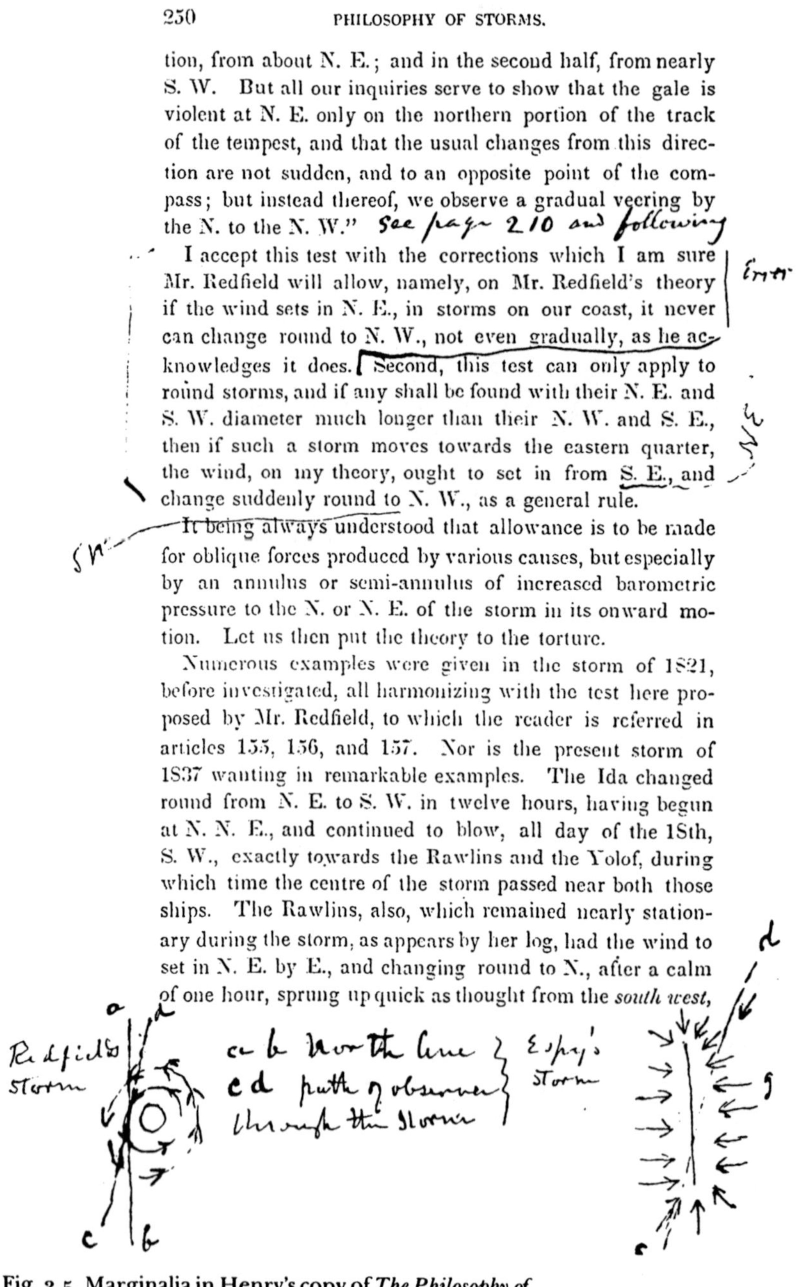

250 PHILOSOPHY OF STORMS.

tion, from about N. E.; and in the second half, from nearly S. W. But all our inquiries serve to show that the gale is violent at N. E. only on the northern portion of the track of the tempest, and that the usual changes from this direction are not sudden, and to an opposite point of the compass; but instead thereof, we observe a gradual veering by the N. to the N. W." See page 210 and following

I accept this test with the corrections which I am sure Mr. Redfield will allow, namely, on Mr. Redfield's theory if the wind sets in N. E., in storms on our coast, it never can change round to N. W., not even gradually, as he acknowledges it does. Second, this test can only apply to round storms, and if any shall be found with their N. E. and S. W. diameter much longer than their N. W. and S. E., then if such a storm moves towards the eastern quarter, the wind, on my theory, ought to set in from S. E., and change suddenly round to N. W., as a general rule.

It being always understood that allowance is to be made for oblique forces produced by various causes, but especially by an annulus or semi-annulus of increased barometric pressure to the N. or N. E. of the storm in its onward motion. Let us then put the theory to the torture.

Numerous examples were given in the storm of 1821, before investigated, all harmonizing with the test here proposed by Mr. Redfield, to which the reader is referred in articles 155, 156, and 157. Nor is the present storm of 1837 wanting in remarkable examples. The Ida changed round from N. E. to S. W. in twelve hours, having begun at N. N. E., and continued to blow, all day of the 18th, S. W., exactly towards the Rawlins and the Yolof, during which time the centre of the storm passed near both those ships. The Rawlins, also, which remained nearly stationary during the storm, as appears by her log, had the wind to set in N. E. by E., and changing round to N., after a calm of one hour, sprung up quick as thought from the *south west,*

Fig. 2.5. Marginalia in Henry's copy of *The Philosophy of Storms,* 250. Courtesy of the Bell-Henry Library, Smithsonian Institution.

the observed wind shifts. Smaller-scale objects such as tornadoes and convective updrafts were still fully explained as circular phenomena with centripetal wind patterns.

Espy's Lyceum Circuit

> I need a scientific revival. Here a few years ago we had Courtenay & his astronomy, Espy & his meteorology, Rogers & his geology, you and I electrically and magnetically engaged, meeting to rub each other up. Now where is the club? C dealing in bricks & mortar for a livelihood at Boston, Espy lecturing for the same, Rogers immersed so fully in the details of his business as to neglect communion.
> —Bache to Henry, February 13, 1840, *JHP*

In 1837, Espy began to participate in the popular lyceum movement, offering lectures to the public on his "philosophy of storms" and on rainmaking.[66] Espy's mixture of sound scientific principles and visionary proposals appealed to a wide audience and brought him recognition and income, as well as new volunteer observers. Thus began an intense period of self-promotion and constituency building for Espy. During this period, he experienced a temporary estrangement from his scientific friends. Espy's partial itinerary, reconstructed primarily from newspaper accounts, is shown in Table 2.3.

In May 1837 the American Lyceum passed a resolution requesting its individual lyceum members to make meteorological observations and send them to Espy. On January 23, 1838, the Pennsylvania Lyceum presented a memorial to the state assembly, "praying that James P. Espy may be appointed Meteorologist of the State of Pennsylvania."[67] And on May 2, Espy and the Philadelphia Lyceum petitioned the federal government for aid. This petition contained a proposal for widespread meteorological observations, guidelines

TABLE 2.3. Stops on Espy's speaking tour, 1837–1841

Year	Stops
1837	Philadelphia, American Lyceum (May 4).
1838	Harrisburg, State House Chambers (Dec. 27).
1839	New York City (June 22); Nantucket, Athenaeum (July 13); New York City (Dec. 18); Boston (Dec. 30).
1840	Boston, Hall of Representatives (Jan. 29); Phillips, Massachusetts, Tremont Hall (Feb. 10); Mobile, Alabama (March 14–15); Harrisburg, State House Chambers (May 22–23); Natchez, Mississippi (June?). European Tour: departed Philadelphia June 6, 1840; lectured in Britain, including two papers at the British Association, Glasgow (September); visited French Academy; departed Le Havre (April 6, 1841).
1841	Portland, Maine (Oct. 23); Philadelphia (Dec. 22).

from the work of the Joint Committee, and memorials of the Pennsylvania and American lyceums signed by the luminaries of science in Philadelphia and Boston.[68] In Harrisburg, a committee of state legislators was established to examine Espy's method of producing rain.[69] The committee also received a small appropriation to evaluate his theory of storms.[70] Espy petitioned successfully on two occasions for the use of the Hall of Representatives for his public lectures.[71] His major legislative rebuttal came when the Senate failed to approve a House amendment providing a reward of up to $50,000 for Espy's rainmaking. The amendment, annexed to a bill regulating "hawkers and pedlars," stipulated that to qualify for the reward, Espy was to keep the Ohio River navigable from Pittsburgh to the Mississippi River during the summer season. "Magnificent Humbug," said the *Genesee Farmer.*[72] The Senate's vote was 55 to 24 in the negative.

While Espy lectured the public on rainmaking and petitioned the government for support, his scientific friends turned away—if only for a time. Bache wondered about the "strange course recently taken by my friend Mr. Espy";[73] Sears C. Walker worried over Espy's "mode of making announcements of meteorological phenomena" and remarked candidly about the unrealized expectations of "that worthy but too sanguine meteorologist."[74] Henry believed Espy to be "a very honest man and ardently attached to science" but sometimes exhibiting "a want of prudence."[75] Redfield kept a jealous eye on Espy's lyceum tour and informed Reid of the latest developments: "[Mr. Espy's] late movements, as I am informed, have given offense to the *savans* of Philadelphia, to whose friendship & support he is mainly indebted for the consideration in which his opinions have been held, as few persons have been able to investigate this matter for themselves."[76] The disaffection of the Philadelphians probably came from several sources, one being the wording of the memorial to the government of May 2, 1838. In the memorial Espy claims that the efforts of the Joint Committee "have not been successful," that storms can now be predicted while "yet several hundred miles off," and that "the plan proposed is the only one from which success in making discoveries can be anticipated." Espy's plan for producing rain by setting large fires and his appeal to the general public for support undoubtedly added to the tension.

According to Redfield, Espy's New York lectures were thinly attended, possible because of a saturation of popular lectures on meteorology and other subjects.[77] Nevertheless, many positive popular reviews of his lectures appeared in newspapers around the country. The New York papers reported: "Professor Espy, beyond all question is a man of science. He is enthusiastic but no quack—no pretender—and we have not a doubt would create a profound impression by his theory in any scientific circle of the world. The revolution in society professor Espy will work out, if he establishes his science, is worthy of consideration. Indeed he will be deserving of the Homeric epithet of the cloud-compelling Jupiter . . . if professor Espy can do what he thinks can be done, *make a storm,* at once, man is almost master of the world."[78] The *Bangòr Advertiser* publicized Espy as "one of the *best* lecturers that ever appeared—living or dead." Boston's *Colonial Sentinel* reported a debate between Espy and Denison Olmsted, a staunch supporter of Redfield: "[Espy] stood like a rock in the midst of his storms; and though his manner was firm and undaunted, it was dignified and respectful to his opponents."[79] Olmsted doubted the practicality of Espy's project for artificial rainmaking and lectured publicly against it.[80] He also opposed Espy's theory that shooting stars and aurora were related to the humidity of the air and had "one common origin . . . low in the atmosphere."[81] An anonymous letter to the editor of the *Boston Evening Mercantile Journal*—could it have been Espy himself?—opined eloquently if not unerringly about the Olmsted debate: "The voice of ridicule is at length hushed. . . . The name of Espy may hereafter stand as high upon the list with Galileo, Harvey, Franklin and those other names precious to science and humanity." Two days later, Redfield, who had been keeping close tabs on Espy, sent a letter to the same paper criticizing Espy for confusing noon and sunset observations in charting a storm.[82] Redfield's explanation of Espy's good press involved a combination of Espy's persistence and the editors' patience and naïveté: "In order to account for the present tone of our newspaper press toward him, you must understand that he lays regular siege to the corps editorial and plies them with as many of his puffing missives as their credulity or good nature will endure. A procedure which no one takes the trouble to expose or counteract."[83]

Americans Tour Europe

Between 1836 and 1840, several prominent American scientists with meteorological interests—Bache, Hare, Henry, Loomis, and Espy—visited Europe to "offer their intellectual wares."[84] Espy toured England and France in 1840. Through the earlier trips of his friends Bache and Henry, he made sure that his meteorological theories were well represented abroad. As Bache was about to depart for Europe in 1836, he received a request from his old friend Espy: "Please to present to Dalton, Faraday, Brewster, Forbes, Airy, ApJohn, Daniell, Whewell, Scoresby or if not convenient to them to any others you may find interested in the science, those essays of mine already published on the Theory of Rain, &c. Please to present them also to Gay Lussac, Arago and others in France, telling them that the theory is a result of an investigation commenced with a hope of obtaining the prize offered by the French Academy for a true Theory of Hail. You *only,* know that this theory was written out in time to obtain this prize, but not sent."[85]

Bache recorded in his European diary that, indeed, he staunchly advocated Espy's theories during a visit with John Dalton, Britain's premier meteorologist, chemist, and source of inspiration for Espy's *Philosophy of Storms:* "By appointment took tea with Dr. Dalton to make a farewell visit. . . . The subject of meteorology introduced gave me an opportunity to cite one attempt at Philad. to get correspondents & of our results. Especially of Mr. Espy's law that the wind blows towards a great rain. He thought to invalidate this by describing the action of a gust when the wind blows inwards. This I had guarded against & pointed it out. He would not admit the displacement of air by falling rain to be any thing in explaining the phenomenon. I was so intent upon this idea as hardly to catch his." Bache further recalls that the discussion included the Joint Committee's work in tracing thunderstorms and barometric fluctuations from Tennessee to Maine. At parting, Dalton offered Bache the choice of any of his works, pointing out a volume of the Manchester *Transactions* which contained many of his original papers: "This of course I declined & by preference selected a volume of his last edition of meteorological essays. In this is his autograph."[86]

Joseph Henry's well-documented European tour demonstrates his interest in meteorology and provides an opportunity to make some observations about the status of meteorology in America and Europe.[87] To promote the accomplishments of American science and prepare for his European tour, Henry either sent or promised to hand deliver copies of the New York regents' reports to many prominent scientists in Europe. In 1834 and 1836, for example, he exchanged magnetic needles and meteorological reports with James Forbes of Scotland, perhaps as a way of thanking him for his British Association report "on the recent progress and present state of meteorology," which contained a laudatory reference to the meteorological data collected by the academies of the state of New York.[88] Like Bache, Henry also promised Espy that he would "attend to all [his] affairs in Europe."[89]

After a voyage of twenty days, Henry arrived in Plymouth, England, on March 14, 1837, and remained in England, primarily in London, for two months. Soon after his arrival, Henry was given a tour of Somerset House, where he examined the meteorological instruments of the Royal Society. His critical evaluation was that their location was "badly chosen for observations on temperature." Henry met John Frederic Daniell in his laboratory and lecture room in one wing of Somerset House. Their conversation explored one of the newest theoretical developments in American meteorology—the storm controversy between Redfield and Espy. According to Henry, Daniell "expressed himself much pleased with the views of Mr. Redfield—was unacquainted with the theory of Mr. Espy." The conversation soon turned to taking meteorological observations, and Daniell used the opportunity to compliment the American efforts and to impress on Henry the need for more systematic projects like the Meteorological Society of Mannheim: "Speaking of Meteorology he stated that America was setting the example to the world in reference to observations. He also directed my attention to the labours of a society established many years ago in Man[n]h[e]im which furnished instruments and directions to individuals in almost every part of the world. He regretted that something of the kind was not done at the present time."[90]

Scientific instrumentation was also one of Henry's concerns. At a London gas works, Henry took special note of the self-registering devices used to measure gas flow and noted that "some of them might with success be introduced into the Science of Meteorology." On his second visit with the inventor Charles Wheatstone, Henry was told that the galvanic telegraph "can ring a bell at a distance of 13 miles and transmit any intelligence through a circuit of that length."[91] When Henry visited Charles Babbage, one of his early role models on topics such as observing and handling data, he was shown the analytical engine and noted, "It will also be of great use in calculating the mean results of astronomical and meteorological observations."[92] Clearly, the possibility of automated collection, communication, and reduction of meteorological observations was on Henry's mind during these encounters in Britain.

On May 10 Henry left for Paris. In France he met François Arago, Jacques Babinet, François Magendie, C-S-M Pouillet, and Joseph Louis Gay-Lussac and attended the lectures of the latter two.[93] The Italian exile Macedonio Melloni showed Henry his thermomultiplier or thermal telescope, which Henry subsequently applied to meteorological purposes.[94] Henry also called on Adolphe Quetelet in Belgium but did not find him in.[95] Henry spent August in Scotland, where he visited the shop of Alexander Adie, "inventor of the sympisometer," and examined a chart of the weather projected so as to show the curves of the thermometer and the barometer for each of ten years.[96] Henry called again on James Forbes, whom he had met in London in the spring, but discovered he was on a two-year trip to the Continent to meet scientists and make meteorological observations.[97] Most significantly, Henry met Sir David Brewster. In a discussion on Baconianism, both men repudiated it as a valid scientific methodology. Henry noted in his journal: "Sir David's opinion of Bacon's rules for making investigations the same as my own. . . . No working man of science advocates Bacon's method."[98] Brewster argued "energetically and often" that the philosophy of Francis Bacon had no influence on either the history of science or its practice, although he admitted that nonscientists often thought the contrary.[99] Here Henry departed from T. R. Beck's definition of meteorology as the "natural history of the atmosphere." Although observation and experiment were basic processes for investigating the workings of nature, induction and deduction (analysis and synthesis) verified by prediction and further observational and experimental tests were necessary to discover natural laws. When scientists in the mid-nineteenth century invoked the inductive method, it was merely another way of saying *scientific method.* For example, when Henry stated that "the great system of modern physical science" had been established by the inductive method, he was quick to add that "few physical investigation[s] could be made without the adoption of some provisional hypothesis. . . . Science properly so called does not consist in an accumulation of mere facts but in a *Knowledge of the Laws of Phenomena*."[100] If Henry and his scientific colleagues were not Baconian fact gatherers, neither were they necessarily Humboldtians. Although on occasion his researches paralleled Humboldt's, Henry never explicitly acknowledged such inspiration. True, Humboldt's observations of declination, dip, and magnetic intensity (1798–1803) were the first of their kind in America and elsewhere, but by the 1830s he was not in the scientific mainstream. As John Cawood has argued, "The Humboltean scheme [the derivation of numerical laws by observational methods] became more and more isolated from the mainstream of science as methodologies of hypothesis replaced inductivist approaches." Perhaps Walter Cannon, who suggested Humboldt as an alternative American role model, jumped rather hastily from the frying pan of Baconianism into the caldera of Humboldtianism.[101]

In September Henry attended the British Association meeting in Liverpool, and on October 2, 1837, he left England. His tour reduced the mystique of Europe and its savants. In addition to the "poverty and barbarism" he found in European society, he discovered certain weaknesses in its learned community. Commenting on the library of the Royal Society of Edinburgh, Henry thought it "not better" (and perhaps not as good) as that of the American Philosophical Society. He and Bache had low opinions of the noted scientific instrument maker John Frederick Newman. Following his visit to the Continent, Henry commented on "the little that is now doing in Paris in the way of original science." This is not to say that he was not impressed by European scientifics; he was just not overawed. He encountered controversy and antagonism from some members of the British Association and concluded that it was "rather a hazardous affair for an American to make a communication unless he be well known and his communications addressed to the Section in which his friends are found."[102] On the positive side, Henry was welcomed as a scientific collaborator and conducted hands-on experiments with some of his British colleagues. After returning from Europe, Henry advocated increased scientific exchanges with other nations. He continued to supply the New York meteorological reports to Forbes and others, notably Quetelet and Heinrich Wilhelm Dove.[103]

With the help of Bache, and through the American Philosophical Society, he agitated for government support to extend the Göttingen Magnetische Verein of Carl Friedrich Gauss and Wilhelm Weber. A series of physical observatories, or "watch towers of nature," were requested, which would be supplied with magnetic instruments and the following meteorological apparatus: two barometers (Newman and mountain), four thermometers (standard, maximum and minimum, and wet and dry bulb), a hygrometer (Daniell), an anemometer (Osler), and an apparatus for atmospheric electricity.[104] John Quincy Adams penned a positive report from the U.S. House of Representatives recommending that the secretary of war administer the project with a $20,000 appropriation for five stations and instruments.[105] Although the funds were not forthcoming, several privately funded observatories were eventually established, notably Bache's station at Girard College, active between 1840 and 1845.[106] For science in general, but particularly in the geophysical tradition, Henry, Bache, and their colleagues were creating new opportunities for American partnership with Europe.

James Espy sailed on June 6, 1840, from Philadelphia, bound for Liverpool.[107] Although his itinerary cannot be reconstructed in detail, his two papers read before the British Association meeting in Glasgow in September 1840 and the favorable report issued by the French Academy in March 1841 are well documented.[108] Henry reported that "the communication which attracted most attention" at the tenth meeting of the British Association "was from our countryman, Mr. Espy, on storms."[109] Espy presented an example of a centripetal storm in Britain, gave a lengthy account of his theory, and demonstrated his nephelescope (see Chapter 5). A lively discussion followed. Objections by John Stevelly, James Forbes, and Archibald Smith were balanced by supportive comments by Edward Osler and John Phillips. Stevelly thought that clouds should be colder, not hotter, than the surrounding air, and therefore that Espy's "violent ascending vortex" would not exist. Forbes was of the opinion that the motion of tornadoes was more complex than could be revealed by deductions from the trace of fallen trees. Smith argued that unless Espy's centripetal winds rushed to a single central point, rotation must develop from the conservation of areas around finite moments. In support, Osler claimed that he had examined the same storm as Espy and also found centripetal currents. Sir David Brewster thought that Reid's *observation* of rotation in a waterspout was worth a thousand *inferences,* although he admitted that in the case of a large winter storm, the great want of well-attested facts made it impossible to form any definite opinion.[110] This is perhaps the most astute comment, given Espy's later attempts to expand the empirical horizons of meteorology through governmental systems of data collection.

Although Espy's performance received at best mixed reviews, Henry's report of the meeting ended with a pat on the back for American science: "It is certainly a matter of congratulation to those who are interested in the cause of American science, that so much has been done among us in the way of meteorology. The interesting theories of Espy and of Redfield, contradictory as they may now appear, will probably be found not incompatible with each other; and they will undoubtedly form the most important steps towards the widest generalizations which have yet been attempted in reference to the complex phenomena of the atmosphere."[111] Reid, however, harbored no such sympathies and reported to Redfield that his adversary had received a cool reception in England: "I hear from England that people's minds were satisfied with the revolving theory of Storms; so that few cared to listen at Glasgow, and elsewhere, to Mr. Espy's explanations of his particular theory."[112] The American press, however, reported Espy's victory: he "lectured at the principal towns and cities with much effect, and to large audiences."[113]

Espy's European tour ended in Paris with the approbation of the French Academy. The Paris correspondent of the *Boston Courier* reported, "Mr. Espy has been here for some time, trying to get an opinion from the academy, respecting his theory of storms, &c. and at last has succeeded. The committee has made a very favorable report, which was to have been read in public session yesterday."[114] Attributed to the French savant Arago on this occasion is the cryptic remark, "England has its Newton, France its Cuvier, and America its Espy."[115] This could be a clever insult or an indication that meteorologists were anticipating a synthetic theory to bring order to the data, but in light of the favorable report issued by the French Academy, it was probably no more than a sincere compliment: "The works of Monsieur Espy gave rise in 1841 to a very favorable report made to the Academy of Sciences, in the name of a commission of which I was a member along with Messieurs Pouillet and Babinet, rapporteur. The facts on tornadoes observed by Monsieur Espy were numerous enough and presented well enough for us to recommend that the Academy approve his work."[116]

Espy's *Philosophy of Storms* reprints, in English translation, the report of the French Academy. The conclusion of the report is of particular importance for the subsequent development of government support for meteorology in the United States: "The committee expressed then, the wish that Mr. Espy should be placed by the government of the United States in a position to continue his important investigations, and to complete his theory, already so remarkable, by means of all the observations and all experiments which the deductions even of his theory may suggest to him, in a vast country, where enlightened men are not wanting for science, and which is besides, as it were, the home of these fearful meteors. . . . The conclusions of this report are adopted."[117] With the manuscript version of *Philosophy of Storms* and the support of the French Academy in hand, Espy departed from Le Havre on April 6, 1841, to renew his petitions to the government of the United States.

Enter Robert Hare

> The Doctor is deemed to be a pertinacious antagonist, and will probably persevere in his attack, whether successful or not in throwing any new light on the subject of controversy.
> —Redfield to Heinrich Dove, June 8, 1844, Redfield Letterbooks

Robert Hare entered the agonistic field with a vengeance in 1841 by requesting a retraction of the French Academy's approbation of Espy and demanding its recognition of his priority over Jean Charles Peltier in the electrical theory of tornadoes. He then published a series of attacks on Redfield. Hare's seemingly limitless capacity for polemics is remarkable since he was also involved in controversies with Michael Faraday, William Whewell, Henrich Dove, and Justus von Liebig at this time. Table 2.4 provides the pattern for the storm controversy in the early 1840s.

Before Espy's visit to Paris, the French Academy had reported favorably on Peltier's electrical theory of storms in a debate over insurance claims for damages caused by an electrical storm. Hare claimed priority for the electrical theory and felt that the academy, by now supporting Espy, was being inconsistent if not unjust.[118] In a letter to Arago early in 1840, Hare stated his claim over Peltier, referring to his analysis of the Brunswick spout of 1835.[119] After Espy's successful European tour, Hare went to Paris seeking a retraction: "Dr. Hare has just returned. While in Paris he endeavored to obtain a recommittal of Mr. Espy's theory to the Com. of the Institute to modify the part of their report relating to the electrical theory."[120] Hare's second letter to Arago expressed his surprise: "I was not a little surprised to see your name at the bottom of the report submitted to the Royal Academy of Sciences relative to the work of Mr. Espy on tornadoes . . . without the slightest reference given to my report." He thought Espy had been "pressing" for an immediate conclusion from the academy, and the haste of the young rapporteur Babinet, the "negligence du sup-

TABLE 2.4. Robert Hare enters the controversy (see bibliography for full citations)

Hare v. Espy (1840–1842)

1. *Hare's Letter to Arago, April 24, 1840.*
2. *Of the Conclusion Arrived at by a Committee of the Academy of Sciences of France . . . with Objections to the Opinions of Peltier and Espy.*

Hare v. Redfield (1841–1843)

1. "On the Theory of Storms, with Reference to the Views of Mr. Redfield," *Phil. Mag.*
2. "Objections to Mr. Redfield's Theory of Storms with Some Strictures upon His Reasoning," *Amer. Journ. Sci.*
3. "Additional Objections to Redfield's Theory of Storms," *Amer. Journ. Sci.*
4. "Strictures on Prof. Dove's Essay 'On the Law of Storms'," *Amer. Journ. Sci.*

Redfield v. Hare (1842–1843)

1. "Reply to Dr. Hare's Objections to the Whirlwind Theory of Storms," *Amer. Journ. Sci.*
2. *On Whirlwind Storms; with Replies to the Objections and Strictures of Dr. Hare.*
3. "Notice of Dr. Hare's 'Strictures' on Prof. Dove's 'Essay on the law of storms'," *Amer. Journ. Sci.*

port" of Peltier, and Arago's "grandes occupations" with more pressing duties had led to the mistaken report.[121]

Hare had carried several of Redfield's published papers to Arago and Pouillet.[122] Now he encouraged Redfield to write to them on his own to clarify his differences with Espy. Redfield wrote at length:

> It has not seemed proper for me to ask the attention of the Academy to my humble labors in this department of meteorology, nor to canvass the Report of its able Committee on the claim presented to the consideration of the Academy by my countryman Mr. Espy. In the discussion of Mr. Espy's views at the meeting of the British Association, I have understood it was stated, that the principal facts relied on by Mr. Espy had been questioned in this country as inaccurate: but of this it was not easy for your Committee to have been informed. So far as I understand the several cases relied on by Mr. Espy, I may say, that there has been no instance in this country in which the violent winds which have belonged to the active portion of a great tempest have been found to blow, at one and the same moment, in opposing directions towards a central point or line within the body of the storm. Nor has it been shown by any proper evidence that a great storm has in this region been found moving towards the southeast; a course which is needful for reconciling the observed changes of wind in our storms with the conditions of Mr. Espy's theory. In the matter of the smaller tornadoes or whirlwinds, I have found some of Mr. Espy's observations not reconcilable with my own inspections of the same phenomena; and some of the effects on opposite sides of the track of the tornado, which he thinks to have been of simultaneous occurrence, are best explained by viewing them as the consecutive effects of a moving whirlwind.[123]

This alliance did not last long. Hare requested, innocently enough, one of Redfield's pamphlets "in which you contend against *the long cherished theory which is founded on calorific rarefaction.*" Redfield replied that his theoretical position on this subject had not been developed in detail, but he found heat "an agent which is far more limited in its effects than is generally believed," and he referred Hare to his papers in the *American Journal of Science.*[124] Hare's "objections" to Redfield were published soon after. Initially, he objected to Redfield's "perfect circles," with hurricane winds blasting from every tangent, and thought that the centripetal theory fit the data equally well. Then he attacked Redfield's causal theory of gravitation, arguing that, "in the absence of calorific and electrical reaction, what other effect could gravitation have unless that of producing a state of inert quiescence." Finally, Hare attacked Redfield's logic, registering the impression that his opponent was indulging in circular reasoning by asserting that "*gyration*" could be a cause of "*gyratory*" meteors.[125]

Redfield's immediate reply was restricted to the accuracy of his facts and their collaboration by independent investigators.[126] He claimed no theoretical pretensions and again invoked Baconianism as a defensive strategy: "It seems incumbent on an objector to set aside these facts [of Redfield, Reid, and others] as unfounded and inaccurate, or to show the results which they appear to establish have been deduced erroneously."[127]

Hare responded with "additional objections" attacking the reliability of "facts" gathered by Redfield from ships' logs and newspaper clippings. He believed that Redfield was hiding his "absurd" storm theory behind Baconian rhetoric and the illusion of objectivity: "It strikes me, however, that a fault now prevails which is the opposite of that which Bacon has been applauded for correcting. Instead of the extreme of entertaining plausible theories having no adequate foundation in observation or experiment, some men of science of the present time are prone to lend a favorable ear to any hypothesis, however in itself absurd, provided it be *associated with observations*."[128]

Espy was aware of the Hare-Redfield exchanges and attempted to enlist both Loomis and Quetelet on his side of the controversy: "I wish also you would find time to examine the evidence which Reid's storms afford of an inward motion of the air; and let the world know your opinion."[129] Espy exchanged meteorological observations regularly with Loomis and made inquiries on his behalf about positions at the Naval Observatory. Now Espy freely expressed his low opinion of Redfield: "Is it not remarkable that Mr. Redfield in his controversy with Prof. Hare, persists in the assertion that storms travel towards the N.E. when you show that your great storm commenced in Quebec about the same time it did at C. Hatteras? and sneers at me on the same subject? Speak out on this point to the world. . . . How can men permit themselves to be so blinded."[130]

Benjamin Peirce Attacks Espy

Other authors as well were beginning to address the issues raised by the storm controversy. Benjamin Peirce launched an attack on Espy's theory in which he called into question first his character and then his calculations. Peirce's *ad hominem* begins:

> I shall attempt to examine professor Espy's theory in a spirit of perfect candor and impartiality, but, I fear, without success; for there is an air of self-satisfaction and contempt for the views of other observers in his statements, which irresistibly arouses the demon of obstinacy. Even storm-kings are intolerable in a republic; and not the "*infinite utility*" of this new theory, not the singular modesty with which its author affects to be merely the Newton of Meteorology, can lull the hardened democrat into submission to the tyranny with which he acts the sole monarch of the winds, the veritable cloud-compeller, the modern Ζευς νεφεληγερετης and treats other observers as invaders of his domains.[131]

Peirce continued his assault by focusing on Espy's calculations, which to him, to use Arago's phrase, were certainly not "très suffisamment exact": "It appears that the most important of Mr. Espy's data [on the steam power of uprising columns] is in the highest degree uncertain. . . . But whatever be the result, the merit of first noticing this tendency of the liberated caloric must be conceded to Mr. Espy, and never can be disregarded as unimportant in any correct theory of storms."[132] Espy did not publish a response to Peirce, but we find in his letters to Loomis that one was being formulated: "You will have observed that Prof Peirce in the Boston Miscellany has attempted to show that the main point in my theory of storms is not tenable. He has however made two great mistakes in his calculation. He assumes that the air in an up-moving column will expand into one tenth less pressure—and that the law of cooling is uniformly one degree for every 1/142 that it expands and of course that air in expanding into double the bulk would be cooled 142°.—Both these assumptions are wide of the mark."[133] The *Cambridge Miscellany* maintained the offensive by publishing Dove's paper in which a discussion of the centrifugal (Redfield-Dove) versus the centripetal (Espy-Brandes) theory of storms dismisses Espy's contribution: "We are indebted to the repeated attacks of this author [Espy] for having given occasion to some excellent memoirs from Mr. Redfield."[134]

The American storm controversy set the tone and the content of the meteorological agenda for decades to come. The argument over the *primum mobile* of storms, gravity, caloric, or electricity, was not resolved. Neither was the argument over methodology. Basic distinctions, however, had been established among the phenomena of hurricanes, thunderstorms, tornadoes, and winter storms. A new emphasis on violent, short-lived phenomena in meteorology had emerged.

American scientists touring Europe between 1836 and 1841 helped bring the controversy to the attention of their European counterparts. America was center stage, at least in this particular geophysical science. Espy's lecture tour of America and Europe convinced many that the government should support his meteorological research. The storm controversy was not over. Indeed, its effect on American meteorology was just beginning to be felt as new institutional arrangements and observational programs began to focus on storms.

CHAPTER 3 Observational Horizons in the 1830s and 1840s

I will cheerfully furnish you with my files of selected marine intelligence (alias newspaper scraps) which I have collected since 1830.
—Redfield to Reid, April 17, 1839, Reid's Correspondence

Could not an association be formed of observers at selected places, who would send in their observations to some fixed point say New York, & then publish them, apportioning the expense between them if the subscription list should fail to defray it.
—Coffin to Loomis, September 2, 1840, Loomis Papers

Until about 1834, observers focused on long-term climatic series and investigations relating weather conditions to sickness and mortality. Systematic observations of such short-lived and severe meteorological phenomena as hurricanes, tornadoes, and winter storms were rare. As the storm controversy blackened the theoretical horizon, however, Espy widened the observational horizons of meteorology by turning the attention of the Joint Committee to the violent phenomena of the atmosphere.[1] The U.S. Navy and the Albany Institute also sponsored projects in the 1830s, and several meteorologists expanded their private data collections.

At the beginning of the storm controversy, few institutional opportunities existed for meteorologists in America. Support for atmospheric research was limited to volunteer societies and committees. Rudimentary training in meteorology was available at colleges such as Williams, Harvard, and Yale, but no well-defined career paths existed for meteorologists, nor indeed for scientists in general. American meteorologists presented their papers to general scientific societies and published them in the scientific journals of the day, especially the *American Journal of Science*, the *American Philosophical Society Transactions*, and the *Journal of the Franklin Institute*.

Elias Loomis's memorial in support of a statewide system of meteorological observations in Ohio summarized some of the outstanding questions in meteorological science remaining from the early decades of the century. These questions included illustrating the origin of disease, expanding knowledge of the laws of storms and their practical benefits for commerce, examining the proposition that the periods of the equinoxes are marked by extraordinary storms, testing the influence of the moon, comparing the presumed warmer and wetter climate of the Mississippi Valley with that of the Atlantic coast, inquiring into climatic change caused by settlement, and providing standard barometric height surveys for canals and railroads. The subject matter of meteorology was changing rapidly at this time. Bache and Espy debated with Olmsted in the mid-1830s over whether shooting stars were low or high in the atmosphere and of celestial or terrestrial origin.[2] Espy thought that shooting stars and aurora were related to the humidity of the atmosphere and had "one common origin . . . low in the atmosphere." Joseph Henry thought auroral displays were global phenomena.[3] By the end of the 1830s, although auroral displays were still part of meteorology, meteors were being consigned, with some nostalgia, to the realm of astronomy, at least by Olmsted's associate E. C. Herrick.[4]

In addition to reviewing the content of meteorology, Loomis's memorial reviewed the state of organization for meteorological data gathering in 1842, mentioning the army system, which for twenty years had been investigating climate and disease at military posts and hospitals; the New York system, with thirty-seven academies and annual reports "furnishing the most valuable meteorological document which America has produced"; and the Pennsylvania system, "a praiseworthy example."[5] His plans for an Ohio system called for twenty to thirty stations, an initial appropriation of $1,000 for instruments, guidance by a committee of scientific men, volunteer observers, reports to a central station, and annual reports to the legislature, printed for distribution as in the state of New York.[6] These plans were never funded. Instead, Loomis became an outspoken advocate of a meteorological "crusade"—an American complement to the international magnetic crusade: "I call it *one year's Crusade,* because I believe that one year's observations such as I contemplate would well nigh exhaust the subject, but in practice I expect it would take a year or two to get the army into the field, so that men ought to be provisioned for a longer campaign."[7] The work of Espy with the Army Medical Department and the founding of the Smithsonian Institution with Henry at the helm kept these hopes alive.

The Joint Committee on Meteorology

In 1831, Espy issued a call for meteorological diarists to follow a common plan and invited them to publish their observations monthly in the *Journal of the Franklin Institute.*[8] Later that year he pointed to the "neglect" of meteorological science: "Why is it that this highly interesting and useful branch of human knowledge makes such slow advances, whilst almost every other department of physical science is receiving important additions every year? . . . Let learned and scientific societies answer to themselves and to the interests of mankind for this egregious neglect. Do philosophers think that no more discoveries can be made on this subject? and that the weather never can be predicted?" Still, Espy was optimistic and called for cooperative observations, philosophical studies, and scientific predictions to "solve the problem of the weather."[9]

Within a few years, Espy had made Philadelphia the "center of meteorological activity in the United States."[10] At the Franklin Institute Espy established a committee on meteorological observations which began publishing monthly reports in January 1831. In the same month Joseph Lovell, the U.S. Army surgeon general, offered the American Philosophical Society a complete series of meteorological observations made at military stations around the country. Within a few years, the two meteorological projects had merged. On September 9, 1834, a Joint Committee on Meteorology of the American Philosophical Society and the Franklin Institute was established with Espy as its chairman. Representing the Franklin Institute were Henry D. Rogers, Sears C. Walker, Paul B. Goddard, Bache, and Espy. Members of the committee from the American Philosophical Society were Charles N. Bancker, Gouverneur Emerson, Bache, and Espy (after 1835).

Although Walter E. Gross maintains that Espy's name is "rarely associated with the American Philosophical Society," Espy was an active member of the society. Its library contains numerous Espy letters and manuscripts, a significant example being his twelve-page proposal for ocean and land observations of clouds, storms, and barometric fluctuations at high altitudes and at sea level. Gross also refers to the scientists connected with the American Philosophical Society (Bache, Henry, Loomis, Hare, Espy, Redfield, and Emerson) as forming "only a list, not a group," and claims that they "went their separate ways." Anyone reading the correspondence of these individuals, examining their role in the storm controversy, or studying the development of meteorological science or American science in general will see that nothing could be further from the truth.[11]

Motivated by Espy's storm studies, the Joint Committee issued circulars requesting credible witnesses to provide detailed information about specific storms.[12] Likely candidates included government officials, college presidents, judges, and newspaper publishers. One of the committee's broadsides, printed and issued the day following a major storm, clearly links the observations to the storm controversy:

> The prime object of this circular is to obtain a complete knowledge of all the phenomena accompanying one or more storms of rain or hail, not only where the violence of the storm is felt, but at and beyond its borders, its beginning and its end. . . . Particularly inquire the course of the wind at the commencement of the storm, and at its termination; the width of the storm; its direction; its velocity; the direction of the wind at its sides; how the wind veers round—whether in different directions at its sides, or not; whether, in case of hail, there are two veins, or only one; where there is the greatest fall of rain, near the borders or near the centre of the storm—and whether this fall takes place near the beginning, middle or end of the storm; whether the clouds are seen moving

> with the wind, or against it, and whether differently among themselves; and every thing else which you think may tend to an explanation of this most interesting phenomenon.[13]

In response to the first circular, issued in 1834, fourteen persons scattered across the country mailed in their private meteorological journals. They had received few guidelines and no forms on which to record their observations; they had no standardized instruments; many had no instruments at all. By 1838, 250 circulars were sent to different parts of the United States and Canada and between forty and fifty answers were received.[14] Although these correspondents of the Joint Committee formed a very small meteorological instrument with a rather poor resolution (see Figure 3.1), the project was significant not for its size or sophistication but because it was the beginning of widespread, simultaneous meteorological observations focused on storms and other transient atmospheric phenomena. It was the beginning of Espy's "volunteers."[15]

The Joint Committee published regular reports in the *Journal of the Franklin Institute*. The first (1835) began hopefully: "Only four months have elapsed since the reception of the earliest of these journals, and already some valuable facts have been deduced from a comparison of the simultaneous observations which they contain." Although the committee attempted to confirm Benjamin Franklin's conclusions on the motion of storms—"Do these barometric fluctuations of great magnitude travel north-eastwardly?"–most of the inquiries were designed to support Espy's theoretical position in the storm controversy: "Are rains caused by an upward motion of air, commencing where the dew-point is

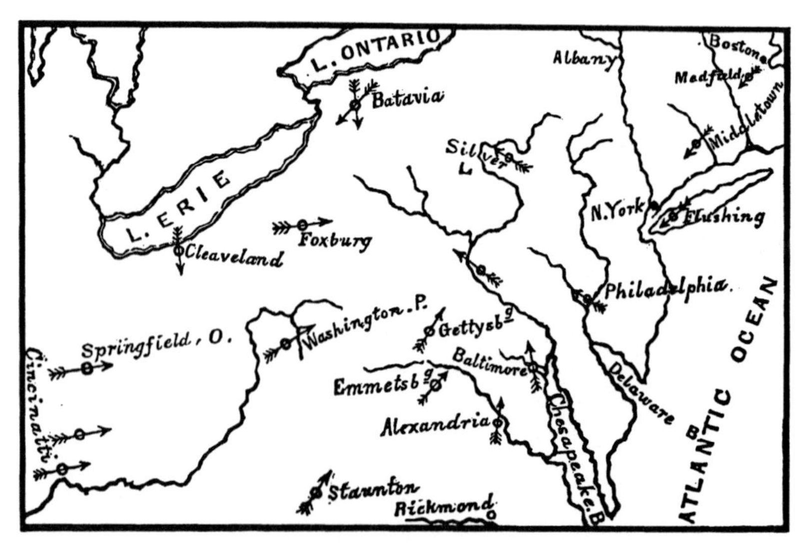

Fig. 3.1. Weather map of the Joint Committee, 1837, showing converging wind patterns during a storm. From *Journ. Frankl. Inst.* n.s. 19 (1837): 19.

highest, or where the barometer is lowest? Do storms in the temperate zone generally travel from some westerly point? And are those storms which so travel preceded by an easterly wind, and also followed by a westerly? . . . Is it the upper current [of air] which gives direction to the storms in [the mid-] latitude? . . . What are [the] two diameters [NE-SW and NW-SE] of storms generally?"[16]

The second report of the Joint Committee (1836) began with "regret that the extent of our correspondence north-west and south-east is not great enough to enable us to ascertain the boundaries of our great storms in those directions." It went on, however, to report observations of seven storms. The most "remarkable" generalization confirmed Espy's centripetal theory of storm winds: "*In all the seven storms examined, the wind blew towards the point where it was raining.*" The report concluded with a call for government assistance and a warning: "And if our present attempt should fail to stimulate men of science to engage in the undertaking of investigating the dynamical laws to which the movements of the atmosphere are subject, and this can only be done by simultaneous observations over a wide extent of territory, then this committee, *unless aided by Government,* will have to leave the work unfinished, and reluctantly close their labors, with perhaps one more report."[17]

The third report of the Joint Committee (January 1837) is notable because it contains the first map published in the United States that displays meteorological (as opposed to climatological) data collected over a wide area (see Figure 3.1).[18] The centripetal wind patterns on the map support Espy's storm theory. The report reiterated the need for an appropriation from the government to purchase meteorological instruments and to support the work of a national meteorologist, Espy. If the government would provide the funds, academies, schools, and colleges that pledged to keep a journal of the weather for five years, follow a prescribed plan, and send a monthly statement to the meteorologist would receive free instruments.[19]

To support this initiative, Espy asked Bache to petition the managers of the Franklin Institute to agitate for government support: "I wish to get scientific societies and Colleges to unite in petitioning Congress to establish and provide for Meteorological observations to be made all over the U.S., one observer to each 100 miles square, so that we may know when every shower commences and terminates, the course it moves, &c. Believing that this is the only information now wanted to predict rain[,] please . . . leave a written request to the Board of managers of [the] F[ranklin] Inst[itute] or to any society you please to set an example on this point and other societies and learned bodies will probably follow."[20] Bache responded to Espy's request, and in 1837 the Board of Managers of the Franklin Institute sent a memorial to Congress requesting government support for the Joint Committee. One year later John Sergeant of Pennsylvania introduced a motion to the U.S. House of Representatives to appoint a committee "to inquire into the expediency of encouraging and aiding meteorological observations throughout the United States." The motion was not approved.[21]

Simultaneously, Espy sought the support of the Commonwealth of Pennsylvania. On April 1, 1837, primarily because of Espy's personal lobbying efforts in Harrisburg, a grant of $4,000 from the legislature of Pennsylvania was provided to the Franklin Institute for statewide meteorological investigations. The legislature appropriated $2,000 immediately and reserved $1,000 for expenditures in 1838 and 1839. Approximately $1,000 was budgeted for sets of instruments for each county in the state, fifty-two in all. A set consisted of a barometer with attached thermometer, a twelve-inch thermometer graduated to single degrees, an eight-inch thermometer graduated to two degrees, a self-registering thermometer with a metallic scale for registering low temperatures, and a rain gauge furnished with a graduated glass tube. L. C. Francis of Philadelphia contracted to supply the instruments and agreed to manufacture them for $16 a set. Gouverneur Emerson recalled that completing the order and standardizing the instruments was extremely difficult: "We had a great deal of trouble to get instruments finished, probably from the contract having been made rather too low. In regard to the construction of the Barometers, we were desirous of having them all graduated as nearly as possible with a standard we have constructed for the purpose. When the deviation from this standard exceeds 2/100 the instrument has to be corrected. There are however causes which operate to change the graduation and prevent it from having that perfect uniformity at first contemplated. We followed a similar rule in the inspection of our Thermometers, etc." The Franklin Institute claimed the right to regulate the distribution and disposition of instruments in each county. The right of property, however, resided in the county commissioners. Inevitably, this arrangement led to disagreements among observers and county officials. The institute preferred as observers "the principal of a College, Academy or Lyceum, where these were to be found," and in the absence of such, entrusted the instruments to some "intelligent person" who could volunteer to make the necessary observations and report monthly by mail. Finding competent and reliable volunteer observers and distributing the instruments to them, unbroken, was not an easy task.[22]

According to Emerson, part of the appropriation was used for collecting and compiling the observations: "Our Met[l] Report rec[d] from the various counties are addressed to the Actuary of the Franklin Institute who pays the postage, files them according to a plan best adopted for their presentation and convenience of reference, computes the averages, prepares tables, &c. for which he is allowed compensation of $250 per an."[23] Funds were also used to print and mail to the observers directions for placing, reading, and calibrating the instruments and methods for measuring temperature, pressure, dew point, clearness of sky, and precipitation.[24] Espy, now the meteorologist of the committee, also wrote a more detailed handbook, *Hints to Observers on Meteorology* (1837), to alert the observers to the fundamental goal of the research: "*to find out the course that storms take over the surface of the earth in all the different seasons of the year,* and consequently, their velocity and *shape,* . . . the force and direction of the wind in their borders, both at the surface of the earth, and in the region of the clouds."[25] Espy asked the observers to pay special attention to visual techniques for estimating upper-air wind velocities and the heights of cloud bases and tops.

After the appropriation was deposited with the Franklin Institute, the Joint Committee, "at a loss to determine to which body they were accountable," was dissolved in June 1838. The American Philosophical Society ended its meteorological work.[26] Emerson's letter to Loomis provided further details:

> The reason why the Am. Phil^{l} society seceded from the Joint Committee, was chiefly the embarrassment sometimes created by one circumstance or another. The Joint Com^{tee}, for example had often to procure authority from the bodies from which they respectively emanated. This was a source of delay and trouble. Then again, there was a sum of money to be taken charge of, and this perhaps was the strongest reason, why the committee should emanate solely from the institution having the care of the funds. On these grounds chiefly, the Phil^{l} Society, with the most amicable dispositions, thought proper to recede and leave the whole subject of a Meteorological Com^{tee} to the Franklin Institute, which embraces so many of its prominent members.[27]

With the demise of the Joint Committee, the Franklin Institute formed a new meteorological committee under the chairmanship of Robley Dunglison. For a short time Espy served as a member. The new Pennsylvania meteorological system, however, with its limited geographical coverage, was best designed to investigate the state's climatology. Although all fifty-two counties in Pennsylvania were supposed to have received instruments, only about sixteen reported monthly to the Franklin Institute. By 1849 the institute informed Henry that only six stations were still sending reports.[28] More important, its appropriation was too small to support Espy as a meteorologist. Espy turned to the lyceum lecture circuit, advocating the artificial production of rain and seeking national support for his projects. Soon he issued "in his individual capacity" a circular "To the Friends of Science in the United States."[29]

Meteorological Reports from Navy Yards and Ships

Simultaneous with the work of the Philadelphia savants, the U.S. Navy began a program to collect meteorological data at navy yards and on board its ships. On March 27, 1834, the senior surgeon of each command was ordered to "keep or prepare a journal of the changes in temperature and weather, as indicated by the Thermometer and Barometer." The orders continued: "The journal is to contain also, notes of gales, storms and hurricanes, with their direction, time of commencement and termination; and a brief statement of the effects, immediate or remote, supposed to be produced by any of the important changes, on the health of the Officers and men."[30] This initiative to collect statistics on weather conditions and health was undoubtedly copied from the Army Medical Department. Keeping notes on gales, storms, and hurricanes, however, was an innovation that soon attracted the attention of the Joint Committee in Philadelphia.

Following a talk by Bache on meteors at the Academy of Natural Sciences in Philadelphia, naval surgeon W. W. Ruschenberger suggested that the observations being collected at the coastal stations of the navy might provide useful data for the investigations of the Joint Committee.[31] Bache agreed and, with Ruschenberger's help, provided the Medical Bureau of the navy with circulars requesting information on storms and meteor showers. In several cases, information was returned to the Joint Committee.[32] Between 1834 and 1837, the navy received reports from nine surface stations. The most notable is a continuous series taken by Charles Wilkes between June 1834 and March 1837 at the Depot of Naval Instruments in Washington. Several ships at sea also contributed reports from remote areas. The surviving reports from this system of observation are fragmentary. It does not appear that large amounts of data were supplied to scientists on a regular basis.[33]

The Albany Institute's Term Days

The Joint Committee and the navy were not the only organizations collecting meteorological data in the third decade of the nineteenth century. The Albany Institute formed a committee to compile observations on the state of the atmosphere on the twenty-first of June, September, December, and March—the dates of the equinoxes and solstices.[34] These observations on "term days" were first initiated by Sir John Herschel at the South African Literary and Philosophical Institution. Herschel's program was designed to investigate the diurnal changes taking place in the atmosphere and, by collecting measurements from widely dispersed stations, to obtain knowledge of the "correspondence of its movements and affections over great regions of the earth's surface, or even over the whole globe." Although Laplace's speculations on atmospheric tides had stimulated continued interest in uncovering solar and lunar influences, especially on the observed diurnal fluctuations of the barometer, Herschel's program was not, as it might seem, primarily an attempt to find direct causal links between astronomical bodies and the weather. Rather, it was an attempt to delineate, in considerable detail, the entire global atmospheric system on a regular basis. Herschel recommended that henceforth four days in each year—at the equinoxes and solstices—should be set aside by meteorologists in every part of the world and devoted to "a most scrupulous and accurate registry" of the state of the atmosphere for each of the twenty-four hours of those days and for six hours of the days preceding and following—a total of thirty-six hours. These "somewhat harassing observations" were recorded and dutifully reported at up to a dozen different locations in North America.[35]

Although some observers searched for "equinoctial" storms and cold spells until about 1850, the term day observations did not constitute a storm research program. Nevertheless, a large winter storm passed through the Ohio Valley and into the Northeast during the term day observations of December 20–21, 1836. This serendipitous occurrence provided Elias Loomis with an opportunity for a detailed investigation. Loomis was convinced that more would be learned from a complete analysis of a single storm than from the partial analysis of several storms. He hoped his storm analysis would provide the *observationis crucis* between the theories of Redfield and Espy. This is distinctly non-Baconian behavior for an investigator branded by Daniels as a prototypical Baconian.[36]

Loomis gathered barometrical and other observations from 27 private meteorological journals. Temperature, precipitation, and wind data were available from 28 military posts, 42 academies in New York, and 5 other observers—a total of 102 locations. No standards of comparison, however, were established among the different sources of data, and the observations covered only one-half the area of the storm. Loomis concluded that the storm was not circular but that the winds blew from opposite sides toward a central line of minimum pressure.[37] Loomis received wide acclaim for his model study and influenced Espy's later efforts to chart lines of minimum pressure. Strictly comparable observations, however, were still needed. The search for new observational standards and new institutional sponsors for meteorology became the major issue of the 1840s.

Collegiate Associations

The Meteorological Association of Williams College, founded in April 1838 by Albert Hopkins, professor of astronomy and natural philosophy, consisted of members of the two higher classes. Their agenda consisted of investigating "the whole subject of atmospheric changes," taking frequent observations at the college, and collecting and collating daily observations "over as great an extent of country as possible." They also maintained a meteorological observatory with self-registering instruments on Gray Lock peak of Saddle Mountain.[38] The association had no clear research program until a new tutor, James Henry Coffin, a veteran meteorological observer of the New York academy system and inventor of a self-registering wind vane, came to the college. Coffin wrote to Loomis soon after his arrival: "A meteorological association has existed in the college here for two or three years; but till lately nothing was done but to record facts. Not long ago at the request of the society I proposed a series of topics for investigation, & the society resolved itself into committees to consider them, each taking one. I hope they will effect something. At any rate they furnish matter for interesting reports at the meetings of the society, & thus keep up the interest.[39] Coffin's list of thirty-nine topics for investigation, published in the *Berkshire County Whig*, demonstrated his awareness of meteorological theories "now under active investigation both in this country and Europe." His list included the Espy and Redfield storm models, the role of atmospheric electricity, the relations of weather and health, climate change, lunar atmospheric tides, and auroral studies.[40] Obviously, the students he tutored received a full dose of meteorology.

Although Williams College had the oldest association, others soon followed. The Harvard Meteorological Society was formed in 1841 under the leadership of William Cranch Bond, Jr., the astronomer's son. The society was established by students in response to the "growing interest" in meteorological science, especially as conveyed by "Mr. Espy's theory."[41] At Yale College, Olmsted presented a special series of lectures on meteorology.[42] At Princeton, however, Henry incorporated meteorological topics into his general lectures on natural philosophy.[43] This pattern was more typical of widespread instructional practice.

Private Compilations

William Redfield, like other meteorological theorists, took observations himself; sustained an extensive network of correspondence with other observers, theorists, and institutions; and advocated plans for simultaneous observations from accurate observers in different parts of the world.[44] For example, he thought the U.S. Exploring Expedition should "obtain the temporary loan" of the journals of the whaling ships it met and copy from them the daily direction of winds in the Pacific.[45] In addition to his extended collaboration with Sir William Reid, his meteorological correspondents included Dove in Germany, Henry Piddington in India, and Charles Darwin and Herschel in England.[46]

The correspondence network of James Coffin included such notables as Redfield, Espy, Loomis, and Henry. By 1842 his private collection of wind data embraced 171 locations "scattered over all the states & territories except Arkansas & Kentucky."[47] His research program, to delineate the prevailing winds of North America, the Northern Hemisphere, and eventually the globe, "& the connection between them & barometric fluctuations," required large numbers of reliable observations. Coffin's early results, on the winds of North

America, were presented to the Association of American Geologists and Naturalists.[48] He then expanded his work to include data from European observers, to "make it embrace the entire northern hemisphere."[49]

While working as a tutor in Ogdensburg, New York, Coffin had attempted to publish a specialized journal on meteorology, the *Meteorological Register and Scientific Journal.* He hoped the journal would help improve the system of observations at academies in the state of New York and encourage others to begin observations and persevere, but the initiative faltered after one issue from lack of support.[50] Redfield received a copy which he sent to Reid, and Reid to Herschel. Loomis subscribed as well but soon received a refund from Coffin and an apology:

> The first No. was issued as an experiment, with but little hope of its succeeding; but, as I remarked in the prospectus, it seemed to me that so much labour was lost by the meteorological observers in this state [New York] through want of correspondence, & thus the benefit of each others' observations, by which alone they could hope to arrive at any thing like *general* conclusions in relation to the changes occurring in the atmosphere, that I felt willing to make an effort to bring about united action. There appears however to be but little interest felt on the subject by most of our observers, which I attribute mainly to three reasons: 1st the observations required by the state are too limited; 2d they are in a measure compulsory on the part of the state, & 3d as I have said, they are isolated experiments, affording separately no data for philosophical investigation. Out of between 50 & 60 institutions where observations are required to be taken I had but 3 or 4 subscribers for the Register, & less than 40 in the whole number, not one third sufficient to defray the expense of publishing. The subscribers consisted principally of professors in colleges & other scientific gentlemen in the eastern and middle States.[51]

Although his *Meteorological Register* was unsuccessful, Coffin never lost sight of his goal to see a cooperative system established for gathering and publishing meteorological data:

> Relying on the interest you manifest in the advancement of science, I take the liberty to suggest to you some ideas I have had of trying to establish a system of correspondence between selected places in different points of the U. States on the subject of meteorology. In the present state of the science, it is comparatively of little use to prosecute it without some such arrangement. The object would be partially accomplished if all observers were to send their registers to the Joint Com[mittee] on the subject at Philadelphia, but still they only would be in possession of the data thus obtained, & consequently we should lose the aid of men of science elsewhere in the investigations growing out of them. I can think of no other way by which the end can be attained without great trouble & expense except by publishing the registers consistently in a periodical. It was with this view in part that I started my paper at Ogdensburg, which you are aware was abandoned for want of support. Could not an associa-

> tion be formed of observers at selected places, who would send in their observations to some fixed point say New York, & then publish them, apportioning the expense between them if the subscription list should fail to defray it.[52]

Espy Moves to Washington

Espy maintained an extensive private correspondence network in support of his storm studies and artificial precipitation experiments. Among prominent meteorologists, however, Espy was the only one experienced in establishing an institutional observational system. His chairmanship of the Joint Committee, success with the Pennsylvania legislature, successful lyceum circuit and European tour, and the public response to his new book convinced him that the time was right to seek federal support for his meteorological investigations. The French Academy's approval of Espy's theories and its advocacy of his employment as the official meteorologist of the United States were widely quoted after Espy's return from Europe.[53] Arago recalled in his memoirs that the academy report played a prominent role in Espy's acquiring a government position:

> We have proposed that the academy approve the quite interesting investigations of Monsieur Espy and encourage this learned physicist to continue in his work. Monsieur Espy informed us (in June 1843) that the United States government has furnished him with the means to establish a system of simultaneous observations which will serve to complete the natural history of this phenomenon [storms]. . . . It would be desirable for the French government to take the necessary measures so that similar observations on the same plan could be made on board naval vessels. Monsieur Espy is pleased that the French Academy encouraged him to continue his research. As he wrote me in September 1844, the favorable report on his work was very influential in placing him in a position where he can procure the information that he needs.[54]

"Espy watchers" in the popular press advocated his cause and followed his move to Washington to solicit aid from Congress for extended meteorological observations: "Could [Espy] obtain the situation of meteorologist in one of the departments at Washington, or receive in any shape the patronage of the general government, his theory would be thoroughly tested in all points, and valuable results obtained, if not for the confirmation of his views, then for the establishment of sound and universal principles on a subject of such vast importance and primary interest to mankind."[55] The memoirs of John Quincy Adams, advocate of a national observatory, recount the initial difficulties facing Espy in Washington:

> Jan. 6, 1842: Mr. Espy, the storm breeder . . . left me with a paper exposing his three wishes of appropriations by Congress for his benefit—about as rational as those of Hans Carvel and his wife. The man is methodically monomaniac,

> and the dimensions of his organ of self-esteem have been swollen to the size of a goitre by a report from a committee of the National Institute of France, endorsing all his crack-brained discoveries in meteorology. . . . I told him, with all possible civility, that it would be of no use to memorialize the House of Reps. in behalf of his three wishes. . . . If the Senate should pass such a bill I would do all that I could for him in the House.[56]

Two weeks later Espy requested support for meteorology at a meeting of the select committee on the Smithsonian bequest of which Adams was a member: "Jan. 19, 1842: Habersham presented a letter from James F. [sic] Espy, proposing that a portion of the fund should be appropriated for simultaneous meteorological observations all over the Union, with him for a central meteorologist, stationed at Washington with a comfortable salary."[57]

Although Espy did not receive a paid government appointment immediately, he successfully eliminated the chief expense in maintaining his network of correspondents—postage—when the franking privilege was extended to him by members of Congress and the Navy Department appointed him professor of mathematics.[58] As he wrote to Loomis early in 1842: "I send you herewith all the documents which I have been able to collect on the storms which you are now investigating; You will please to return them to me as soon as you can conveniently. Enclose to the Navy Department, Washington City, Care of J. P. Espy and thus my letters will come free." The postscript to the letter encouraged Loomis to provide a list "of all the persons in the U.S. and Canada &c who, you think, will keep journals of the weather for me, and I will send them blank forms when I begin operations again."[59] Espy's widely published circulars "To the Friends of Science in the United States, Canada, West Indies, Bermuda, etc.," recommended that all observers send in their meteorological registers so that "[we] will be able to solve the question of storm movement in a few years by the system I have proposed, if it is carried into complete operation."[60] In his second circular of 1842, Espy expressed his resolve "to prosecute this subject to the full extent of my means (and my ventilator will probably furnish me the means)—whether Government aids me or not."[61]

As a professor of mathematics in the navy, Espy had gained important credentials, and his work on the ventilation of ships and buildings won him powerful friends in the government: "Mr. Espy has just returned from Norfolk, which he was requested by the secretary of the Navy to visit, for the purpose of putting into practice one of his simple yet useful inventions [the conical ventilator]. It is intended to clear the ships of the United States from their foul air. . . . It has lately been put on one of the chimneys of the U.S. senate chamber, and it appears when all the other chimneys puff down smoke and ashes into the chamber, as they do sometimes when the wind blows, the one furnished with the ventilator produces a draft upward, stronger as the wind increases in force."[62] Espy's conical ventilator (some members of Congress called it his "comical" ventilator) was a metal cone that redirected the airflow around it and diminished the pressure, enhancing the updraft in a chimney. In the naval experiments the ventilator was connected to a canvas bag two feet in diameter and thirty feet long to reach into the ship's hold. In a later experiment, the bag was eighteen inches in diameter and forty feet long. According to Espy's calculations, fifty thousand cubic feet of foul air could be discharged from a ship's hold in one hour. Espy claimed that President John Tyler ordered the ventilator installed "on every chimney on the president's house, not only to cure smokey chimneys in winter, but to ventilate the rooms in summer."[63]

Army Medical Department: Forry and Espy

The Army Medical Department, of course, had a long tradition of observing airs, waters, and places for the protection of the health of the troops. Surgeon General Thomas Lawson, appointed in 1836, continued the meteorological work begun by his predecessors by employing Dr. Samuel Forry to compile the observations of the post surgeons and discuss the medical implications of the data. Between 1840 and 1842 Forry compiled the *Army Meteorological Register* for 1826 to 1830 and the *Statistical Report on the Sickness and Mortality of the Army* and authored *The Climate of the United States and Its Endemic Influences.*[64] In the first work, Forry tabulated the army weather observations for five years; in the second he promised to "furnish some general laws towards the basis of a system of *medical geography.*" In the third work Forry used medical statistics, defined as "the application of numbers to the natural history of man in health and disease," to give percentages for deaths and for particular diseases in each region. Forry called these percentages "numerical ratios," which he regarded as "fair expressions of the general laws of this system of climate in relation to diseases." For example, Forry's data for the Great Lakes region covered 6,377 individuals in seven different localities over a period of ten years. Based on army medical records of deaths and sickness, he calculated a mortality rate (m) between 0.9 and 1.3 percent per year and a sickness interval (s) of five and a half months before the average soldier in this region required some form of medical treatment. The effects of cholera were excluded. In Forry's system the northern region exhibited the greatest "salubrity of the seasons"; the interior stations of the middle region had a value of m between 3.0 and 4.5 percent and s of 3.5 months; while East Florida in the southern region had the potentially lethal indexes of m = 6.1 percent and s = 7 days. Forry's expanded definition of climate included the atmosphere's humidity, wind currents, electric tension, and barometric pressure as determining factors in regions previously distinguished solely by their temperature:

> DEFINITIONS: The term *climate,* which is limited, in its rigorous acceptation, to a mere geographical division, and in ordinary parlance to the temperature only

> of a region, possesses, in medical science, a wider signification. It embraces not only the temperature of the atmosphere, but all those modifications of it which produce a sensible effect on our organs, such as its serenity and humidity, changes of electric tension, variations of barometric pressure, the admixture of terrestrial emanations dissolved in its moisture, and its tranquility as respects both horizontal and vertical currents. Climate, in a word, as already defined, constitutes the aggregate of all the external physical circumstances appertaining to each locality in its relation to organic nature.[65]

Before leaving the army to return to private practice, Forry plotted a climate map with imaginary lines of equal temperatures joining at a point in the Far West. The map is shown in Figure 3.2.

Surgeon General Lawson, who had learned of the work of the Joint Committee in Philadelphia through a pamphlet prepared by J. N. Nicollet, was anxious to make improvements in the army system and had asked Espy and Bache to serve as consultants in the process.[66] When, on August 23, 1842, Congress appropriated $3,000 for "extending and rendering more complete

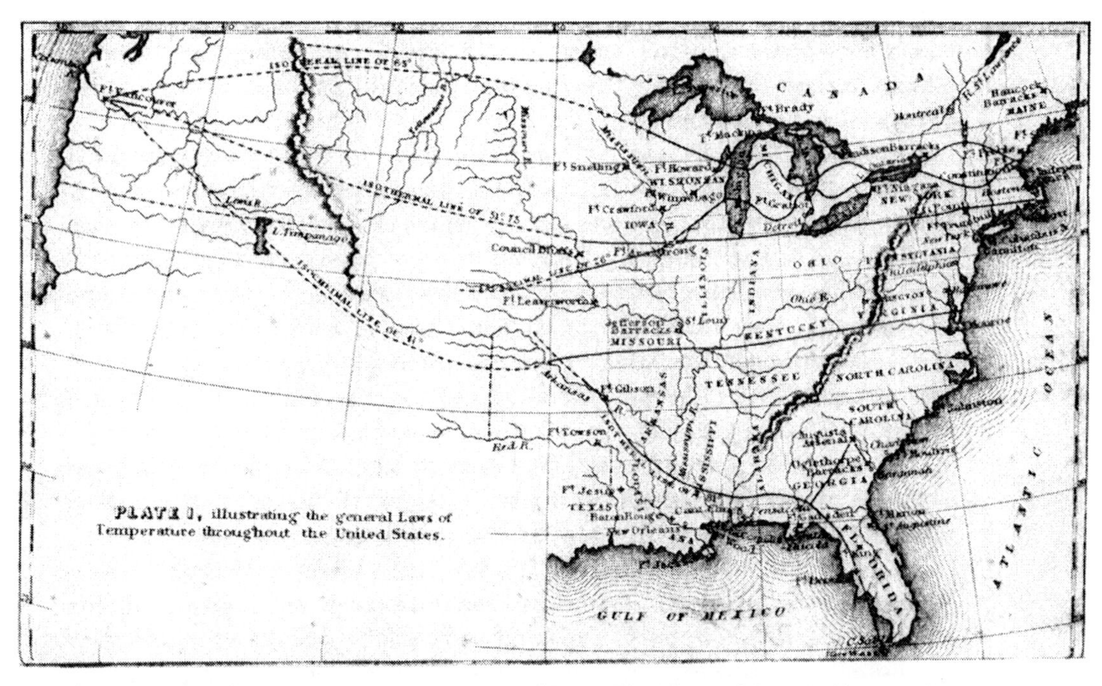

Fig. 3.2. Forry's climate map, 1842. From Samuel Forry, *The Climate of the United States and Its Endemic Influences* (New York, 1842), Plate I. Forry's map shows isothermal lines (lines of equal annual temperature), isotheral lines (lines of equal summer temperature), and isocheimal lines (lines of equal winter temperature). His data was provided by army post surgeons.

the meteorological observations conducted at the Military posts of the United States under the direction of the Surgeon General," Lawson appointed Espy, now in Washington, to take charge of the subject and instructed him to report annually to the War Department, making him "in effect the national meteorologist."[67] Expeditiously, and with considerable authority, Espy issued his first official circular in December 1842: "Journals, according to the common plan, will be kept at all military stations of the United States; and the secretary of the Navy has given orders for the same to be done at the naval stations and ships of war on our coast. Forms also will be sent to all the lighthouses and floating lights. . . . Governor Reid, of Bermuda, has promised to send me journals from that island, and I have the promise of various journals from Canada, Newfoundland, and Nova Scotia." Espy ended with appeals for observations to "the 103 colleges of the U.S. and very many high schools. . . . The number of observers cannot be too great."[68] The circular was reissued four days later with a significant addition stating that the good offices of the secretary of state had been provided for the receipt of international observations: "I am authorized by the secretary of state to request all our ministers, consuls, and other diplomatic and commercial agents of the U. States in foreign countries to whom the form is sent, to transmit to the *department of state* the journals which they may keep or procure from others, that they may be immediately placed in my hands."[69] Moreover, masters of vessels sailing in the Atlantic Ocean or Gulf of Mexico were asked to send in a copy of their logs. The *American Journal of Science* advocated full cooperation.[70]

Although Espy was not the first person hired by the army to examine its meteorological data, he was the first to use that material for storm studies. As funds provided by Congress allowed the army's observational program to expand, Espy's research efforts redirected its focus from medical geography to storms.[71] Twenty-four posts were furnished with "accurate" barometers, times of observation were made to conform with those of the Royal Society of London, and term day observations were requested for equinoxes and solstices. Still, as confirmed by Lawson's annual report for 1844, information on storms was Espy's primary goal: "Mr. Espy has been assiduously employed in completing the meteorological charts, referred to in my last annual report, with a view to the elucidation of his 'theory of storms.' He exhibits, by symbols [facts] which could not be satisfactorily developed by any other mode of illustration."[72]

Espy's *First Report on Meteorology to the Surgeon General of the United States Army*, although dated October 9, 1843, was issued in 1845. In it he reprinted his first official circular and announced with pleasure that "more than fifty observers who note the barometer, and more than sixty others who have no barometers, have commenced sending in their journals." Reveling in the luxury of his newfound financial support, Espy promised his correspondents that they would receive a copy of his report when it was published.[73] In support of his theories,

Espy requested that observers take measurements of fallen trees along tornado paths and send him eyewitness testimony of fire-induced artificial rains. The report concluded with engravings of twenty-nine of the ninety-two maps prepared for January, February, and March 1843; a chart of barometric curves for thirty-three stations; and a list of twenty empirical generalizations taken from the maps and charts. Figure 3.3, a reproduction of one of the maps, illustrates his technique of searching for long lines of barometric minima and their associated wind shifts. This form of analysis represents a departure from Espy's earlier theory of circular, centripetal wind patterns in storms and marks his acceptance of Loomis's published analysis of the storm of 1836.

Other meteorologists took careful notice of Espy's first report. Although Redfield had not seen it, he hoped that Espy would provide "an accurate and useful report of the *facts* which have been observed at the different stations."[74]

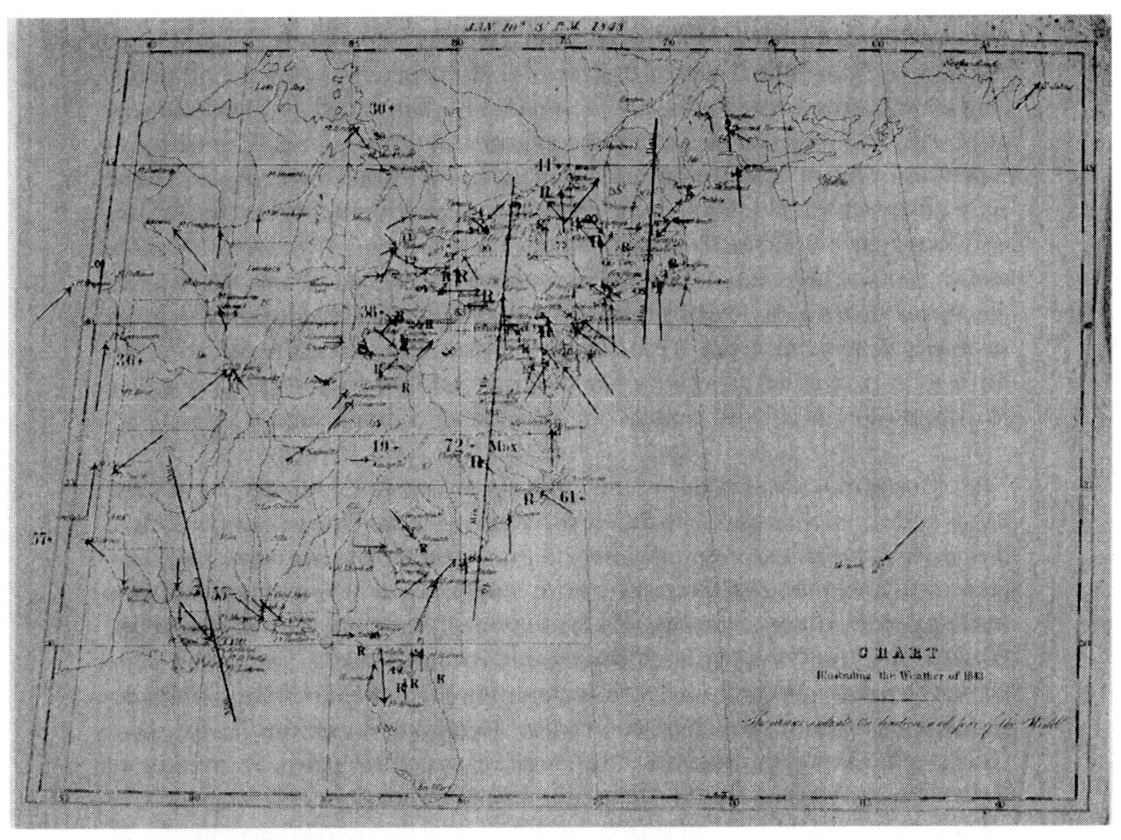

Fig. 3.3. Map from Espy, *First Report on Meteorology to the Surgeon General of the United States Army* (Washington, D.C., 1843), showing advancing lines of minimum pressure and winds directed inward toward the lines.

Henry sent a copy of the report to Forbes in Edinburgh, adding, "You will probably be interested although you may not be inclined to adopt all his views."[75] Although it is not widely known, Espy's "second" report on meteorology was presented to the surgeon general on October 16, 1845. Ninety-two maps prepared for the first three months of 1844 were not published but were transmitted in manuscript form to the secretary of war.[76] As before, Espy continued his controversy with Redfield and Hare, lectured to the public on storms and rainmaking, and maintained a large network of meteorological correspondents. Each year his appropriation came up for renewal before Congress.[77]

Espy's appointment as professor of mathematics in the navy, which initially provided him with franking privileges, was concurrent with his salaried position with the Army Medical Department. In 1843 Secretary of the Navy A. P. Upshur ordered Espy to report for duty under Matthew Fontaine Maury at the Depot of Charts and Nautical Instruments. Espy refused.[78] Two years later, after receiving orders to report for sea duty, Espy resigned: "Having seen in the public prints that an order has issued from the Navy Department for me to join the Sloop of War John Adams as Professor of Mathematics, I avail myself of the earliest opportunity to send in my resignation of that Office. . . . The business in which I am engaged in the Surgeon General's Office, I consider of paramount importance, and [since] it is incompatible with the duty assigned to me as Professor of Mathematics in the Navy, I feel myself bound to resign. . . . [It] has always prevented me from performing any services as Professor of Mathematics, and consequently I have neither claimed nor received any Salary from the Navy Department."[79] Discounting these statements made in the heat of the moment, Espy would later try to claim compensation from the navy for his services during these years as a professor of mathematics. Espy's talent for surrounding himself with controversy continued undiminished.

The observational horizons of meteorology expanded dramatically in the 1830s and 1840s in response to the storm controversy and the growing needs of the nation. The Joint Committee, the U.S. Navy, the Albany Institute, the Army Medical Department, and several theorists with private correspondence networks collected observations nationally, and state climatological projects served Pennsylvania and New York. Although opportunities for training were very limited, several colleges established meteorological associations and others instituted special courses or lectures. Coffin attempted to publish the first specialized meteorological journal. Espy, one of the most active promoters of observational systems, became the national meteorologist in Washington. The transforming effect of expanded and improved observations on meteorological theory, however, could not be felt until an institution beyond the direct influence of the protagonists in the storm controversy could establish a national system to collect and compile observations according to a common plan. The meteorological project that would attempt this task was established by the Smithsonian Institution under the direction of Joseph Henry.

CHAPTER 4 The Structure of the Smithsonian Meteorological Project

Permit me to suggest a thought that I have entertained, that being in a sense a national Institution, [the Smithsonian] might serve as a medium through which to concentrate the scientific observations in different parts of the country, & by its publications afford to the various observers, the benefit of each others labors. It seems to me that something of the kind is greatly needed. . . . We want a *central point* to which we can resort with the confident assurance that the embodied knowledge of the country is to be found there & no other scheme appears to me so feasible.
—Coffin to Henry, December 21, 1846, Henry Papers

The storm controversy of the 1830s and 1840s and the appointment of Joseph Henry as the first secretary of the Smithsonian Institution brought new impetus to building meteorological systems in America. Coffin, whose sentiments were representative of those of many scientists, rejoiced at Henry's appointment; he saw it not in personal terms but as a great opportunity for the advancement of science—a dream he had worked for throughout his life. Henry thanked Coffin for his unselfish support.[1] A new phase of development for meteorology in America had begun.

The Smithsonian meteorological project was the first major scientific undertaking of the new institution. In pursuit of theoretical goals related to the storm controversy, it charted the path of storms across the nation. In pursuit of practical goals related to western settlement, it explored the climate of the North American continent. In pursuit of institutional goals related to the Smithsonian charter, it soon became, as Coffin and others had hoped, the "national center" for atmospheric research in the mid-nineteenth century.

The heart of the project was an extensive system of volunteer observers who kept weather journals by a common plan and submitted their reports monthly by mail. At its greatest extent the system had over six hundred correspondents located across the country and in Canada, Mexico, Latin America, and the Caribbean. The Smithsonian developed and distributed calibrated instruments, standardized blank forms, uniform guidelines, and tables for the reduction of observations. It also facilitated the publication and international exchange of meteorological results.

Henry established cooperative ties with numerous organizations interested in meteorology (see Chapters 5 and 6). Instruments were supplied to several telegraph companies with the provision that the operators clear their lines each morning with a brief description of the weather (see Chapter 7). The Smithsonian received observations from railroad and canal companies, missionaries, lighthouse keepers, and official exploring expeditions under the direction of the departments of State, War, and Interior as well.[2] Henry also collected public documents, private weather journals, and other manuscript records bearing on the historical climatology of the American continent.[3]

The Smithsonian system provided a mechanism for individuals to pursue new scientific research interests as they advanced themselves professionally. During his long tenure as secretary, Henry served as a prominent spokesman for and promoter of meteorological research. Smithsonian data, equipment, publications, and other resources contributed to the work of scientists such as Arnold Guyot, Charles Schott, William Ferrel, Espy, Loomis, and Coffin. Spencer Baird, Henry's successor as Smithsonian secretary, and Lorin Blodget, author of one of the earliest texts on climatology, became professional scientists in part because of their involvement as observers in the Smithsonian meteorological project.

In 1848, the Board of Regents of the Smithsonian Institution allotted $1,000 "for the commencement of the meteorological observations."[4] Between 1849 and 1874, the meteorology budget averaged 20 percent of the institution's budget for publication, research, and lectures, or 6 percent of the Smithsonian's total budget. Expenditures for meteorological research peaked in 1860 at $4,431.07. This amount was 30 percent of the research budget and 10.5 percent of the total Smithsonian budget. Supplementary funds were available from other sources as well. Initially, the Navy Department provided franking privileges and paid Espy's salary. From 1855 to 1861 the Patent Office entered into a partnership with the Smithsonian, provided funds for data collection and analysis, and published the observations in its agricultural series. In 1863, the Smithsonian entered into a similar cooperative arrangement with the Department of Agriculture. Figure 4.1 shows the amounts budgeted for meteorology between 1849 and 1878. Over three decades, the amount spent on the meteorological project averaged 20 percent of the total research budget.

The "'Programme' of Organization" and the "Problem of American Storms"

The storm controversy was influential in Henry's decision to initiate a meteorological project at the Smithsonian Institution. In 1847, when asked by the Board of Regents to outline a plan of action for the new institution, Henry called for a "system of extended meteorological observations, for solving the problem of American Storms. . . . Of late years, in our country, more additions

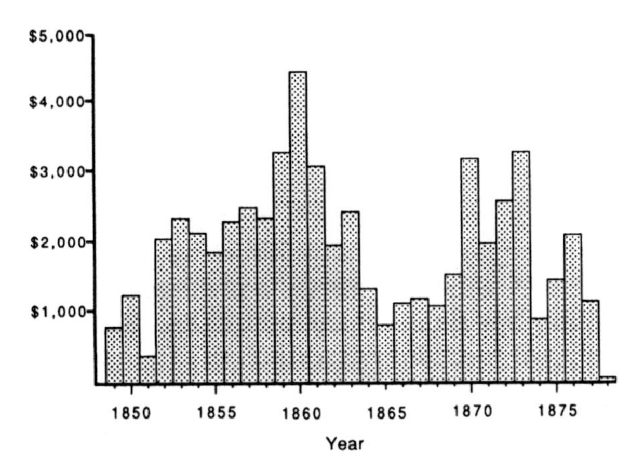

Fig. 4.1. Smithsonian meteorology budgets, 1849–1878. The precipitous drop in funding was attributable to the Civil War. The budgetary peak in the 1870s represents the Smithsonian's final attempt to compile, reduce, and publish all the collected data.

have been made to meteorology than to any other branch of physical science. Several important generalizations have been arrived at, and definite theories proposed, which now enable us to direct our attention with scientific precision, to such points of observation as can not fail to reward us with new and interesting results."[5] By "American Storms" Henry meant the debates of the storm controversy, which centered around the phenomena of hurricanes, tornadoes, thunderstorms, and winter storms common to North America. Henry thought the present time "peculiarly auspicious" for commencing such an enterprise. "The citizens of the United States are now scattered over every part of the southern and western portion of North America, and the extended lines of telegraph will furnish a ready means of warning the more northern and eastern observers to be on the watch for the first appearance of an advancing storm."[6]

Loomis, one of Henry's consultants on the proposal, believed that a "grand meteorological crusade," similar in zeal and scope to the British magnetic crusade, would be consonant with the provisions in the will of James Smithson. His "Report on the Meteorology of the United States," a masterful contemporary survey of the status of the science, presented the advantages to be derived from the study of meteorology for the mariner, the farmer, and the physician: "If the laws of storms can be discovered, this knowledge must be of the highest importance to mankind, particularly to those who are employed in navigating the sea. If the prevalent character of a season can be anticipated, it would save the husbandman much bitter disappointment from the failure of his crops. If the influence of climate upon disease could be detected, it might add years to mean duration of human life." He believed that the time was ripe for "a rich harvest of important results." According to Loomis, a "grand meteorological effort," along the lines of investigation suggested by Redfield, Espy, and Hare, would lead to the discovery of the laws of storms and the means of predicting them: "We are justified, then, in inferring that storms are subject to laws; that these laws are uniform in their operation, and that they may be discovered. . . . When we have fully learned the laws of storms, we shall be able to *predict* them." He believed that one year's observations—certainly no more than three—would be sufficient to determine these laws, the storms of each year being "probably but a repetition of those of the preceding."[7] Such optimism was undoubtedly bolstered by Loomis's earlier success in rediscovering and computing the orbit of Halley's comet. He was making the assumption that storms, like many astronomical phenomena, were subject to natural laws and that their periodic behavior could be both explained and readily predicted. In reality, storms were much more complex.

In a shorter communication, Espy suggested that the Smithsonian meteorological project be based on the model of the Joint Committee in Philadelphia twelve years earlier and follow the line of investigation outlined in his book *Philosophy of Storms.*[8] The advice of Loomis and Espy provided the blueprint which Henry followed for twenty-five years in the Smithsonian meteorological project.

The Navy Department and Espy

Espy's meteorological work in the office of the surgeon general was not going smoothly in the late 1840s. His second report was published in 1845 as a brief addendum to the report of the secretary of war but appeared without any of the ninety-two maps he had prepared. Espy reported "to the friends of science" in 1847 that his work on meteorology had been delayed because the Mexican War had resulted in a lack of stable observers in the southern and western regions.[9] In fact, the army system was in relatively good shape in 1847 (see Figure 1.2), but Espy had not published any new findings. When Espy heard that Henry was moving to Washington to direct the new Smithsonian Institution, the happiness he felt for his old friend was magnified by the new opportunity to advance his own meteorological experiments: "My head is full of imagined experiments on the electricity of steam. . . . I have long had a desire to experiment on the specific caloric of Atmospheric Air and other gasses as measured by the latent caloric of aqueous vapor condensed in them. Were you in full operation here these experiments might be made at once."[10]

In April 1847, Espy's salary was stricken from the appropriation bill for the War Department, and he planned to leave Washington in July to care for his invalid wife in Harrisburg. Henry, as Smithsonian secretary, expressed the desire to bring Espy back to Washington: "We will invite [Espy] to furnish for publication in the Smithsonian contributions any researches he may hereafter make. I should however if it were in my power be pleased to do something in the way of restoring Mr. Espy to his former position. I consider him a man of excellent character who has done much for the science of meteorology and who in a country of so much wealth as ours should not be thus deprived of the pittance to which he was before thought entitled." After all, Espy had been a member of the "club" in Philadelphia, and the two families were very close.[11]

Henry's first step was to write a letter to the secretary of the navy requesting that Espy be reinstated as the government's meteorologist: "I am informed that a proposition has been made to review the Meteorological researches of Mr. Espy under the charge of the Navy Department and I beg to express my opinion in favor of these researches. I have always considered them as tending to develop principles of the highest interest both in a practical and a scientific point of view and I am certain that the reinstatement of Mr. Espy into his former offices would meet the approbation of all persons interested in the progress of science."[12] After Henry wrote this letter, he met with the secretary of the navy. As he confided to Bache, "I found him not well disposed to our old friend . . . but before I left he assured me that *he would be happy to turn Mr. Espy over to the Smithsonian,* and would [do] anything in his power to advance our meteorological scheme."[13] Espy's salary bill passed both houses of Congress, and the navy subsequently ordered him to cooperate with the Smithsonian meteorological project.

Espy wanted to issue "Circulars to the friends of Meteorology" as the national meteorologist acting jointly with the Smithsonian and collate the private journals of weather diarists, much as he had done a decade earlier with the Joint Committee. Henry thought him too anxious to begin again on his "*old plan*" and advised him to slow down and be willing to take directions from others.[14] Henry decided that one circular, issued by the Smithsonian but signed by both Espy and him, would be more appropriate. A list of likely recipients included scientific societies, members of Congress, Espy's volunteers, and the correspondents of Coffin. A total of 412 persons from 30 states and territories received the circular. Approximately 150 meteorological observers offered their assistance and began sending in returns in March 1849 (see Figure 4.2).[15]

The Smithsonian system gave Espy the extra observations he needed from the South and West. He also gained new inspiration through a closer working relationship with his well-appointed and levelheaded scientific friends. He issued two reports to the secretary of the navy in quick succession, one on meteorological observations and one on experiments.[16] Henry and Bache reciprocated by lobbying strenuously in Congress for the renewal of Espy's appropriation—an annual event.[17] Indeed, Henry's diary for January 15 to April 4, 1849, records a remarkable three-month campaign to keep Espy on the government payroll:

> *1849, Mon. Jan. 15:* Letter from Mr. Espy—attend to his appropriation—Speak to Mr. Vinton—attend to his report left at the Navy Department to be given to the committee on printing. . . . *Tues. Jan. 16:* Saw Mr. Brown of Penn.

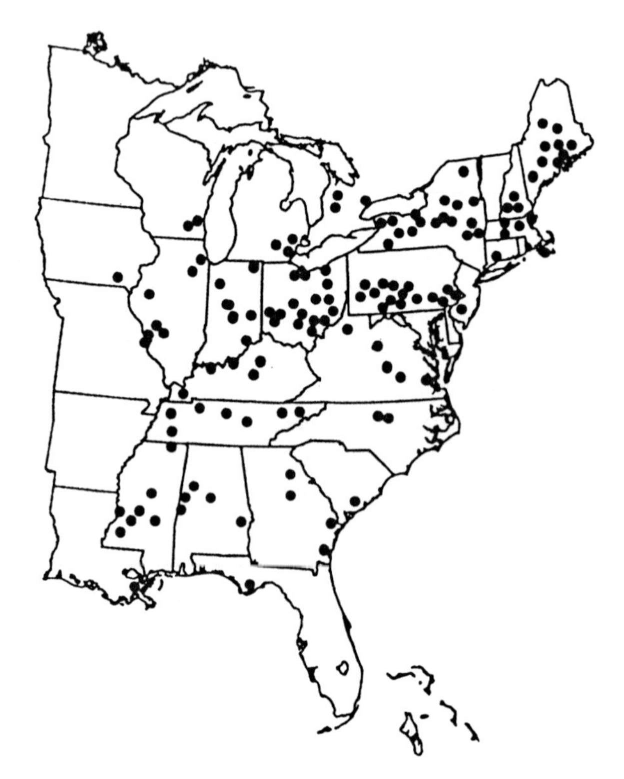

Fig. 4.2. Locations of Smithsonian observers, ca. 1850.

he promised to look after Mr. Espy's appropriation. . . . *Wed. Jan. 17:* Brow[n] about Mr. E's appropriation—Letter from C. W. Brown fears Mr. E's appropriation left out must be put in the Senate. . . . *Fri. Jan. 19:* Saw Mr. Brown of Penn. who promised to get the appropriation of Mr. Espy on the supply bill. . . . *Tues. Jan. 30:* Called upon Mr. Vinton relative to appropriation for Meteorology (Mr. Espy)—Saw Mr. Cameron; was informed that Mr. Espy's report had not been ordered printed. . . . *Thurs. Feb. 1:* Engaged this day the House & Senate endeavoring to get Mr. Espy's appropriation through—call upon chairman of executive printing committee—Mr. E's report had not been ordered to be printed—Got promise from the Secretary of the Navy that he would recommend the appropriation. . . . *Fri. Feb. 2:* Got letter of recommendation from Sec. of Navy in favor of Meteorological observations; gave the same to Mr. Vinton who promised to get the appropriation through. . . . *Tues. Feb. 6:* Wrote to Mr. Vinton urging the case of appropriation for Meteor[ology]; also to Mr. Brown—Saw Mr. Brown; will endeavor to get meteorological report upon firm basis under clerk here. . . . *Sat. Feb. 10:* Committee of Senate struck out Mr. E's appropriation. Informed Mr. Brown who has promised to have it put in again. . . . *Mon. Feb. 19:* Saw Mr. Brown of Penn. Promised to attend to Mr. Espy appropriation. . . . *Mon. Mar. 19:* Wrote to Mr. Espy to inform him of the history of his appropriation for Meteorology—

> Called at the Navy Dept.; Chief clerk promised to speak to the Secretary about Mr. Espy's appropriation. . . . *Thurs. Mar. 22:* Called upon the Secretary of the Navy to inform him of the arrangement previously made relative to Mr. Espy. . . . *Fri. Mar. 23:* Concluded to send letter blanks and other instructions to all correspondents of Mr. Espy. . . . *Wed. Ap. 4:* Mr. Espy—has been retained in the office of Meteorologist by the Secretary of the Navy.[18]

Beyond the genuine feelings of friendship and respect between the two scientists, it was Henry's practice to seek specialized consultants such as Espy for each of the Smithsonian's major research projects. Moreover, the navy's annual appropriation of $500 for Espy's expenses could be used to pay for the services of Smithsonian meteorological clerk Edward Foreman.[19]

The Observers: Science by Volition

> Every man is a valuable member of society who by his observations, researches, & experiments procures knowledge for men.
> —James Smithson, quoted in Henry Pocket Notebook, 1848, Henry Papers

In the nineteenth century, amateurs were able to make significant contributions to meteorology; indeed, groups of volunteer observers were indispensable for meteorology's progress. John Herschel stated the point thus: "[Meteorology] *can* only be effectually improved by the united observations of great numbers widely dispersed . . . [it is] one of the most complicated but important branches of science, . . . [and] at the same time one in which any person who will attend to plain rules, and bestow the necessary degree of attention, may do effectual service."[20] Henry made a related point by choosing Smithson's aphorism for the motto of the *Smithsonian Contributions to Knowledge.* Thousands of volunteer observers thus contributed to the increase and diffusion of knowledge. Figure 4.3 shows the rise and fall of the number of observers in the Smithsonian system between 1849 and 1874.

The Smithsonian had three classes of observers. Those of the first class had the most instruments, typically a barometer in addition to the ubiquitous thermometer, wind vane, and rain gauge. Some also had hygrometers. They reported on "Number one" forms originally devised by Espy. Observers of the second class had no barometers but followed the general guidelines of the first group. The third class had no instruments. All observes were requested to provide estimates of the direction and speed of the wind, type and amounts of clouds, and time and duration of precipitation. Following the practice of the army medical officers, observations were taken at "sunrise," 9:00 A.M., 3:00 P.M., and 9:00 P.M. Observers were instructed to use the "mean time" at their station. Some thirty years before the advent of "standard time," this was the best available option—close enough for climatological work but of limited value for

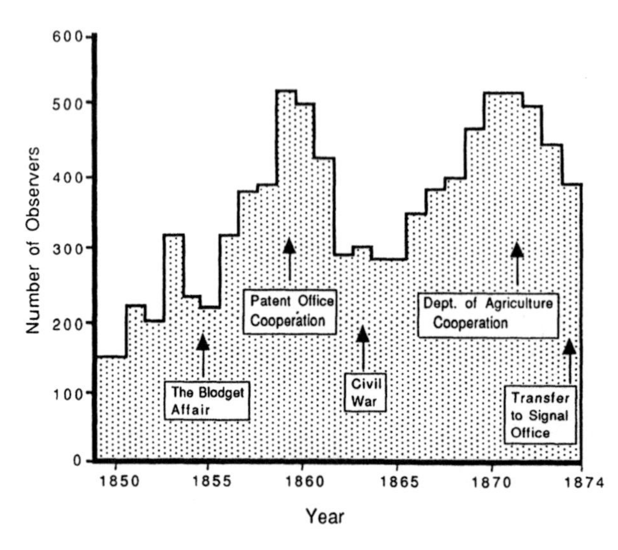

Fig. 4.3. Smithsonian meteorological observers, 1849–1874. Additional observations were available from cooperating agencies. The decline in 1854 and 1855 resulted from Henry's controversy with Lorin Blodget; the subsequent rise is attributable to Smithsonian partnership with the Patent Office. The sharp decline beginning in 1861 was caused by the Civil War. The decline after 1870 reflects the rising fortunes of the new storm warning system under the U.S. Army Signal Office.

precise comparisons or storm studies because the observations were not taken simultaneously.[21] Following the advice of Arnold Guyot, these instructions were superseded by a new form in 1853. Observations at sunrise were dropped because the varying time made reductions especially difficult. The new hours of observation became 7:00 A.M., 2:00 P.M., and 9:00 P.M. local time. Revisions were made in 1855 and again in 1860. National simultaneous observations were instituted by the U.S. Army Signal Office in the 1870s. In addition to their daily weather observations, correspondents were asked to collect other information on natural phenomena. At the request of naturalists, they frequently collected specimens of local flora and fauna and took notes on "periodical phenomena" such as the blossoming of plants, the migration of birds, and the hatching of insects. Some found specimen collecting more interesting than observing the weather (see Chapter 6).

Correspondents were also asked to observe "casual phenomena" such as thunderstorms, tornadoes, lightning, hailstorms, aurora borealis, meteors, solar and lunar halos, frosts, depth of frozen ground, opening and closing of waterways and their extreme rise and fall, the temperature of wells and springs, and the occurrence of earthquakes. Figure 4.4 is a reproduction of one of the few remaining completed forms in Record Group 27 in the National Archives. Most of the forms were microfilmed in the 1950s and the originals were destroyed. This example, by observers of the first class, was completed in May 1857 by the Reverend J. M. Hensley and Professor Henry How at King's College, Windsor, Nova Scotia. Under "casual phenomena," the observers noted occurrences of frost, aurora borealis, thunder and lightning, and so on. Observers sometimes sent in meteorological observations during solar eclipses.

Although the Smithsonian issued detailed guidelines to the observers, the reports of correspondents varied greatly in reliability, duration, content, and even format. Many returns were not complete, and others, as Coffin noted, were filled with "hieroglyphics," which were intelligible only to the observer:

> I have looked over the directions to observers in the blank forms, & I do not think of anything to suggest, unless it be a note calling the attention of observers to the importance of uniformity in the mode of recording their observations. The diversity now existing, each observer supposing that his own peculiar marks & mode will be understood by others, is a source of endless perplexities in reducing the observations as it renders it impossible for me to give general directions to my assistants without exposing them to run into frequent errors—such as it has taken a large share of the time I have been able to devote to the work to ferret out & correct. For example, in the barometer, some omit to record the *whole inches entirely*, others record them only when they differ from 29, others when they differ from the number *next above in the same column*, others when they differ from the next preceding *in the order of time*, & one set of registers only *the difference*, positive or negative from 29 inches is recorded. So also the ditto marks thus ("), some use it to denote the entry is the same as that next above in the same column, others that it is the same as that next preceding in the order of time, & others only to fill up a blank without any signification at all. There is nearly the same diversity in the use of the dash, besides numerous abbreviations, new-coined characters & hieroglyphics intelligible only to the person who makes them. It would be far better to make the record of each observation complete in itself without reference to another, & either to avoid private marks altogether, or to insert a *key* to them on each sheet that they return.[22]

Each observer received twenty-four blanks a year, one for each month for the observer and one to be returned to the Smithsonian. *Results of Meteorological Observations, 1854 to 1859*, a comprehensive two-volume compilation of observations on periodical phenomena, climate, and storms, was published by the commissioner of patents as special reports in 1861 and 1864 (see Chapter 6).

In a circular issued in 1849, the Smithsonian announced that instruments would be supplied to capable observers at certain important locations. The demand for instruments, however, quickly outstripped the supply on hand. The flood of applications forced the Smithsonian to change to a policy of defraying only half the cost of instruments for individual observers in remote locations. Colleges and other institutions were told to look elsewhere for funds, as were individuals in areas that were thickly settled. Telegraph offices and exploring expeditions were exceptions to this rule, but in general the instruments were merely placed on loan. Shipping the delicate instruments was also a problem as it had been earlier for the Joint Committee in Philadelphia. Many instruments were broken in transit. For example, Increase Lapham offered an express company three times its usual rate if it would use extra care and assume liability for shipping two Smithsonian barometers from the instrument maker

REGISTER OF METEOROLOGICAL OBSERVATIONS UNDER THE DIRECTION OF THE SMITHSONIAN

Place of Observation King's College Windsor **County of** Hants **State of** Nova Scotia B. N. America.

Latitude 44° 58' 34" N. **Longitude** 64° 7' 30" **Height above the sea** 200 feet circa.

Day of Month.	BAROMETER. Observed height. 7 a.m.	2 p.m.	9 p.m.	Thermometer attached to barometer. 7 a.m.	2 p.m.	9 p.m.	Barometer height reduced to freezing point. 7 a.m.	2 p.m.	9 p.m.	Mean.	THERMOMETER IN THE OPEN AIR. 7 a.m.	2 p.m.	9 p.m.	Mean.	PSYCHROMETER, OR HYGROMETER. Dry bulb. 7 a.m.	2 p.m.	9 p.m.	Wet bulb. 7 a.m.	2 p.m.	9 p.m.	RAIN AND SNOW. Time of beginning of rain or snow.	Time of ending of rain or snow.	Amount of rain or melted snow in gauge in inches.
1	30.52	30.475	30.46	46°	51.5	43.5	30.477	30.416	30.425	30.439	33.5	55.5	41.5	43.5				33°	49°	36°			
2	30.375	30.27	30.22	48°	51.8	41.6	30.236	30.210	30.170	30.235	40°	56°	45.5	47.16				39.5	50.5	43.5			
3	30.06	30.10	30.04	53.8	60.2	55.3	29.9946	29.915	29.970	29.9591	50.5	63.75	55.5	56.58				47°	59.5	54°			
4	30.16	30.18	30.16	57.2	58°	55.3	30.084	30.1017	30.0899	30.0911	49.5	60°	54°	54.5				49°	58	53°	1.45 p.m.	—	
5	30.05	30.03	29.94	58°	64.2	58.6	29.997	29.933	29.8605	29.9300	55.5	61°	57	60.16				55°	64°	54.5	—	9 p.m.	0.57
6	29.88	29.95	29.96	59°	56.8	51.5	29.799	29.8759	29.900	29.8583	58.5	52.2	44.5	51.7				57.5	50.5	44°			
7	29.99	29.995	30.09	55°	59.5	58°	29.927	29.9127	30.0119	29.9485	53.5	56°	50.5	54.1				52.5	55°	52.5			
8	30.03	29.97	30.10	56	58.3	57°	29.957	29.891	30.0249	29.9[illegible]9	51°	59.5	46.5	52.3				49.5	53.5	42°	6 a.m.	1 a.m.	0.11
9	30.22	30.14	30.07	50.5	63.6	60.3	30.164	30.045	29.985	30.0646	46.5	67.5	57.5	57.16				44.5	57°	50°			
10	29.855	29.64	29.59	55	67.3	57°	29.776	29.585	29.576	29.625	52	71.5	53°	58.8				47.5	57°	49°			
11	29.555	29.62	29.83	57.8	44.5	40.5	29.479	29.583	29.774	29.612	48.2	33°	35°	38.7				47.5	32.2	33.5	8 a.m.	5 P.M.	0.06
12	29.99	29.98	29.95	43.6	46.8	46.5	29.955	29.935	29.8761	29.9221	34°	43°	35.5	37.5				30	36°	32°			
13	29.95	29.83	29.81	46.8	54°	52.5	29.884	29.704	29.749	29.799	42	54°	47.5	48.5				35.5	44°	41.5			
14	29.800	29.72	29.73	53°	56.9	48°	29.737	29.646	29.6125	29.6885	48	54°	39.5	47.16				40	43.5	36°			
15	29.375	29.72	29.75	49.5	56.4	48°	29.662	29.647	29.7024	29.6704	39.5	55.5	45.5	46.8				37.5	44°	42.5			
16	29.75	29.79	29.87	44.2	52°	57.5	29.699	29.750	29.8117	29.7536	47.5	54.5	44°	48.66				45.5	49°	38.5			
17	30.01	30.05	30.12	50°	54.3	52°	29.952	29.983	30.0651	30.0016	39.5	51.5	43.5	44.5				42.5	42.5	40°			
18	30.23	30.23	30.22	52°	57.5	48°	30.169	30.153	30.1717	30.1545	41	57.5	44°	47.5				40	48°	39.5			
19	30.22	30.19	30.10	50.3	56.5	47.5	30.143	30.162	30.0536	30.1202	44.5	61°	48.5	51.3				42.5	57.5	46.5	1 P.M.	—	
20	30.035	29.98	29.68	54°	52.5	50.3	29.967	29.9180	29.6257	29.8371	47°	55.5	45.5	49.3				45	50.5	45°	1 P.M.	5 a.m.	0.1
21	29.67	29.67	29.71	55.3	55°	52°	29.571	29.5619	29.550	29.5076	49°	59°	50°	52.66				48.5	54°	47.5	—	1 a.m.	0.6
22	29.71	29.75	29.89	57°	57.8	54.7	29.635	29.673	29.822	29.710	52.5	54°	46°	50.8				49.5	50.5	44.5			
23	29.94	29.93	29.93	55.2	67°	55.5	29.876	29.825	29.8598	29.8570	50°	67.5	49°	55.5				49	57.5	46°			
24	29.925	29.92	29.98	57.5	66°	63.7	29.849	29.818	29.885	29.8576	53.5	65°	56.5	58.3				57.5	59°	53°			
25	29.98	29.92	29.89	53.8	76.4	66°	29.905	29.787	29.790	29.8270	61°	77°	65.5	68.5				60	64	57.5			
26	29.94	30.075	30.21	58.5	61°	51.7	850 29.337	29.988	30.156	30.0010	53.5	58.5	45.5	52.5				50.5	5[illegible]	41.5			
27	30.335	30.32	30.23	52°	56.2	49.5	30.274	30.2401	30.177	30.2316	43°	67.5	47.5	50.66				40.5	52°	43.5			
28	30.19	30.13	30.08	53°	59°	58.2	30.1177	30.0789	30.0013	30.0459	58.2	75°	58°	63.7				57°	57.5	57°			
29	30.00	29.93	29.86	58°	67°	60.3	29.922	29.8256	29.7757	29.8411	61°	68°	59°	62.60				59	63.5	58.3	2 P.M.	—	
30	29.7	29.7	29.75	62	65.2	64.8	29.611	29.6017	29.652	29.6215	62°	68°	60°	63.3				67	56°	57°	—	1.30 P.M.	0.33
31	29.88	29.88	29.915	60	65°	62.5	29.798	29.781	29.824	29.8012	56°	66°	56°	59.3				52	56°	57.5			
Sums.	929.615	929.100	929.135	1673.3	1829.1	1666.8	4691 927.6263	926.5945	927.0591	927.0933	1526.4	1852.95	1531.5	1634.2				1615.5	1630.2	1433.8			1.71
Means.	29.989	29.971	29.972	53.98	59.	53.76	29.9234	29.8907	29.9051	29.9062	49.24	59.71	49.4	52.7[illegible]				49.95	52.6	46.23			0.285

EXPLANATION OF THE

BAROMETER.

Observed Height.—Under this head are entered the records of observations made on the barometer at three hours daily—7 a. m., 2 p. m., and 9 p. m.

Thermometer attached to Barometer.—Under this head are the records of the thermometer attached to the barometer, observations of which are necessary to correct the barometer for variations in height of the mercury on account of difference of temperature.

Barometer reduced to Freezing Point.—In this space are placed the figures which indicate the observations on the barometer reduced to the height at 32° Fahrenheit, by means of a table; and also the mean or average of the three reduced daily observations.

THERMOMETER IN THE OPEN AIR.

This is intended for the register of the standard thermometer, and for the daily mean or average of the three observations.

PSYCHROMETER, OR HYGROMETER.

Under this head are entered the records of the dry and wet-bulb thermometers, from which the "Force of Vapor" and "Relative Humidity" are obtained by means of tables.

RAIN AND SNOW.

Under this head are entered the times of beginning and ending of the fall of rain or snow, and the amount, in inches and hundredths, of rain or melted snow collected in a gauge at the surface of the ground: also the depth of the snow.

CLOUDS.

Under this general head are entered three daily observations on the aspect of the sky, &c. 1st. The "Amount of cloudiness," designated by figures, 10 being entire cloudiness; 5 half cloudiness; and 0 entire clearness. 2d. "Course of the higher clouds," which pass directly over the head of the observer, as given to eight points of the compass. This observation is important, as the course of the higher clouds is sometimes different from that of the surface wind, which is given in another column. 3d. The "Velocity," or rate of motion, 10 being the highest and 0 apparent rest. 4th. The description or "Kind of clouds," to be entered by means of the following abbreviations: St. Stratus; Cu. Cumulus; Cir. Cirrus; Nim. Nimbus; Cir. st. Cirro-stratus; Cu. st. Cumulo-stratus; Cir. cu. Cirro-cumulus.

Fig. 4.4a (above) and b (overleaf). Completed Smithsonian "Register of Meteorological Observations," 1857. The form was oversized with printed instructions (not included here) on the back.

32 9

For the Month of May 185 7.

Name and address of Observer Rev. J. M. Hensley M. A. King's College Windsor N. S.
Henry How Esq. Prof. King's College Windsor N. S.

Depth of snow, in inches.	CLOUDS. 7 A. M. Amount of cloudiness.	Course of higher clouds.	Velocity.	Kind of clouds.	2 P. M. Amount of cloudiness.	Course of higher clouds.	Velocity.	Kind of clouds.	9 P. M. Amount of cloudiness.	Course of higher clouds.	Velocity.	Kind of clouds.	WINDS. 7 A. M. Direction.	Force.	2 P. M. Direction.	Force.	9 P. M. Direction.	Force.	FORCE OR PRESSURE OF VAPOR IN INCHES. 7 a. m.	2 p. m.	9 p. m.	RELATIVE HUMIDITY, OR FRACTION OF SATURATION. 7 a. m.	2 p. m.	9 p. m.
	0	c	0	0	0	0	0	c	0				N.	1	N. E.	1	W.	3	·182	·262	·140	95	59	53
	8	W	1	Cir. st.	5	W.	1	Cir. cum. Cir. st.	10				S. W.	2	S. W.	3	S. W.	2	·236	295	·255	95	55	84
	10	S. W.	2	Cat. Clou.	10			Cir. Clou.	10				S. W.	2	S. W.	3	S. W.	1	·328	·452	398	89	76	90
	Fog				10			Nim.	10			Nim.	0		S. E.	1	S. W.	1	·341	·436	390	90	88	93
	10			Nim.	10			Cum. Clou.	10			Nim.	S. W.	2	S. W.	2	S. W.	1	·427	543	·392	97	79	84
	2	W	2	Cir. st. Clou.	10			Nim.	Fog.				S. W.	2	N.	2	0		·460	349 (343)	·282 (368)	94	89	96 (100)
	9			Nim.	1	S. W.	2	Cir. st.	7	W.	1	Cir. cum.	S. W.	1	N.	3	W.	1	·382	·363	362	98	81	84
	10			Nim.	7	N. W.	3	Cir. Nim.	0				S. E.	2	N. W.	3	N. W.	1	·335	·330	208	89	65	66
	2	N. W.	1	Cat. Clou.	5	W.		Cir. st.	0				N. W.	1	W.	2	N. W.	1	·268	·326	·262	84	47	55
	2	W.	2	Cir. st.	1	W.		Cirri.	0				W.	2	W.	4	W.	1	·270	·273	·295	69	35	73
	10			Nim.	10			Nim.	0				N.	2	N.	4	N.	2	325	·188 (172)	·173	96 (84.5)	89 (81)	85
	0				0			0	0				N. W.	4	N. W.	3	N. W.	1	·120	·121	·136	61	43	65
	0				7			Cirri	10				W.	4	W.	5	N. W.	1	·128	·157	·184	46	38	55
	0				1			Cum.	0				N. W.	3	W.	4	0		143	·145	·167	43	35	58
	6	N. W.	1	Cir. st. Nim.	4	N. W.	3	Cir. st.	6				S. W.	1	N. W.	3	E.	1	·199	·137	·233	82	31	76
	10			Nim.	10			Nim.	16				N. E.	1	E.	2	N. E.	1	·279	·275	·162	85	65	55
	10			Nim.	7	W.	1	Cir. st. Clou.	6			Nim.	N.	3	N.	3	N.	1	·190	·152	202	82	40	71
	0				3	W.		Cat. Clou.	0			Nim.	E.	1	N.	2	0		·235	·216	·185	91	44	63
	3	W.	1	Cirri.	8	W.	4	Cir. Nim.	10			Nim.	E.	1	W.	4	S. E.	3	·245	255	291	84	47	85
	10			Nim.	10			Nim.	10				S. E.	2	S. E.	4	S. E.	2	273	·307	293	85	61	96
	10			Nim.	10			Nim.	10			Nim.	S. E.	2	E.	2	E.		·335	·352	·246	96	70	82
	10			Nim.	8			Nim.	0			Nim.	S.	1	N. W.	3	N. W.	1	·314	·321	·275	79	77	88
	0				6			Cir. Cum.	0			Nim.	S. E.	1	S. W.	4	0		·335	341	·272	93	57	78
	0				1	S. E.	1	Cirri.	2				N. W.	1	S. E.	1	N. W.	1	·355	·420	·357	86	68	78
	0				1			Cirri.	6				c		S. W.	4	S.	1	·305	·399 (6)	·367	94	40	58
	10			Nim.	4			Cir. st.	1				N. E.	4	N. E.	4	N. E.	2	328	·275	·276	80	56	58
	1			Strat.	6			Cir. st.	10			Nim.	N.	2	S. W.	4	N. E.	2	·227 (226)	·263	·304	93 (79)	47	92
	10			Nim.	7			Cum.	16			Nim.	S. W.	2	S.	3	S.	1	·450	·873	·452	93	65	94
	10			Nim.	10			Nim.	10			Nim.	S.	2	S. W.	5	S. W.	1	·473	·526	·481 (478)	88	77	96
	10			Nim.	9			Nim.	4			Cum.	S. W.	1	S. W.	1	N. W.	1	·523	·612	·426	94	90	82
	0				3			Cum.	0			C	N. W.	3	N. W.	3	N. W.	1	·336	·316	·322	75	49	72
																			9·520 (30)	9·965	8·773	2617	1876	2385 (2401)
	5.4				5.7				4.9										·307	·321	·283	84	60	77

5.3

ABOVE COLUMNS.

WINDS.

This is for the record of the direction of the wind as indicated by a vane, and its force by estimation. The direction is entered in eight points of the compass: N., N.E., E., S.E., S., S.W., W., N.W. The force is to be estimated and registered by the following table, in figures, from 1 to 10:

1. Very light breeze	2	miles per hour.
2. Gentle breeze	4	" "
3. Fresh breeze	12	" "
4. Strong wind	25	" "
5. High wind	35	" "
6. Gale	45	" "
7. Strong gale	60	" "
8. Violent gale	75	" "
9. Hurricane	90	" "
10. Most violent hurricane	100	" "

FORCE OR PRESSURE OF VAPOR.

In this space are entered the numbers which indicate the inches of mercury of a barometer which the force of vapor of the atmosphere alone will sustain. The whole pressure of the atmosphere, as given under the head of "Barometer," is due, conjointly, to the weight of the air and the moisture which it contains. The figures under this table indicate the separate pressure of the moisture, and are obtained from the records of the wet and dry-bulb thermometers by means of a table based on experiment.

RELATIVE HUMIDITY, OR FRACTION OF SATURATION.

The quantity of water in a state of vapor which a given space can contain, depends on the temperature. When a given space has as much water as it can hold at a given temperature, it is said to be saturated with vapor. If the temperature be lowered, a portion of the vapor will be condensed into water, but the space will still be saturated. If the temperature be raised and no more vapor be admitted, the space will only be partially saturated. The numbers under the head of "Relative humidity" therefore denote the per-centage of saturation; full saturation being indicated by 1, and half saturation by 0.5. These figures are deduced from the corresponding forces of vapor by means of a table in the Smithsonian Collection.

[☞ SEE OTHER SIDE.]

CASUAL PHENOMENA.

Note observations of the following:

THUNDER STORMS—Time of occurrence and direction of motion. TORNADOES—Time of occurrence, width and direction of path, effects produced, and whether attended by electricity or hail. LIGHTNING AT A DISTANCE—Time of occurrence, direction from observer, whether zigzag, forked, or diffuse. Objects struck by lightning, as trees, buildings, &c. HAIL STORMS—Time of occurrence, direction and width of path, size and quantity of stones, and amount of injury. AURORA BOREALIS—Time of appearance and disappearance; time of the formation of arch, beams, and corona, and whether there is a dark cloud below the arch. Direction and time of occurrence of METEORS, SHOOTING STARS, SOLAR and LUNAR HALOS, PARHELIA, and PARASELENES. Time of early and late FROSTS, particularly first and last. DEPTH OF GROUND FROZEN, in feet and inches; disappearance of frost from the ground. Time of closing and opening of RIVERS, LAKES, CANALS, and STREAMS, and their extreme rise and fall. TEMPERATURE of wells and springs at least once each season. EARTHQUAKES—Time of occurrence, direction of impulse, number of shocks, and effects produced.

May. 1st. White frost during the night.
May. 10th. Brilliant Aurora Borealis Commenced soon after dark. At 9 p.m. the arch nearly reached the zenith, touching the horizon nearly in the N. E. and N. W. points. The dark cloud underneath was about 12° in altitude. At this time there were brilliant undulations, but little radiation. The light was sufficiently strong to allow the reading of small Capitals. The Moon rose about 10 p.m. and interfered with the spectacle, but at 11.30. p.m. in spite of the moon-light there could be seen at times rays and streamers of intense brilliancy, shooting up towards the zenith.
May. 11th. Squalls of snow.
May. 12th. Thick ice in the morning.
May. 30th. Thunder and lightning attended with heavy rain. The storm travelled from W. to E.

Minimum reading of Thermometer on 11th day. 33°.
Maximum — " — 25th day. 79°.
Greatest difference between wet and dry Bulbs. 25th day. 15°.

A remarkable depression of temperature occurred on the 11th and few following days, agreeing with the long observed phenomenon of European Meteorology, noticed in the Cosmos.

Fig. 4.4b

James Green in New York City to his home in Milwaukee, Wisconsin.[23] Even if the instruments arrived safely, they were not always installed properly, and observations were not always recorded with precision. Figures 4.5 and 4.6 show the locations in 1860 and 1870 where at least a barometer was included in the set of instruments. All other observers had only thermometers or were entirely without instruments. Some encountered unexpected problems. The report of one military survey team showed an abrupt break in the thermometric record and indicated that the thermometer, which had been placed in the shade of a wagon, had been "stolen by the Indians."[24]

Responses from the Field

> When I began to keep an abstract of the weather for the Smithsonian Institution, I had no idea I should attend to it more than one year, but the more I watched the Clouds and the changes in vegetation & the various Seasons of the year as they rapidly came round, the more I became interested in everything relating to the object of your Institution.
>
> —Rufus Buck to Smithsonian Institution, August 5, 1853, SIMC-NA

The Smithsonian project linked amateurs and professionals in a task that encompassed the entire United States for twenty-five years. Meteorological observations, especially those from remote locations and those taken faithfully

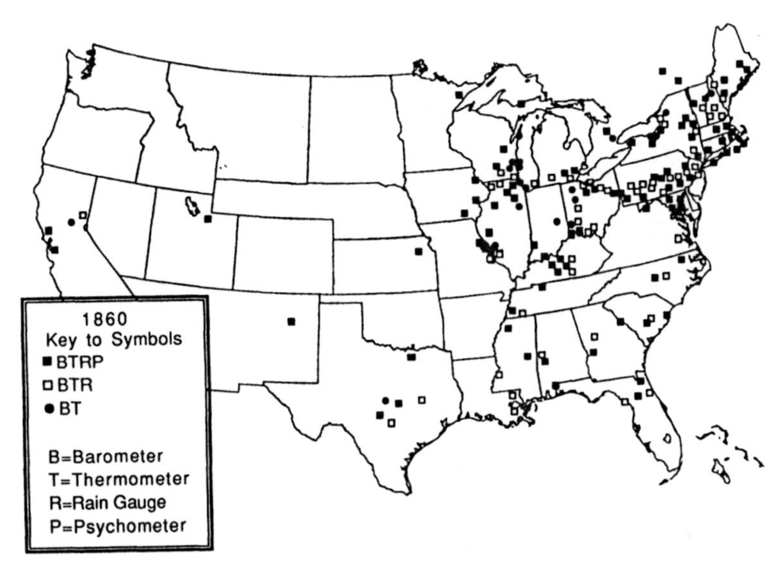

Fig. 4.5. Locations of Smithsonian observers having barometers and other instruments in 1860.

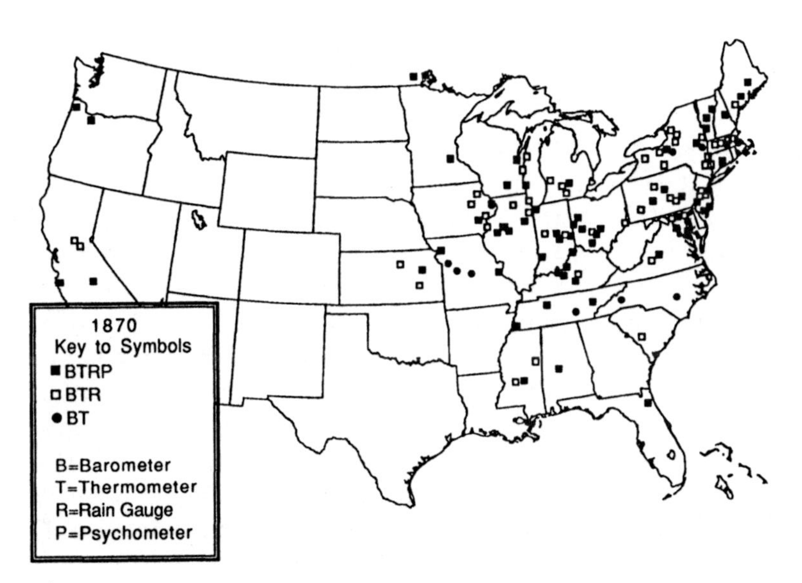

Fig. 4.6. Locations of Smithsonian observers having barometers and other instruments in 1870.

over many years, were a way for nonelites to contribute to the advancement of science. Perseverance was valued more than profound thought, cooperation more than creativity.

Volunteers joined the Smithsonian meteorological project for various reasons. Some had kept weather journals over long periods for their own amusement and now found an outlet in which their contribution was valued. Ovid Plumb, a physician in Salisbury, Connecticut (see Appendix) wrote that he had long kept a journal of observations taken three times a day "for the Amusement of my self & family, & I believe very accurately."[25] Another, a former student of Henry at Princeton, now a successful minister in rural Mississippi, was seeking a philosophical complement to his "soul invigorating profession" and was convinced to join the system by Henry's "beautiful exposition of Prof. Espy's beautiful Theory of Storms."[26] Professors, principals, and college presidents interested in stocking their libraries and laboratories requested government reports and free instruments.[27]

The observers were accorded special status in Henry's eyes. Although they were not paid for their efforts, they were placed high on the list of those eligible to receive the Smithsonian *Annual Reports* and *Contributions* and other government publications. They could also receive free advice from Washington on a wide range of scientific questions. Some even were given instruments free or at a reduced price. The project served to spread the aura and the apparatus of science to isolated areas far from the intellectual centers of the East Coast. Individual contributors gained the pleasure and prestige of participating in a national scientific undertaking. Some became known locally as "Smithsonian observers"; a few, who kept reliable, long-term records in one location, were rewarded with publication of their journals in the *Contributions*.[28] Others, such as John Chapplesmith, wrote articles on local phenomena.[29] Chapplesmith, a wealthy Englishman and resident of the utopian community of New Harmony, Indiana, was an eager observer who reduced his own data and attempted to make his own correlations. On April 30, 1852, a tornado passed near New Harmony, and Chapplesmith, following the great American tradition of surveying tornado damage, set out to test the centripetal and rotatory storm theories of Espy and Redfield: "I forward for the SI a map and memoir relating to the tornado that passed near this place in April last. . . . Although the facts presented are opposed to some of Mr. Espy's statements respecting the phenomena, still my opinion is that they abundantly support Mr. Espy's theory, that the devastation in tornadoes is produced by an ascentional column of air."[30] Chapplesmith's article, published in the *Contributions,* contains numerous illustrations, among them the one reproduced in Figure 4.7.

A Collective Biography of Observers

> The observers are generally persons engaged in occupations which admit to some extent of their being present at the place of observation at the required hours of the day all the year round. . . . The classes to which the observers belong, are professors in colleges, principals or

Fig. 4.7. "Monarchs of the forest felled by the meteor," quarto plate from John Chapplesmith, "Account of a Tornado near New Harmony, Indiana, April 30, 1852," *Smithson. Contrib.* 7, Article II (1855). "The quarto plate is intended to represent a group of trees, on the track of the tornado, about three hundred yards a little west of south of the five mile post on the New Harmony Plank Road."

> teachers of academies, farmers, physicians, members of the legal and clerical profession, and a few engaged in mechanical and mercantile pursuits.
> —Edward Foreman, *Smithsonian Report,* 1851

The documentation preserved in the Smithsonian Institution Archives and the National Archives allows a closer look into the nature of the meteorological system. The names, locations, and dates of activity of all the Smithsonian observers are on record. In some cases the instruments they used are known. What can be said about their education, vocation, and social standing? How did the group of observers change over time? To address these questions, a collective biography, presented in Appendix A, was prepared for a random group of Smithsonian observers. In the words of Lawrence Stone, "collective biography . . . has developed into one of the most valuable and most familiar techniques of the research historian."[31] In the history of science, however, collective biographies often have been restricted to successful or powerful populations. The collective biography presented here is meant to provide an impressionistic glimpse into the mundane affairs of the "little actor who belonged to no powerful élite" and in so doing to uncover dimensions of the meteorological project heretofore neglected.[32]

This prosopography, however, is not an end in itself. It is meant to complement the theoretical and institutional dimensions of this study. Given the choice between ignoring the issue of the role of the individual observers (standard practice for most institutional histories) and abandoning history of science for collective biography (more than two thousand individual observers represent a daunting task), a representative sample was taken of observers active in the years 1851 (the first year with reliable information), 1860 (the system's antebellum peak), and 1870 (just before the end of the project). This was deemed an acceptable compromise and a means of reducing an intractable problem to the realm of possibility. The names of observers for whom biographical information was compiled were chosen at random using a regional weighting system. For the year 1851, a sample of 36 observers was selected from a total of 155; for 1860, the sample was 42 observers drawn from a total of 500; and for 1870 a sample of 43 observers was selected from a total of 507. These samples represent 23.2, 8.40, and 8.48 percent of the total number of observers active in those years.

What assurance do we have that the chosen samples reveal significant features of the entire population of observers? For example, using the sample data to determine the average number of years that individuals participated in the system produces a relative error (D) that is calculated using equation 1.

1. $$D^2 = 3.84\, v^2\, (1/n - 1/N),$$

where v^2 is the relative variance, n is the sample size, N is the population size, and

2. $$v^2 = \frac{s^2}{\bar{x}^2} = \frac{\sum_{i=1}^{n} (x_i - \bar{x})^2}{(n-1)\,\bar{x}^2}.$$

Here s^2 is the sample variance and $\bar{x}$ is the sample mean.[33] The sample mean for 1851 was 5 years with a relative error of 30 percent. This means that with a confidence level of 95 percent, the sample mean plus or minus 1.5 years is the interval that contains the true population mean. To reduce the uncertainty of this estimate by 0.5 year, the sample size would need to increase from 36 to 63—almost double. Following the same procedures for 1860 gives a sample mean of 6.4 years and an uncertainty of plus or minus 1.6 years—a relative error of 25 percent. The figures for 1870 give a sample mean of 8.2 years and an uncertainty of plus or minus 1.9 years—a relative error of 23 percent. To reduce the uncertainty to plus or minus 1 year, the sample size would have to increase from

43 to 121—almost triple. Table 4.1 summarizes these statistics. The distribution of observers' participation in the system is presented in Figure 4.8.

In the 1851 sample we find Coffin, author of "Winds of the Northern Hemisphere" (1853) and "Winds of the Globe" (1875) and meteorologist in

TABLE 4.1. Statistical summary for sampled observers, 1851, 1860, and 1870

Sample Year	Population Size N	Sample Size n	Relative Variance v^2	Relative Error D (%)	Sample Mean $\bar{x}$ (yrs.)	Uncertainty[a] (yrs.)
1851	155	36	1.115	30	5.0	±1.5
1860	500	42	0.745	25	6.4	±1.6
1870	507	43	0.624	23	8.2	±1.9

[a]95% confidence level.

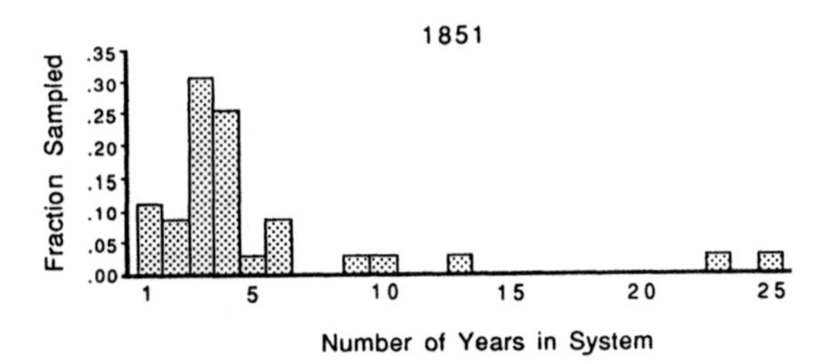

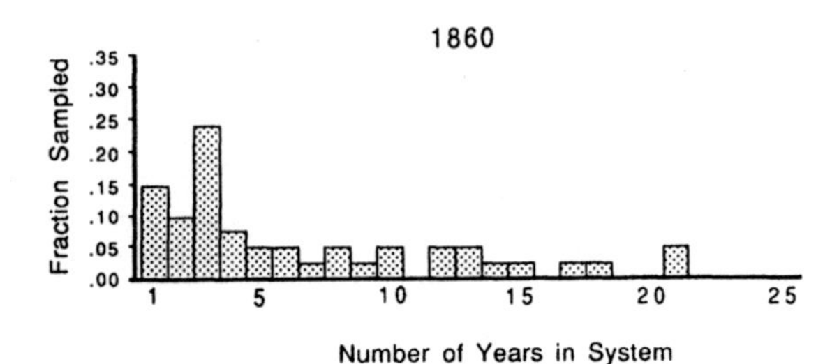

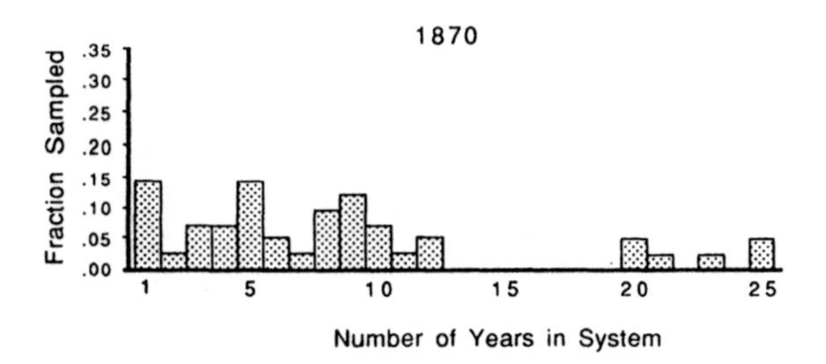

Fig. 4.8. Number of years individuals participated as observers in the Smithsonian system, 1851, 1860, and 1870.

charge of data reduction for the Smithsonian Institution (see Chapter 6). In the field of education, notables include the Reverend Levi Washburn Leonard, who organized the lyceum, library, schools, and town government of Dublin, New Hampshire. Professor William Williams Mather, "the celebrated geologist and mathematician," was both a meteorological observer and acting president of Ohio University during the college's "dark ages" of underfunding.[34] Josiah Little Pickard, after nine years as an academy principal and observer, had a distinguished career as the superintendent of public instruction of Wisconsin, the superintendent of public schools of Chicago, the president of the University of Iowa, and the president of the National Educational Association. In the 1860 sample notables included Robert Hallowell Gardiner, founder of the Gardiner Lyceum and member of the Maine House of Representatives, and Oliver Hudson Kelley, founder and first secretary of the Patrons of Husbandry (the Grange). Dr. Zina Pitcher was president of the American Medical Association during the time he participated in the Smithsonian meteorological project. In 1870, the only active scientist sampled was Asa Horr, president of a local scientific society in Iowa. He conducted experiments in 1861 using the telegraph to warn shippers on the Great Lakes of impending storms.

Table 4.2 below summarizes the occupational categories and status indicators of the sample. Initially the Smithsonian meteorological project attracted a rather eminent group of observers. Almost half of the 1851 sample (47 percent) were engaged in scientific, technical, or educational occupations; 28 percent appeared in the standard biographical dictionaries; 14 percent were listed as authors in the *Royal Society Catalog of Scientific Papers;* and 15 percent were AAAS members. By 1860, these percentages had decreased to 23, 17, 7, and 4. In 1870 they were even smaller: 16, 9, 2, and 1 percent. During the same period the percentage of farm workers in the sample rose from only 8 percent in 1851 to 37 percent in 1870.

TABLE 4.2. Occupational categories and status indicators of sampled observers

		Percentage of the sample		
		1851	1860	1870
Occupational Category	Scientific, technical, and educational	47	23	16
	Agriculture	8	17	37
Status Indicator	Published biography	28	17	9
	Royal Society Catalog listing	14	7	2
	AAAS members	15	4	1

The steady decline in status indicators from 1851 to 1870 may be attributed to several related factors. Certainly, the Smithsonian meteorological project attracted greater attention within the scientific community during its formative, experimental years when a solution to the storm problem was actively being sought. Moreover, the number of farmers in the system increased with time, largely because of Henry's cooperative programs with the Patent Office and the Department of Agriculture. The developing infrastructure of American science also played a major role in the decreasing status indicators. Observer-professors active in 1851 were replaced at their institutions by individuals who typically had more formal education, more professional opportunities for research and publication in their specialties, and less time to spend as passive and poorly rewarded weather observers. The chemistry faculty members at Beloit College illustrate this point. From 1849 to 1854 professor of chemistry and natural science S. Pearl Lathrop, M.D., was an observer for the Smithsonian system. His successors on the chemical faculty all had Ph.D.'s in chemistry from Göttingen: Henry Bradford Nason (1858), organizer and later president of the American Chemical Society and the editor of two manuals on chemical methods in mineralogy; Elijah Paddock Harris (1866), author of a very popular chemistry textbook; and James H. Eaton (1868), the only Ph.D. on the faculty at the time.[35]

The Smithsonian meteorological project provided standardized instruments, uniform procedures, free publications, and a sense of scientific unity which extended beyond the normal reach of colleges and local scholarly societies. Data compilations shifted from private diaries to published volumes, from local series lasting several years to more universal collections spanning the continent and the century. The Smithsonian project formed a "seedbed" for the continued growth of theories rooted in data.[36] To *increase* knowledge of the atmosphere it sponsored original research on storms, climatic change, and phenology; to *diffuse* knowledge it published and distributed free reports, instructions, and translations. As Loomis had hoped, it became the center of America's "grand meteorological crusade."[37]

CHAPTER 5 Stormy Relations among Theorists and Administrators

Consultation with Prof. Bache on the subject of meteorology—agrees to the proposition that: 1. The Army take the West, 2. The Smithsonian the East, 3. The British government the North, 4. & The observatory the sea, 5. The returns of all given to the S.I.
—Henry Desk Diary, January 26, 1852, Henry Papers

Although "cooperation—not monopoly" was Henry's motto as he attempted to "supplement and harmonize" the efforts of various groups involved in the collection of meteorological statistics, harmonious arrangements were sometimes established more in Henry's imagination than between various organizations and government agencies.[1] Competition between strong-willed and jealous bureau chiefs was the rule as often as it was the exception. The storm controversy, now in its third decade, continued to rage with Henry in the role of informal moderator. Maury, at the Naval (or National) Observatory, and Surgeon General Lawson, head of the Army Medical Department, found reasons to contest the hegemony of the Smithsonian meteorological project. Since no single group had the resources or the authority to cover the entire continent, controversy was inevitable.

The Storm Controversy Revisited, 1848–1860

> The great object . . . of the Smithsonian collection of meteorological observations is, to settle definitely the question as to the origin, progress, and character of the winter storms of our continent. On this subject it is well known there are various opinions.
> —Henry, *An Account of the Smithsonian Institution*, 1854

Henry's plan for a meteorological project at the Smithsonian to "solve the problem of American storms" was not universally welcomed among meteorologists. Espy was an eager collaborator, but Redfield felt that the question of storms had already been settled in favor of his whirlwind theory.[2] Hare, however, was just beginning a new assault on Redfield and Espy. In published salvoes, at gatherings of the AAAS, and in formal requests to the government, the public disagreement over storms continued throughout the 1850s. Table 5.1 lists the dates of published exchanges during the third phase of the controversy.

TABLE 5.1. The American storm controversy revisited, 1849–1857 (see bibliography for full citations)

Hare v. Redfield (1850–1853)
1. "On the Whirlwind Theory of Storms," *AAAS, Proceedings.*
2. *On the Law of Storms.*
3. "Dr. Robert Hare on the Cause of Storms," *Merchants' Mag.*
4. *The Whirlwind Theory of Storms.*

Hare v. Espy (1851–1852)
1. "Strictures on Professor Espy's Report on Storms to the Secretary of the Navy, as Respects the Theoretical Inferences," *AAAS, Proceedings.*
2. *Queries and Strictures by Dr. Hare, Respecting Professor Espy's Meteorological Report to the Naval Department.*

Redfield v. Espy (1850)
1. "On the Apparent Necessity of Revising the Received Systems of Dynamical Meteorology," *AAAS, Proceedings.*

Espy v. Redfield (1849–1857)
1. *Second Report on Meteorology to the Secretary of the Navy.*
2. *Fourth Meteorological Report.*

Espy v. Hare (1851)
1. *Third Report on Meteorology to the Secretary of the Navy.*

Redfield, the reluctant chairman of the inaugural meeting of the AAAS in 1848 in Philadelphia, used the occasion to present to the assembly a copy of Henry Piddington's *Horn Book of Storms* in support of his rotary theory of hurricanes. Hare "immediately rose and delivered himself of an argument against the *possibility* of the rotation" and, after a brief intervention by Maury, proceeded with "allegations and remarks" which, Redfield noted, "had long since been presented to the public." Redfield registered a "complaint and caveat" on Hare's behavior and proceeded with other business, but not before Hare had promised to "come forth again" at the next annual meeting.[3]

At the meeting in Cambridge, Massachusetts, the following year, Hare urged the committee on meteorology to allow a more general hearing on the theory of storms but, instead of attacking Redfield, introduced an experiment to demonstrate that rarefaction by heat could not produce Espy's up-moving columns of air.[4] The three controversialists met face-to-face at the 1850 meeting in New Haven. There Redfield attacked "the received systems of dynamical meteorology," and Hare, "in an off-hand speech, fluid and animated, assailed the views of Mr. Redfield, who was all the while a quiet and silent listener."[5] Espy, by this time a collaborator with Henry in the Smithsonian meteorological project, avoided a direct confrontation and presented a noncontroversial report of his experiments on the expansion of moist air at low temperatures.[6]

The controversy erupted again in Albany in 1851, when Hare presented his "strictures" on Espy's annual report to the navy. When Hare mentioned the electrical effects of the Brunswick spout in 1835, Arnold Guyot replied impatiently that the storm controversy, now in its sixteenth year, was caused by "theorizing instead of resorting to received observations of unquestionable accuracy." He suggested that the proper way to resolve such conflicts was through a national observing program like that at the Smithsonian.[7] Hare missed the point and responded by noting that observations unassisted by generalization were often the source of error:

> The observations of Redfield are irreconcilable with those of Espy and Loomis though never were there more zealous observers occupied in an inquiry. . . . Amid these discordant results of practical observation generalization comes in to try them severally by the well ascertained laws of nature founded on facts universally admitted by men of science. . . . There is manifest deficiency in the hypotheses both of Espy and Redfield that the part performed by Electricity is wholly neglected although its dazzling and thundering manifestations ought to make the blindest observer both see and hear them. No explanation of storms can be satisfactory which fails to account for the presence of this all powerful agent.[8]

Overall, the AAAS meetings created heated exchanges but shed little light on the fundamental issues of the storm controversy.

In 1849 Redfield sent a memorial to the secretary of the navy requesting official recognition as the theorist of whirlwinds in the Atlantic and proposing that the navy purchase copies of Piddington's *Horn Book* and Reid's *Progress of the Development of the Law of Storms* to be used as guides for navigators on navy vessels and as textbooks for officers at Annapolis.[9] The Navy Department declined Redfield's proposition.[10] About the same time, Espy submitted his *Second Report on Meteorology* to the secretary of the navy. The report was critical of Redfield, Reid, and the other rotary theorists and was issued as a government publication.[11] Because of Espy's official position with the navy, this represented a severe rebuff for Redfield and necessitated a further response. Redfield was worried that the government patronage Espy had been receiving since early in the Tyler administration would be construed as "*the official endorsement by government*" of Espy's theory. Yet, Redfield sniped, "we have overwhelming evidence that this alleged theory of storms and gales is a positive error; and that its practical application must often and necessarily be attended by injurious and even fatal consequences." Redfield was upset that his memorial to the navy had been rejected "while *thousands* had been expended, perhaps properly, for meteorological labors and services which have led to no such important results. . . . To such expenditures, however, I ~~do~~ will not object: although the effect of its connexion with Mr. Espy at least, is to mystify the public mind as regards a true theory of storms, and thus shut out from practical use, in our national and commercial marine, the great truths of the *rotative character* and *determinate progress* of storms, which are acknowledged to have been first clearly made known in our own country; which other navigators, in various seas, are deriving important benefits therefrom."[12]

While Redfield was complaining to the navy, Espy was extending his observations to naval stations, on ships at sea, and at lighthouses, using Smithsonian blank forms and instructions.[13] He also conducted new experiments with an instrument he developed called the double nephelescope (literally a cloud-scope) to determine the properties of "atmospheric air," oxygen, hydrogen, and carbonic acid.[14] The later experiments, apparently related to medical inquiries, were designed "to ascertain whether the quantity of carbonic acid generated by respiration varies with the dew point—as the quantity of [water] vapor generated in the lungs certainly does."[15] In 1859 he claimed to have used the double nephelescope to prove that James P. Joule's unit for the mechanical equivalent of heat was equal to his results for the specific caloric of air. When applied to the case of air under and above the base of clouds, the experiments supported his convective theory of storms (see Figure 5.1).

In the single nephelescope, a condensing pump *a*, is used to force air into the glass vessel *b*, and its degree of condensation can be measured by the barometer gauge *c*. When the stopcock was opened, the compressed air rushed out suddenly and cooled by expansion, the amount of cooling being indicated by the change shown on the barometer gauge. In this instrument air had to be compressed before it could expand and cool. The double nephelescope obviated this requirement and provided more flexibility. The improved instrument consisted of two vessels, the larger one, *a*, made of copper and the smaller one, *b*, made of glass; their relative sizes were as twenty-five to one. The vessels were connected by a pipe with a stopcock, *c*. Each vessel was connected to a barometer gauge, *b* and *e*. An air pump connected through a leaden tube, *d*, could evacuate or compress the gas in either vessel.

Other experiments performed by Espy demonstrated that water vapor did not behave as an ideal gas in the free atmosphere and, in the absence of convection and advection, its diffusion was extremely slow. That is, the meteorological conditions in the free atmosphere, such as vertical and horizontal wind currents, and temperature changes were much more important than Dalton's law of partial pressures in distributing water vapor. Espy also designed an experimental demonstration of the effect of updrafts in a waterspout. With Henry and members of Congress present as witnesses, Espy used a powerful fan to induce a vacuum equal to several inches of mercury in a vertical glass tube several feet in length. Air rushed into the mouth of the tube with a velocity proportional to the square root of the diminished pressure in the tube—about 240 feet per second for a pressure reduction of one inch of mercury. A basin of water placed near the mouth of the tube was drawn up rapidly and carried off

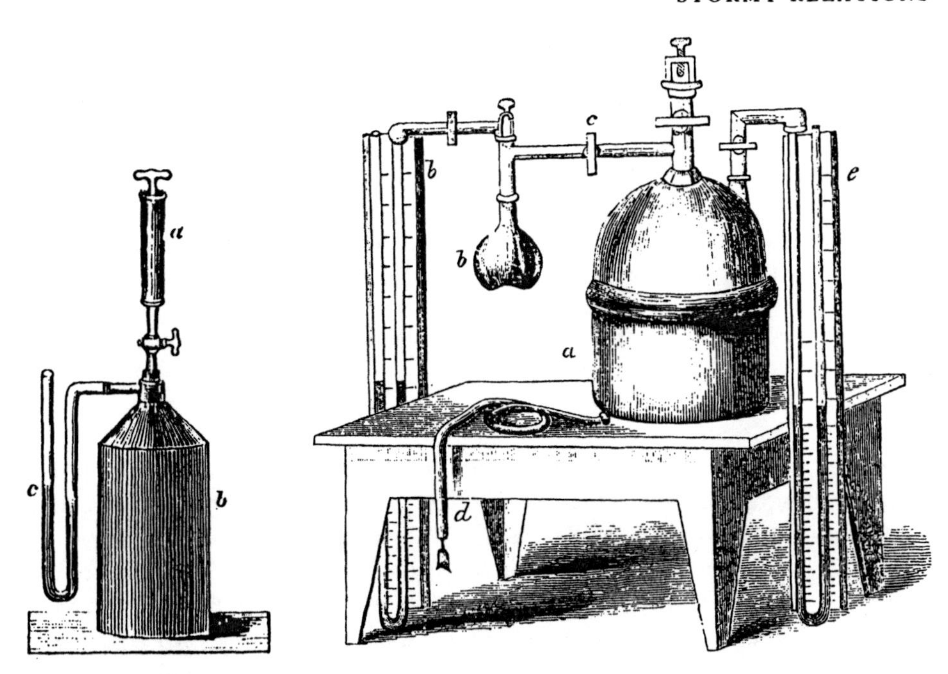

Fig. 5.1. Espy's nephelescope (1830s), which he took on his European tour (*left*), and (*right*) double nephelescope (ca. 1849) used at the Smithsonian in the 1850s. From James P. Espy, *The Philosophy of Storms* (Boston, 1841) viii; and *Fourth Meteorological Report*, U.S. Senate, Ex. Doc. 65, 34th Cong., 3d sess., 42.

up the tube, imitating a miniature waterspout and spreading out above in an ascending inverted cone of spray.[16]

Espy also continued his attempts to generate artificial rain by setting fire to large tracts of forest. In 1849 he contracted for twelve acres of timber in Fairfax County, Virginia, "grown up with pines as thick as a man's leg or arm," to be cut and burned in the hope of producing an intense column of heated air, clouds, and artificial rain.[17] He asked Henry to guarantee the owner of the woodlot $60 for supervising the experiment in his absence.[18] Henry did not expect much: "The conditions necessary to success are too many to occur simultaneously unless by unhoped for good luck."[19] Nevertheless, Espy tried the experiment in the last week of July 1849 but made sure that if rain was not produced, the failure could be attributed to unfavorable ambient conditions and not to any deficiency in his convective theory.[20] The experiment failed.[21] Henry warned Espy not to make extravagent claims about artificial rainmaking, especially in his reports to the secretary of the navy. He thought such impractical suggestions were unscientific and could only fuel the controversy between Espy and Robert Hare:

> Mr. Espy has shown me his copy of his Report to Congress and I have informed him that I regret very much that he had not shown it to me before the impres-

> sions to supply the Senate were struck off. Had he done so, I would have insisted upon his modifying some passages. The extra copies are however still in press, and I have endeavored to change the language, so as to render the parts, above alluded to, less objectionable. He brings forward again the old proposition of producing rain by firing patches of woods, along a line seven miles in length: a proposition in my mind so entirely impractical that, as I informed him, should one of his enemies get hold of it, the influence of yourself, myself and all his other friends, would not be sufficient to sustain him. Dr. Hare has been busily engaged in preparing an attack upon some parts of this Report of Mr. Espy; I shall endeavor to cool him down and prevent a war of words.[22]

Indeed, Espy's *Third Report* did offend Hare by calling his electrical theory of storms "unphilosophical." Adding injury to insult, Espy used some of Hare's chemical apparatus, donated to the Smithsonian in 1848, in his experiments without authorization and in a careless way.[23] During a visit to Washington, Hare found his donation in a state of "utmost confusion." Many of the chemicals were unlabeled, and empty glass vessels were strewn in the Smithsonian cellar "in all degrees of destitution." Hare discovered one vessel of particular value on an open shelf in use by Espy to hold "common salt."[24] Hare, until now a supporter of Espy's centripetal patterns, although advocating electrical causation, exploded with rage and called for a debate with Espy in the halls of Congress. Henry's diary records: "*1852, Wed., May 26:* Dr. Hare wishes to lecture before members of Congress on his theory of storms—He gave me an exposition of his theory of electrical polarization—very ingenious but indefinite and not minutely worked out. *Sat., May 29:* Mr. Espy consents to discuss in a lecture before the members of Congress with Dr. Hare his theory of storms—Mr. E. to speak first Dr. H. to answer him Mr. E. to close."[25] There is no evidence that the debate was ever held. Instead, Hare published a series of "strictures" on Espy's reports to the secretary of the navy.[26]

In scientific journals and the popular press Hare announced that it was time for Americans to declare their "intellectual independence" from the tyranny of the French Academy, which had supported Espy. Hare accused the French of allowing authors to write reports on their own communications and of making scientific pronouncements carrying the force of a "Papal Bull" but giving little or no study to the issues involved. According to Hare, "Arago assured me he had nothing to do with the report [on Espy], and Pouillet let it pass, because Arago could not attend, and he could not agree with Babinet; while Espy was very urgent to have the report before leaving Paris."[27] Hare was also angry that the French Academy had ignored his own priority claim against Peltier for an electrical theory of tornadoes. He sent his criticisms, in French translation, to members of the academy and to all of the learned societies, foreign and domestic, in correspondence with the Smithsonian Institution.

When the Navy Department asked Hare for his opinion "on the propriety of continuing Mr. Espy's labours," Hare's retort was immediate and instinctive.[28] He printed one thousand copies of *Queries and Strictures* at his own expense and distributed them to every member of Congress and the cabinet. This pamphlet was part of a memorial asking that Espy's government appropriation be shared with other participants in the storm controversy: "Is it not then reasonable that under the same patronage the results of further investigations shall be held up to view, so that the government shall not continue [to be] the instrument of a one sided enquiry . . . the government cannot consistently avoid lending their aid to additional enquiries which tend to confirm what may be true, and expose that which is erroneous." Hare was careful in his memorial to distance himself from Redfield. He included reasons why "no whirlwind storm consistent with the definitions given by the advocates of that hypothesis can exist in nature" and why the directions to mariners according to the rotary theory "can only perplex or mislead."[29]

While Redfield and Hare complained to the government, Espy petitioned the savants of Europe for support, as he did in the 1840s. He hoped their influence would help decide the issue of the storm controversy in his favor. He thanked Quetelet for the "kind notice" of Espy's three reports on meteorology in the *Annuaire météorologique de France:* "The expression of your favorable opinion of my labors has been very useful to me, in continuing for me the aid of the United States Government."[30] However, because of strong support for Redfield and Herschel's negative assessment of Espy's work, his theory still lacked support in Britain. Espy asked Quetelet's help in changing Herschel's opinion:

> Sixteen or seventeen years ago, Sir John Herschel pronounced before the British Association, that "my Theory could not be true, for if the wind in storms blew to the central regions, the barometer would rise there above the mean which is contrary to the fact." This dictum of Sir John, I have reason to believe, has prevented a careful examination of the doctrine I teach in Great Britain. I have tried in various ways to prevail on Sir John to examine the subject *de novo,* and declare again whether he was right or wrong in that hasty decision. Sir John has reputation enough not to suffer loss, by acknowledging that in this case he was in error. You know him well; a word from you to him on the subject would bring him out.[31]

Closer to home, Henry and Bache continued to support Espy's annual appropriation and recommended that their scientific colleagues do so as well. As Henry confided to Stephen Alexander, "You can safely recommend his appropriation to the committee. In any other country [Espy would] receive a pension for what he has done."[32] Espy continued to receive his salary until 1859.[33]

While Espy sought approbation, Redfield gathered evidence to support his notion that the natural whirling motions of "fluid matter" were the *vera causa* of tempests, tornadoes, and hurricanes.[34] He quickly adopted Henry Piddington's new term *cyclone,* which designated "any considerable extent or area of wind which exhibits *a turning or revolving motion,*" and incorporated it into his hurricane theory. Yearning to appropriate Espy's Continental storms, and indeed *all* storms, in his theory, Redfield cited Coffin's work on the relation between wind direction and barometric pressure as independent proof of "the cyclonic character of the variable winds in the temperate latitudes."[35] A paper by O. N. Stoddard on the cycloidal curvature of a tornado track in Ohio offered additional evidence.[36] He also toured the Great Lakes region in 1854 and encountered lake storms, which he saw as confirming examples of the rotary theory.[37]

To prove Redfield wrong on the issue of cyclones, Hare invented two "cycloidographs": a large one for demonstrating cycloidal curves to an audience and a smaller one for engraving the curves on metal for publication. Hare argued that curves produced by this machine demonstrated both the kinematic and the dynamic impossibility of the rotary theory. Hare's cycloidal curves were meant to show that a traveling whirlwind could never sustain circular winds. If, he argued, the union of rotary and progressive motions in a traveling storm caused the wind to blow in cycloidal curves, parcels of air accelerated along these trajectories would tend to tear the storm apart. Hare argued that Redfield's insistence on circular wind patterns showed "his observations to be unreliable, or his theory untrue."[38] Although Piddington's terms *cyclone* and *cycloidal* conveyed the idea of a curve that is neither a cycloid nor a circle, Hare replied that no such curve existed in nature—a rotary and progressive motion always generates a cycloid. To reinforce his argument, Hare placed one of Piddington's *Horn Cards* next to a diagram produced by his cycloidograph. Both are shown in Figure 5.2.

Hare intended the bottom figure to show the resulting complication of traces when the "travelling motion in proportion to the whirling motion is as 1 : 9" and to demonstrate "the utter impossibility that a fluid mass could retain the form of a disk, and revolve so as to preserve the movements attributed to a cyclone." In addition to criticizing the kinematics of the motion, Hare presented an argument against the fluid dynamics: "A cylindrical mass of water made to rotate by circuitous stirring with a stick is by Redfield, Reid and Piddington, considered as exemplifying, by the consequent depression about the middle, and accumulation towards the border, the analogous consequences of centrifugal influence during the rotation of a cyclone. . . but air has no restraining vessel, no constant stirrer."[39]

Hare gave two talks on his cycloidograph to the AAAS but never published the papers.[40] In 1856 he submitted a long manuscript, also never published, "On the Suppositious Travelling Whirlwinds Called Cyclones," to the *Smithsonian Contributions to Knowledge*.[41] Although Henry wanted a "clear exposition of principles, independent of controversy," he confided to Hare that he was

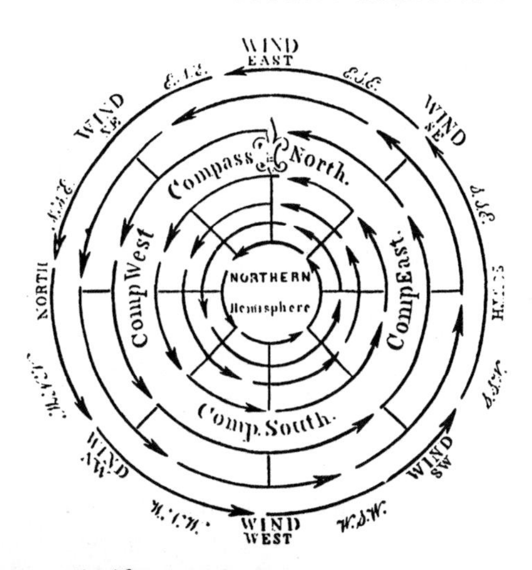

Fig. 5.2. Two illustrations from Robert Hare's unpublished manuscript "On the Suppositious Travelling Whirlwinds Called Cyclones," n.d., Henry Papers. Courtesy of the Bell-Henry Library, Smithsonian Institution. At right, "Fig. 1 Piddington's imaginary Cyclone as represented by his horn card." Below, "Fig. 2 This engraving shows what a cyclone [of] onward motion of only ten miles per hour would be were such a phenomenon possible in nature."

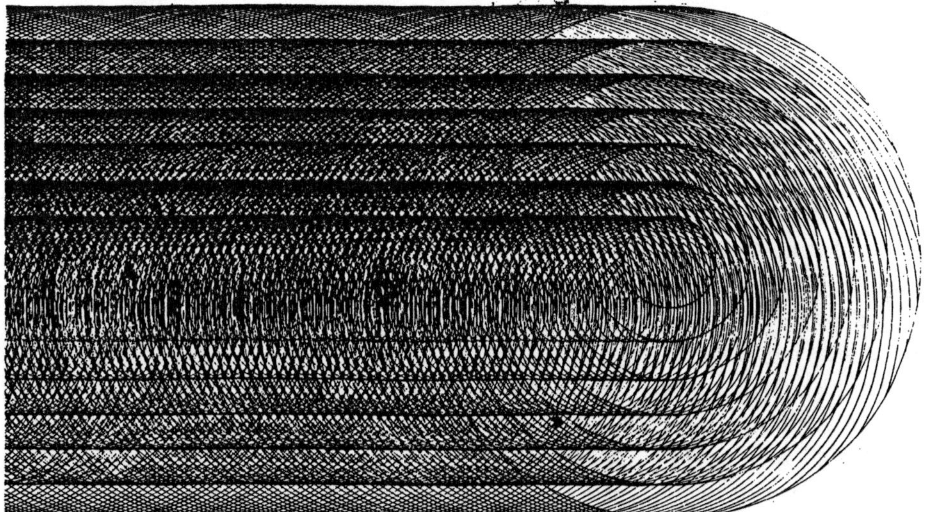

"much amused and instructed" by his "demolition" of the rotatory theory of storms. He added, diplomatically, that the cycloidal curves were "beautiful exhibitions of mechanical principles applied both to art and science."[42]

Redfield did not engage in public polemics after 1852. He published summaries of his hurricane theory, expanded his coverage to include other cyclonic whirlwinds, and, with data provided by Matthew C. Perry's Japan expedition, extended his results to include cyclones in the Pacific Ocean.[43] Redfield's last public appearance among scientists was in August 1856 at the meeting of the AAAS in Albany. He died on February 12, 1857. Redfield's eulogy was written by his long-term supporter Denison Olmsted. Although giving Redfield credit for his role in the storm controversy, Olmsted's remarks fell short of praise. He "regretted" that Redfield had deviated from his original position not to explain the reason "*why*" the winds blow but only to show "*how*" they blow. According to Olmsted, "His facts are impregnable: his hypothesis doubtful." The eulogy provided Olmsted with an opportunity to reiterate his doubts about all the storm controversialists: "Each of the combatants appears to me to be more successful in showing the insufficiency of the other's views than in establishing his own."[44]

Soon after Redfield's death, Espy's massive *Fourth Meteorological Report,* addressed to the president, was published as an executive document of the U.S. Senate. Although much of the report is a reprint of his earlier work, it is noteworthy because it gave Espy the last—and official—word against his original rival. Espy used the occasion to argue that Redfield was mistaken in his explanation of the dynamics of tropical hurricanes. Redfield had maintained that the centrifugal force in whirlwinds draws the air out from the central parts of the storm and brings down the cold air of the upper regions. The cold air then mingles with the warm air below to produce clouds and rain. This is in line with the old Huttonian theory of rain. Espy replied that neither the outward nor downward motion could be correct. His theory of cooling and expanding moist updrafts implied that descending air will be warmed and dried by compression: "*Air can never come down from a great height without being very dry when it reaches the surface of the earth.*" Espy further noted Redfield's inconsistent treatment of centrifugal force; Redfield had argued in another paper that air blows *inward* spirally and *upward* in the middle of a hurricane.[45]

Espy also had the last word in the battle over sailing directions for vessels encountering storms. His *Fourth Report* reprinted "Rules for the Mariner," in which he rearranged Reid's data for hurricanes to support his own theory: "I have examined many other of Reid's hurricanes, and I find that the wind in all of them blew in towards the centre; and the only reason that Colonel Reid did not discover this fact is that he did not draw arrows showing the course of the wind at some particular moment."[46] As he had three years earlier in 1855, Espy again asked Quetelet and other European savants to examine his "Rules" and send a letter recommending their adoption by the U.S. Navy. Espy ended his letter to Quetelet with a final plaintive request: "I am now seventy three years old, and it would give me much pleasure to see, before I die, my labors begin to be useful to mankind."[47]

While Espy savored the final victory over his now silent rival of twenty-six years, Hare was planning a memorial to Congress condemning Espy's *Fourth Report.* Henry intervened, however, with his opinion that further polemics would not be fruitful: "This is the short session of Congress and I do not think there is the least probability that you can succeed in drawing any attention to your memorial relative to storms during the present winter. Besides this Mr. Espy is no longer under the pay of the government no appropriation having been made last session for his salary."[48] Within the year, Hare had abandoned science for spiritualism, and the "war of controversy" ended with Espy the sole survivor. Henry wrote of these developments in a letter to Herschel three months before Hare's death: "The subject [of meteorology] has been rendered somewhat repulsive, by the personal feelings which have been exhibited in this country in the discussions to which it has given rise. Redfield, Hare, and Espy have kept up for many years a war of controversy which has only just now been brought to a close by the lamented death of the first—the abandonment of physical science for spiritualism of the second, and the advancing years of the third."[49] Two days after Henry penned these words, Espy—now unopposed—began a series of public lectures at the Smithsonian on his meteorological theories.[50]

Following his Smithsonian lectures, Espy petitioned the committee on science and the arts of the Franklin Institute to "give an impartial opinion of the merits of his theory," especially its "fundamental" aspect, "that the wind blows inwards to the central parts of all great storms, hurricanes and tornadoes."[51] Although Espy pressed in vain for an early decision from committee member and Lazzarone John Fries Frazer, his petition was discharged without action by the institute in 1862, two years after his death.[52]

Most authors have given but faint praise to Espy's efforts and have added caveats to his work. Bruce Sinclair wrote, "Espy's contributions to meteorology were outstripped in his own lifetime"; Gregg De Young insisted that Espy blocked the introduction of the method of hypothesis into America.[53] They were probably echoing Bache's unflattering eulogy of Espy: "His views were positive and his conclusions absolute, and so was the expression of them. He was not prone to examine and re-examine premises and conclusions, but considered what had once been passed upon by his judgment as finally settled. Hence his views did not make that impression upon cooler temperaments among men of science to which they were entitled."[54] It is clear, however, to anyone who reads the meteorological literature of the time that Espy had convinced Henry, Loomis, and Coffin of the importance of the steam power of the atmosphere. Espy's ideas on latent caloric and his theory of the convective nature of storms were widely accepted in his day. Henry considered Espy's theory basically sound, and wrote in 1858: "Though in subordinate particulars modifications will be required, yet we think the general propositions of his theory will stand the test of time."[55] In fact, Henry believed that the "only adequate cause" of the effects of a storm was explained by Espy's theory of the infusion and condensation of water vapor, which releases latent heat and causes an accelerated upward motion. The Smithsonian investigations also demonstrated that storms crossing the continent were "not of a rotatory character," although violent tornadoes were. He noted that Loomis's views on the origin of storms agreed with Espy's. Coffin concurred, adding that the existence of belts of high and low barometer around the globe "[has] been satisfactorily accounted for by Prof. Espy."[56]

Maury and the Depot of Charts and Instruments

> You are at the focus towards which, since the establishment of the Smithsonian, all the rays of information on the subject [of meteorology] from different parts of the country tend to converge.
> —Coffin to Maury, February 2, 1852, NOR-NA

Matthew Fontaine Maury arrived in Washington in 1842 as superintendent of the Depot of Charts and Nautical Instruments in the Bureau of Ordnance and Hydrography under orders from the secretary of the navy to "continue observations in astronomy, magnetism and meteorology."[57] Although Maury was in charge of the new "national observatory" after 1844, oceanography, not astronomy, was his primary interest.[58] In 1847 Maury and Henry were involved in a misunderstanding over the announcement of orbital elements for the newly discovered planet Neptune. Sears Cook Walker, a member of the Philadelphia "club," one of America's foremost astronomers, and a new civilian employee at the depot, discovered that Neptune had been observed and recorded as a fixed star in 1795 by J. J. Lalande in his *Histoire céleste française* (1801). Using this information, Walker calculated a more precise orbit for the planet.[59]

Disgusted with Maury's administrative shortcomings, Walker left the observatory and gave Henry an abstract of his findings, which Henry sent to the *Astronomische Nachrichten.*[60] Unfortunately, Walker's abstract disparaged the leadership of the observatory. Maury was "embarrassed" that the Smithsonian would announce research results produced by another agency and felt that credit belonged to him and the depot staff for having "*treed* the planet Neptune."[61] He asked Henry, "Was the account freely given by Mr. Walker or at *your invitation?*"[62] Henry, not yet aware of his mistake, claimed he had not read Walker's contribution carefully before sending it off. He regretted Maury's "unpleasant categorical form, unusual in scientific correspondence," and saw no "cause for controversy." To Henry's credit, he apologized to Maury as soon as he realized what had happened and assured Maury that it was merely a regrettable oversight.[63] Maury felt that the wrong done to the observatory by the Smithsonian announcement was "unintentional but, nevertheless, real."[64] However, because of Henry's fair-mindedness and his recognition of Maury's rights as the leader of a research group, the misunderstanding went no farther, and relations were normalized soon after. Henry even served as a mediator in Bache's turf battle with Maury over the scientific exploration of the Gulf Stream.[65]

Promises of cooperation in meteorology were made on both sides, yet little of note was accomplished. For example, Maury acknowledged the receipt of the Smithsonian meteorological circular of 1848 and promised his cooperation "at all times and on all occasions" in scientific matters related to the observatory.[66] Early in 1849 Henry received a proposal from Maury for using the telegraph for meteorological reporting. Henry already had in mind "a fine scheme with the telegraph" for simultaneous observations of the aurora and storms, but funds were short because of construction of the Smithsonian building. Henry noted in his diary, "If [Maury] can put into the [coffer?] 1000 dollars [he] is to be taken in—cooperation and assistance desirable."[67] But Maury's office, which received only $24,000 of an anticipated $58,000 appropriation, was experiencing a budgetary crisis of its own.[68]

Maury invested most of his efforts in "Pilot Charts" of ocean winds. Old navy logbooks and reports from ship captains provided data which Maury's calculators plotted on a map of the ocean in grid squares measuring five degrees of latitude by five degrees of longitude. The wind was recorded for sixteen points of the compass. Barometric measurements were not requested.[69] The charts—widely advertised at home and abroad—were sold through special agents George Manning and A. G. Seaman.[70] Officers of the British navy, scientific institutions, and others were deluged with copies of Maury's maps and his pamphlet *To Mariners*.[71]

Initially Maury conceded the realm of meteorology on land to Henry,[72] but he intended ultimately to make the Smithsonian project a subset of his grandiose plan for a "Universal system" of meteorological observations for sea and land.[73] This idea was sparked when Maury received a request from the British Royal Engineers for cooperative meteorological observations at foreign stations. Maury responded that indeed cooperation at the observatory and at U.S. navy yards was possible and that the system of observation should be extended throughout the world.[74] Maury announced his plan in letters to the foreign ministers of numerous countries, to scientific societies, and to private meteorologists:

> The object is to enlist in this great work, the public and private meteorological observatories, the good will of the friends of science, the labors of amateur meteorologists, and the cooperation of the Navigators, both of the naval and commercial marine of all countries; and by consultation and conference with them, to devise plans and methods of observing and recording, which by being common, effective, and of easy execution, may be followed by Meteorologists and Navigators generally. For this undertaking the Government of the U.S. desires to secure the friendly cooperation of the government and people of all countries; and for the purpose of giving effect to this wish, I have been authorized to confer with the proper authorities on the subject.[75]

In reality, the British request had been sent through routine channels from the State Department, to the Navy Department, and finally to Maury.

Maury took it upon himself to call for international cooperation when no unity of plan yet existed in his own country. His peers called into question both his administrative ability and his scientific competence. Maury's contributions to meteorology consisted of the pilot charts and his theory that the general circulation of the earth's atmosphere was controlled by a mysterious force generated by terrestrial magnetism. He also theorized that the trade winds from the Northern and Southern Hemispheres crossed each other at the equator, presumably maintaining their separate thermal and hygroscopic properties in their interpenetration.[76] Naturally, the scientific community regarded Maury's proposals with suspicion.

In 1852 Maury sent his plans to Henry and to other leading figures in meteorology, asking them to "*cooperate* please."[77] His proposal seemed to ignore the Smithsonian project and led to considerable confusion among those who were familiar with its meteorological work. For example, J. J. Abert, chief of the Bureau of Topographical Engineers, wrote to Maury: "I have been under the impression that this matter of a system of meteorological observations had been taken up, digested, and carried into extensive operation by the Smithsonian Institution, as exposed in its 5th annual report page 15 &c &c, and that you were aiding the efforts by appropriate sea observations, but your letter rather induces to a different impression."[78] One of the Smithsonian observers, Noah B. Webster, when asked by Maury to make meteorological observations for him, noted the conflict of interest: "It will afford me much pleasure to contribute my 'mite' toward the great object of your laudable labors. It may be proper for me to say that I am engaged in making observations for the 'Smithsonian Institution', and if in addition to the observations they require, I can be of any service to the cause of Science, please give me the requisite direction."[79]

Cautious positioning was needed on both sides. Maury wrote Henry to "fix a time to call here, if not convenient I will call there."[80] Henry replied, "If you have business in the city I should prefer [you call on me] at the Smithsonian, if not I will call at the observatory." Although the two men traveled in the same Washington social circle—they dined together the next day with the secretary of war—it was six weeks before Maury finally called on Henry and promised to "cooperate" in meteorology.[81] Three months later, Henry wrote to Maury to remind him of his promised cooperation.

One of Maury's biographers cast Henry as the villain in this power struggle and argued that the Lazzaroni were attempting to isolate Maury from the scientific community. This interpretation, however, neglects Henry's offer to place Maury on the meteorological committee of the AAAS:

> We think of making a movement to carry out the system of observing the phenomena of storms relative to which we have previously corresponded. You may recollect that the proposition was that you should take charge of the

> dominion of the Sea, that the Smithsonian Institution would collect observations from the Eastern portion of the U. States, the Surgeon General of the Army from the Western; and the British Government from the northern part of this Continent. By a letter from Col. Sabine I am informed that Her Majesty's Government will cooperate both with reference to observations on sea and land.... I have requested that a meeting of the Meteorological Committee of the American Association be called. Are you not a member of this Committee? If you are not we will elect you as one.[82]

Maury immediately replied, "Whether I be off or on, my services in helping to carry [it] out are at your command." He cautioned Henry: "The storm cooperation is not intended to supersede or interfere with the government proposition for a general Meteorological Conference, and the establishment of a series, or set of forms which should embrace both sea and land, and become universal."[83]

Maury's universal plan was soon reduced in scope to a "conference upon the subject of a uniform system of observations on board of vessels of war at sea." The conference was held on August 23, 1853, in Brussels.[84] The U.S. Navy subsequently kept its logs by the plan suggested at the Brussels conference, but the British and French failed to cooperate.[85] It seems that "the members of the conference returned home, every one followed his own plan and did what he pleased."[86] Maury may have had one shining moment as a diplomat, but he had tried to organize a meteorological system from the top down with no support from the grass roots.

There is, however, a second episode to the story. Upon his return to America, Maury again began to agitate for a system of observations on land to include agricultural as well as commercial meteorology.[87] On January 10, 1856, Maury addressed the United States Agricultural Society, meeting in the rooms of the Smithsonian Institution. The subject of his speech, given in Henry's presence, was the establishment of a land meteorological system. The substance of Henry's objections was that compiling wind maps for sailors was a "simple matter," involving "no great scientific attainment" and implying "no particular ability" or qualification on Maury's part, "to apply a meteorological system for the seas to the *land* for *agricultural* purposes."[88] Indeed, the Smithsonian had just entered into a partnership with the agricultural branch of the Patent Office for this very purpose.

Through the Agricultural Committee of the Senate, Senator James Harlan of Iowa drew up a bill to support Maury's suggestions. The princely sum of $20,000 was to be appropriated to supply farmers with instruments. Observations were to be transmitted by telegraph to the Hydrographical Office.[89] With stiff opposition from Henry and the scientific community, the bill was defeated in 1857. Maury attributed the defeat of his measure to the opposition from Henry and the Patent Office: "Some years ago I proposed, you recollect, a system of agricultural meteorology for farmers, and of daily weather reports by telegraph from all parts of the country for the benefit of mankind. The Smithsonian Institution and the Agricultural Bureau of the Patent Office stole this idea and attempted to carry it out, but with what success let silence tell. Take notice now that this plan of crop reports is 'my thunder', and if you see some one in Washington running away with it there, recollect if you please where the lightning came from."[90]

Throughout his life, Maury ignored the meteorological work of the Smithsonian. He continued to advocate a congressional appropriation for a central meteorological bureau and claimed as his "vested right" all of the land and sea.[91] Maury resigned his commission in 1861 to serve the Confederacy and, although he continued to advocate federal funding of meteorology for the rest of his life, his influence in Washington had come to an end. For Henry, one of the few "good results" of the war, among many evils, was the departure of Maury from the Naval Observatory: "For the first time in ten years I have lately visited the Observatory. Maury, sustained by the puffing which he was constantly receiving from England, arrogated to himself all the science of the country and did all in his power to interfere with the operations of the Coast Survey and the Smithsonian Institution."[92]

The Blodget Affair and the Surgeon General

Before the Smithsonian meteorological project was announced to the public, the office of the surgeon general had promised its cooperation. Dr. A. S. Wotherspoon, the officer in charge of the army weather records, sent Henry a list of military posts where weather observations were being taken and a list of locations where troops were soon to be established so that Henry could "form some idea of the country occupied."[93] Full cooperation was expected, and the Smithsonian records show that in 1851 a total of 907 registers, or an average of 75 registers each month, were received from the Army Medical Department.[94] Surgeon General Lawson reported this cooperation in his report for 1852: "The registers for 1850 and 1851 are in the hands of Professor Henry, superintendent of the Smithsonian Institute, who is engaged in tracing the course of the storms which have swept over the United States during those years, and endeavoring to discover the laws which govern their progress, &c."[95] The army medical system, however, did not have a formal cooperative agreement with the Smithsonian. The army merely shared its observations with Henry and continued to preserve, publish, and extend its own health-related weather records.[96]

Controversy erupted between the two institutions when Lawson hired Lorin Blodget, a disgruntled clerk in the Smithsonian meteorological project, to prepare a map and essay on the climate of the western United States for the army's meteorological register.[97] A bitter dispute between Henry and Lawson ensued over the fair use of the shared data and proprietary claims made by Blodget to its copyright. Blodget added to Henry's problems in 1854 by attacking his administration of the Smithsonian meteorological project and siding with Henry's opponents, librarian Charles Coffin Jewett and regent Rufus Choate, in the "big library" versus "active operations" dispute.[98]

Lorin Blodget had begun to take observations for the Smithsonian from his home in Sugar Grove, Pennsylvania in 1849.[99] For two years he contributed observations made with instruments supplied by the institution and planned his career in meteorology. Blodget came to Washington in December 1851 and, with the help of his congressman, was hired by the Smithsonian as a temporary clerk at $1.50 per day. He prepared maps of the cloudiness over the county, collected information on storms, compiled old weather diaries, and calculated seasonal and yearly averages from the data.[100] As Henry's desk diary for 1852 shows, Blodget wanted a more permanent situation: "*Fri. Feb. 27:* Conversation with Mr. Blodget—Wishes a definite engagement. Informed him that we could engage him for 6 months at the rate of 600 doll[ar]s per year. See what he will get from the observations on meteorology—To speak again on the subject. *Tues. Mar. 2:* Arrangement with Mr. Blodget. He is to continue for 3 months at the rate of 600 per year. If he deduces matter of consequence to remain at the same rate for the year and to be permitted to make a report subject to my revision."[101]

In the next three months, Blodget prepared a series of maps showing the course of a storm in January 1851 and set out for Cleveland to display them to the AAAS.[102] En route, Blodget visited several academies in western New York to ensure that meteorological observers followed Smithsonian guidelines. Although the AAAS meeting was canceled because of a cholera outbreak, Henry was pleased with Blodget's work and sent copies of his maps to the Royal Society and the British Association as examples of the Smithsonian's plan for investigating storms. Henry also offered Blodget a 50 percent pay raise to $900 a year and a greater role in scientific planning for meteorology. These offers were made verbally, however, and there was no written contract.[103] Blodget had quite another view of the situation. He claimed that he did not receive proper scientific recognition, "either as an officer or assistant in the Institution, or as an independent scientific collaborator," for his communications prepared for both the BAAS and AAAS. Moreover, Blodget thought Henry had promised him at least $1,100 a year and was using the institution's relationship with Espy as an excuse for not promoting him: "The reason assigned for failure to make ample compensation, so far, was the relation of the institution to Prof. Espy—I was unwilling, though I was several times solicited to do so, to place myself in any attitude of effort to displace him."[104]

After only eight months, in October 1852, Blodget left the Smithsonian, "without arrangements or acceptance of terms."[105] He claimed that his expenses for the Cleveland trip had not been repaid and that his report on the climate of the United States, intended for the Smithsonian *Annual Report,* had been intentionally delayed at the printer. Through his friends in Congress, Blodget pressed for aid, setting himself up as a competitor for the meteorological appropriation annually given to Espy. Under strong opposition from Henry, however, Blodget withdrew his request for congressional funding.

During his estrangement from the Smithsonian, Blodget put his scientific experience to work elsewhere. He prepared tables of temperature, pressure, and rainfall to accompany the mortality tables of the *Seventh Census of the United States,* published in 1853.[106] Blodget also attended the rescheduled AAAS meeting in Cleveland and presented three papers "on his own responsibility." Blodget gave no credit to the Smithsonian for the data that had been collected. At the meeting Henry issued a mild rebuff to Blodget for his lack of discretion and presented a disclaimer on behalf of the Smithsonian. But Henry was in a conciliatory mood and offered to reemploy Blodget as a clerk. With greater zeal than discretion, Blodget rushed back to Washington and, without Henry's permission, sent letters to all of the Smithsonian meteorological correspondents informing them that he was now "in charge of the meteorological department."[107]

Blodget's second tenure at the Smithsonian was short and confrontational. In less than a year, by May 1854, Blodget had become an accomplice of Jewett and Choate in the stormy controversy over the proper mission of the institution.[108] Jewett led the "local objects" or "big library" faction against Henry, who favored "active operations"—pilot projects in fundamental research that could not be sponsored elsewhere.[109] Blodget was not really part of the "big library" faction, but he was useful to Jewett because of his attack on the credibility of the meteorological project—one of Henry's most visible "active operations." His attack coincided with several proposals to Congress in late 1853 calling for the transfer of the Smithsonian meteorological project to the Department of the Interior, the War Department, the Navy Department, or the National Observatory.

Blodget's affidavit, presented through Jewett and Choate to the Board of Regents, requested their judgment on "the measure of support proper to accord to [meteorology] on the part of the Smithsonian Institution."[110] He told the regents that because of the institution's strained finances, the meteorological project created "responsibilities which could not here be met." Blodget asked the regents to publish his manuscript on climate, to award him $1,500 for research expenses and official duties, and to pay him for his time and scientific services. He further suggested that Henry be removed as secretary of the institution and self-appointed head of weather research in America—an action he deemed "pressingly necessary to save private and public interests."[111]

Henry expressed his concerns to Guyot: "The Institution is just now encountering a very severe storm. I am not however much concerned as to the final result. If the enemies triumph I shall retire, if not, the Institution will be purified and placed upon a more secure foundation."[112] He expressed his frustrations to Lapham: "Mr. Blodget has no permanent connection with the Institution, but has been employed *by me* to compare and reduce the meteorological observations. I regret to say, though the observations have cost the Institution at least $10,000—he considers the results that have been deduced from them as belonging to himself, & that letters [addressed to the meteorological department] as a communication to him individually."[113]

As the storm clouds gathered around the Smithsonian, Blodget found other opportunities to undermine Henry's authority. He rushed into print an article on agricultural climatology for the *Patent Office Report for 1853*.[114] The article contains no new scientific results; it is merely a rehash of data collected from the army, the Smithsonian, and New York State. It is worthy of note, however, because of two official letters published with the article: one from Commissioner of Patents Charles Mason asking Henry's permission to publish the article and Henry's reply indicating that the data in Blodget's article belonged to the agencies that collected it and "ought first to be published in a general report under their sanction." Because of the Jewett controversy, however, he was not "at liberty" to object to its publication.[115] The battle for control of meteorological data had moved into the public realm.

Blodget's sedition placed a continuing strain on Henry's relationship with Surgeon General Lawson. Lawson blamed Henry for allowing a Smithsonian clerk to publish data collected by the Army Medical Department without proper acknowledgment: "The meteorological registers have been freely loaned to the Smithsonian Institution . . . and it is to data obtained from this office that the paper on 'agricultural climatology of the United States', appended to the report of the Commissioner of Patents for the year 1853, owes much of its completeness."[116] Henry responded with a scientific criticism of the army system: "As first approximations the simple observations at the different posts, which have thus far been published, are acceptable additions to knowledge; but whatever is worth doing at the expense and under the direction of the general government, ought to be as well done as the state of science and the circumstances under which the work is commenced will allow."[117]

In April 1854 Blodget accepted an invitation from Lawson to write an essay on climate to accompany the latest army meteorological register. His first essay was rejected because it contained data copied from Smithsonian records, but the army issued him a written contract to revise the report.[118] When Blodget's essay appeared in the army's *Meteorological Register (1855)*, Henry protested that it had "done injustice to the Smithsonian Institution and to its meteorological correspondents." According to Henry, charts were presented from materials gathered at Smithsonian expense, produced from insufficient data, and "projected on a wrong principle." Moreover, they were injurious to the plan of "harmonious cooperation" between the two offices.[119] Formal cooperative arrangements between the Smithsonian and the Army Medical Department, however, apparently existed only in Henry's mind, for Lawson responded: "Your letter conveys the first intimation I ever received of a plan for 'harmonious co-operation' between the Smithsonian Institution and this Office. Nothing could be further from my every thought and feeling on this subject."[120] Lawson was critical of one-way "co-operation." He had loaned the Smithsonian the army meteorological registers, distributed Smithsonian circulars to the medical officers, and asked them to collect specimens of natural history, but he felt that nothing had come in return: "This Bureau has been passed by in the distribution of the Smithsonian Contributions to Knowledge, one Volume of which was prepared in part from observations made under its direction; and although the museum of that Institution has (as I am credibly informed) been very materially enriched by the contributions of Army Medical Officers, its Secretary has never made an acknowledgment of that fact to me." More important, Lawson was angered by Henry's "offensive and dictatorial language" and his "discourteous" manner. As Lawson saw it, the army system had been in existence for thirty-five years, employed medical officers of the army to take measurements with standard instruments, and possessed the only data of value from the western half of the continent. It had long had its own unique purposes for supporting meteorology: the "advancement of Medical Science and the elucidation of the effects of the climate upon the health of troops." The collected data were also useful to the quartermaster general in answering questions on the transportation of supplies, sources of food for the troops, and the clothing and accommodations required by the army in the field. The Smithsonian system, on the other hand, had been operational for only four years, "and although standard instruments may have been furnished to a few persons, it is believed its observers were for the most part amateurs who supplied themselves with such instruments as they could conveniently obtain. . . . This office did not need co-operation with anyone. . . . I would [not] for one moment consent to play a subordinate role in the affairs of the Smithsonian Institution." Lawson published Blodget's essay in the *Meteorological Register* because, in his words, "[I was] tired of waiting for the long expected 'contribution' to Meteorology from the Smithsonian."[121]

Blodget's bid to wrest control of meteorology from Henry was ultimately frustrated by his bad timing. Two weeks after Blodget publicized his complaints, a special committee of seven regents formally rebuked Jewett. Several weeks later, the full Board of Regents empowered Henry to remove Jewett from office. Rather than criticize Blodget publicly, Henry used James Coffin as a special consultant to settle the score. Coffin suggested that Blodget's work had little scientific merit, his analysis of data was often in error, and with few exceptions Blodget "did not discover any thing but that which a clerk might do."[122] Blodget was dismissed on October 11, 1854, and his rooms at the institution were barricaded. Choate resigned from the Board of Regents in January, and a congressional investigation vindicated Henry by March 1855.[123]

Late in the year 1856 Henry expressed his frustration to Asa Gray, writing that meteorology "has given me the greatest disquietude."[124] According to details provided in Henry's candid letter to Guyot, the main culprit was still Lorin Blodget:

> I hope you will pardon me for the feeling manifested in my note. It was written on the spur of the moment after looking over Blodget's prospectus in which his covert allusions to this Institution aroused by indignation. He is a person of so little regard for moral principle that it is unsafe for anyone to have dealings with him. I cannot give an account of his conduct or make any allusions to it in public without subjecting myself to the charge of persecution and jealousy and since he has no regard to truth he constantly finds new persons to espouse his cause. He is just at this time making a desperate attempt to get the ear of members of Congress in order to make a new assault on the Institution and myself. He has procured a series of large charts for lectures and these have been suspended on the walls of the Hall of the House of Representatives. At the last session of congress he induced a member to prepare a violent attack on myself which was only prevented being made by a private exposition of his character and conduct. His whole course in the Institution as you know was in the highest degree reprehensible. He refused to give up the material for publication which had been collected at great expense to the Smithsonian fund, unless he was paid a large sum and was allowed to publish the results in the manner he alone should judge proper.[125]

The "prospectus" to which Henry referred was Blodget's *Climatology of the United States*, published privately in 1857 and based on pilfered Smithsonian data.[126] To protect the Smithsonian meteorological project, Henry forged a stronger alliance with the Patent Office for cooperative data collection and published his views on meteorology in a series of essays titled "Meteorology in Its Connection with Agriculture" in the annual reports of the Patent Office.[127]

The storm controversy was driven by both personal disagreements and differences in theory. Institutional working relationships were determined more by the personalities of bureau chiefs than by official policy. Agreements were usually verbal, seldom contractual, and subject to constant change. Moreover, no single agency had the resources or authority to cover the entire continent. Clashes were inevitable. Maury, Lawson, and Blodget were convinced that Henry was attempting to gain a monopoly on meteorological results for the Smithsonian. Nevertheless, these frustrating and unproductive confrontations did not preclude cooperative arrangements among theorists and administrators.

CHAPTER 6

Cooperative Observations and Contributions to Knowledge

Let a meteorological department of the institution be organized, under the direction of the Secretary, with a suitable assistant. Let a united effort be made to secure for a limited period, and to the greatest possible extent, the co-operation of the general government, the several State governments, scientific societies, and the friends of science throughout the country.
—Loomis, "Report on the Meteorology of the United States," 1848

Numerous organizations and individuals with weather- and climate-related interests cooperated with the Smithsonian meteorological project. The Navy Department provided franking privileges and paid Espy's salary; the states of New York and Massachusetts, the Canadian government, the Army Topographical Engineers, the Agricultural Branch of the Patent Office, the Coast Survey, and the Department of Agriculture either collected data or contributed other resources—funds, personnel, postage, or printing—to the Smithsonian initiative. Arnold Guyot formulated and published new standards for the observers and their instruments. James Coffin reduced, compiled, and published tables of Smithsonian data as well as two studies of the winds of the Northern Hemisphere and the winds of the globe. Alexander Dallas Bache lent prestige and personnel to Henry's projects. Even Spencer Baird and the community of natural historians found inspiration in the meteorological project, but often for their own reasons. Finally, William Ferrel used data collected by the Smithsonian in his pioneering theoretical studies of the general circulation of the atmosphere and oceans.

Arnold Guyot and the States of New York and Massachusetts

If the system patronized by the Smithsonian Institution seeks to aspire to the needs of science, it has the right, moreover the duty, to be independent and not to descend to the lowest level, but to elevate all others by its example.
—Arnold Guyot to Henry, January 12, 1850, Guyot Letters

At the inception of the Smithsonian meteorological project, T. Romeyn Beck, chairman of the Board of Regents of the University of the State of New York, wrote to his former associate Joseph Henry asking advice on the reform of the New York system. Beck was ready, in his words, "*to fall in as a state detachment* of the U[nited] States Corps of Observers—under your general rules & directions."[1] With Henry's support, Beck issued a circular to the New York academies announcing the reforms and convinced the regents to hire Arnold Henry Guyot, a recent émigré and friend of Louis Agassiz, as a consultant to coordinate and improve the meteorological system.[2] In the winter of 1849–50 Guyot traveled through New York State inspecting, comparing, and positioning the instruments. He instructed the observers on the proper care of instruments and on techniques to minimize observational errors. A unique correction factor was assigned to each instrument to bring the results into agreement with a common standard. Barometer readings were to be further "reduced" to a common standard temperature (32° Fahrenheit) and altitude (sea level).

On his itinerant tour of New York Guyot found deplorable conditions among the observers in the field. Indeed, he had reason to believe that reliable instruments did not exist in the state: "I have not seen one station, even less a coordinated ensemble of stations . . . which operate under circumstances and with instruments much poorer than those which I have seen in my many travels." Guyot did not doubt the sincerity of the "estimable" observers in New York, but he was not able to consider their meteorological journals and results as anything but "temporary approximations."[3]

Guyot urged the Smithsonian to undertake a "complete reform" of the instruments and procedures of the New York observers. At Henry's request, he prepared a model meteorological register and a set of guidelines in an attempt to infuse uniformity into their diverse practices. Among the points made by Guyot were the following:

> 1. Each sheet ought to be a complete and independent document . . . [giving date, location, and signature of the observer]. 2. The order in which the columns follow indicates that in which the instruments ought to be observed: the external thermometer before opening the window to moisten the Psychrometer; the thermometer of the Barometer before the approach of the observer has modified its temperature, then the wind, the state of the sky, &c. and in the last place the Psychrometer. 3. The observations must be made according to this invariable sequence and the instrument readings noted *without corrections.* 4. Additional observations are needed during storms.[4]

Guyot examined the hourly observations at the Toronto Observatory to find the best times to take thrice daily observations which would approximate the true mean. He recommended 6:00 A.M., 2:00 P.M., and 10:00 P.M. based on comparisons with the twenty-four-hour observations taken at Toronto, Lake

Athabasca, and (by Bache) at Girard College. The observation at 6:00 A.M. represented the mean of the minimum temperature, the observation at 2:00 P.M. provided the mean of the maximum temperature, and the average of the three observations gave the daily mean. According to Guyot, "The series of 6, 2, 10 gives [monthly and annual means] completely, with the least possible hours, and with the conveniences of observers: everything that could be required from a general system of observations." He concluded that the difference of the means of both this series and the series 7:00 A.M., 3:00 P.M., and 11:00 P.M., with the means derived from a full series of twenty-four-hour observations, were "so small that they may be considered identical." The need for early morning and late evening observations led Guyot to position the instruments in the houses of the observers, not in the schoolhouses. Besides, the schoolhouses were unoccupied two days each week and during vacations. Guyot advised and Henry concurred that these arrangements agreed "at once with the demands of Science and the convenience of the observers."[5]

Guyot's work in New York benefited the entire Smithsonian meteorological project. His report included detailed comparisons of Newman barometers owned by the Smithsonian Institution, Columbia College, and the Harvard Observatory with two standard barometers: a Fortin-Ernst barometer from Paris and a Newman barometer at the Toronto Observatory calibrated to the standard of the Royal Society. Guyot found the Smithsonian's Newman barometer marked too low when compared with the Toronto standard: "Newman should take it back . . . I am astonished that he did not send a bill of the comparison with that of the Royal Society; and I am sorry to say that the division of [the scale of] his instruments is not irreproachable."[6] To improve the design and construction of domestic instruments, Guyot worked closely with the instrument makers Green and Pike in New York City. Green's standard barometer, patterned after Newman's, came to be known as the Smithsonian Barometer, and Pike and Sons received orders for standard wind vanes and rain gauges. To accompany the instruments, Guyot wrote a set of instructions for all Smithsonian observers entitled "Directions for Meteorological Observations, intended for the first class of observers." The pamphlet contained instructions on positioning the instruments, complete with woodcuts, and engravings of cloud types.[7] By 1851, the instruments prepared for the Smithsonian were considered by Henry "not only equal in accuracy to the instruments of the best construction from abroad, but in some respects superior."[8]

Guyot also prepared an elaborate set of *Meteorological Tables,* first published by the Smithsonian in 1852. The first in a long series of similar publications on instruments and physical constants, these tables enabled "every one to choose . . . and to compare and appreciate the results of others."[9] Guyot's publication included tables for comparing barometers, thermometers, and psychrometers from different makers, typically with differing scales. It also contained tables for computing standard altitude and temperature corrections. Since there were no international agreements on instrument construction or calibration, this was not an insignificant accomplishment. In the expanded second edition of his tables, Guyot included six thermometrical tables for converting among three different scales. The hygrometrical series was expanded from five (all French) to twenty-seven (French, English, and German), and the barometrical tables increased from twelve to twenty-eight and included for the first time a comparison with the Russian barometer. A new hypsometrical series provided forty-two new conversion tables. Guyot also included Dove's series of meteorological corrections for periodic and nonperiodic variations for all parts of the world.

The improved New York regent's system, with instruments chosen by the Smithsonian Institution, commenced operations on January 1, 1850, and was fully operational within the year (see Figure 1.4). The New York legislature provided an appropriation of $1,500 a year for two years to supply twelve stations throughout the state with instruments and to provide each observer with $30 per year for expenses. Fifteen barometers at $20 each and fifteen thermometers at $2.50 each were ordered from James Green of New York. The legislature appropriated $800 a year thereafter until 1863. By that time, volunteer observers were hard to find. The Civil War had redirected the initial enthusiasm, and the growth of Smithsonian cooperation in meteorology with the Department of Agriculture had, in a great degree, "superseded the necessity of continuing a separate system."[10]

Fresh from his tour of New York, Guyot urged the legislature of Massachusetts to sponsor a system of observations to extend those made in New York eastward to the Atlantic Coast.[11] He argued that, when the Massachusetts system became operational and was joined with New York's system, "we shall have a beautiful line of compared instruments of 600 miles from Detroit to the Sea shore, which will afford a fair opportunity to study the courses of the storms & of the barometric waves." By December 1850, New York had established thirty-eight stations, but Guyot found the Massachusetts system "buried in a profound sleep." The instruments were ready, but no official direction had been given to take any further step. Indeed, there was some confusion over who had the authority to begin. The American Academy in Boston thought it had the duty to execute the plan and appointed a committee on meteorology made up of Guyot, Benjamin Peirce, and Nathaniel Bowditch for that purpose. State representative Charles Mitchell of Nantucket, however, believed he had been entrusted with the matter and made arrangements to confer with Guyot on the subject. After a six-month delay, Mitchell visited Guyot, detained, as he said, by his unsuccessful attempt to obtain a supplementary appropriation for paying the observers. Mitchell's procrastination more likely stemmed from the uncertain political situation in the state. When the elections of 1850 swept the

Democratic party into power and threatened the reelection of the governor, Mitchell urged Guyot to proceed with the original plan expeditiously and incorporate the Massachusetts system into the Smithsonian project before the appropriation was repealed. Guyot received this information coolly, for it was Mitchell himself who had contributed to the delay. Moreover, the "endless details" of establishing systems of meteorological observation had taken time away from his other scientific and professional duties. Guyot was frustrated by the confusion, politicking, and lack of progress. He concluded that state systems of meteorological observation were too small, too poorly funded, and too vulnerable to local politics to be of much scientific value unless connected to "a powerful center of action." The Smithsonian Institution, Guyot advised, "ought to assume fully this position, which is now willingly conceded to it by public opinion."[12] Guyot thought the Smithsonian should suggest some fixed arrangement to every state disposed to enter the national system so as to avoid the uncertainty, frustration, and duplication of effort Guyot had experienced in Massachusetts.

The vicissitudes of the Massachusetts meteorological system contained fundamental lessons for the course of action taken by the Smithsonian in compiling, reducing, and publishing the data collected by other groups. Guyot's model for a cooperative national meteorological system contained three components, each requiring a separate budget, and each (ideally) covered by state, Smithsonian, and federal appropriations:

> 1. The cost of the establishment of a Station, with a full set of instruments, can be estimated, according to the experience made in the State of New York, at $100. . . . 2. The [Smithsonian] Institution would take entirely to its own charge the reduction of the observations & the preparation for publication. 3. As to the printing expenses, they would be too large for the Institution—According to the estimates I have gathered here on that subject, a quarto volume of 5 or 600 pages, and 800 to 1000 copies, would cost $2000 . . . the States ought to meet these expenses . . . Mass., for instance, could certainly be induced to appropriate $500 for that purpose. But by far [it would be best] to obtain from Congress an appropriation for this publication, to be made under the charge of the Smithsonian Institution.

By following this plan, Guyot argued, the states would know their expenses and their responsibilities in advance, and the Smithsonian Institution, "a responsible agent, helping, supplying, completing, directing," would remain the "centre and Soul of the whole" and maintain the regularity, steadiness, and scientific character of the overall project, "and so we would avoid this scattering of the forces, this uncertainty in the progress & the march of the plan which can not but end in a total loss for scientific purposes.—Five years of such observations well conducted and regularly published would already have a great value."[13] Henry was impressed with these recommendations and asked Guyot if he would take charge of the Smithsonian meteorological project: "What are your plans for the future? and what is the least sum per annum for which you could afford to take charge of the whole system?"[14] Guyot, still hopeful of attaining a permanent teaching position and continuing his own research, respectfully declined Henry's offer.

Observations finally began in Massachusetts at six locations in the second half of 1851: Amherst, Williamstown, Westfield, West Newton, Worcester, and Pittsfield.[15] Four more stations began observations in 1852: Newburyport, North Attleboro, Truro, and Princeton. Stations at Provincetown and Bernardstown did not submit any observations; Cambridge, Bridgewater, and Nantucket were inactive for several years. As of March 1853, only six stations were active: Newburyport, Worcester, Pittsfield, North Attleboro, Truro, and Princeton, "with *Amherst* probably making seven—Westfield dropped off in June '52, Williamstown in Oct. '52."[16]

The cooperation of other states with the Smithsonian Institution is not as well documented. For example, an article in the *Monthly Weather Review* in 1931 lists, anachronistically, "U.S. State Weather services" for Massachusetts (1849), Maine (1855), Illinois (1855), and Texas (1858).[17] The term *State Weather services,* however, was not used until after 1875 and referred to arrangements between the states and the U.S. Army Signal Office. Even the dates are wrong. Observations did not begin in Massachusetts until 1851, and the system was soon incorporated into the general voluntary program of the Smithsonian Institution. In Illinois, a resolution was passed in 1855 by the State Board of Education seeking greater cooperation with the Smithsonian, but no formal Illinois state weather service was forthcoming.[18] In Texas, the state geologist, Benjamin Franklin Schumard, established three stations with standard instruments in 1858—hardly a weather service. A similar effort was initiated by Schumard's friend George Clinton Swallow, state geologist of Missouri.[19]

A relatively successful project began in 1857 when the state of Maine appropriated $75 to each of nine persons to act as meteorological observers in connection with the Smithsonian Institution. One of the observers, Henry Willis, immediately applied to the governor for funds and purchased a standard barometer, thermometer, psychrometer, and maximum and minimum self-registering thermometer from James Green. After receiving his instruments, Willis wrote to Henry requesting a set of the meteorological tables, "or so much of them as is necessary to make a perfect record."[20] Meteorological reports from Willis and others appeared in the annual reports of the Maine Board of Agriculture.[21]

Canadian Cooperation

> The storms of this country cannot be fully studied without simultaneous observations from all parts of the continent.
> —Henry to G. T. Kingston, August 3, 1858, AES Files

In the summer of 1849, Henry visited Captain J. H. LeFroy in Canada, "principally for the purpose of examining the meteorological instruments and the method of using them employed at the Observatory in Toronto."[22] Henry wanted advice on establishing the Smithsonian system and a promise of cooperation from the British provinces. LeFroy pledged his "hearty cooperation" and provided Henry with a list of eight military posts in Canada which had been keeping "observing books" since 1844.[23] He even contributed his own observations from 1849 to 1852. Soon other observations, some by private individuals and some taken at outposts of the Hudson's Bay Company, became available to the Smithsonian meteorological project.[24]

In 1853 an act was passed by the Provincial Parliament establishing a system of meteorology in Upper Canada at "seminaries and places of education," making it the "duty" of every county grammar school to keep a meteorological journal.[25] Regular reports, to be sent to the local newspapers and to Egerton Ryerson, the chief superintendent of education, were required from these "grammar school meteorological stations" in exchange for an appropriation of $15 per month. Students were to be taught habits of observation and an awareness of natural phenomena. The government hoped to receive in return a better knowledge of the climate and meteorology of Canada in service to military, agricultural, commercial, and scientific interests. Indeed, the committee on immigration of the Canadian House of Assembly found the records useful for promoting settlement; the Royal Engineers consulted the records in planning forts, harbor improvements, and other matters of defense. Regulations stipulated a penalty of $30 for "the omission to take and record observations during any one month, or portion of a month."[26] Strict enforcement of these regulations soon would have ended elementary education in Canada. Although the legislation was passed in 1853 and funds for calibrated instruments were available as early as 1854, observations began only in 1858, and then only three of twelve stations returned reports that were "well prepared."[27] The first series of observations ended in 1862. A second series began in 1866. Observations at military posts, by the Hudson's Bay Company, at the public schools, and by private individuals provided the Smithsonian with data from thirty-six widely scattered northern locations between 1849 and 1873. As noted by Henry, these records were considered essential for the study of the storms of North America. A list of Canadian correspondents of the Smithsonian meteorological project between 1849 and 1873 is found in the *Smithsonian Annual Report for 1873*.[28]

Following Canadian unification in 1867, the weather service expanded. In 1870 the Ministry of Marine and Fisheries approved a proposal for lighthouse keepers to take meteorological observations. By 1871, $5,000 had been appropriated and observations began at thirty-seven locations in a thin strip bordering the United States. Exchanges between a few telegraph stations in Canada and the U.S. Army Signal Office began in 1872. Canadians began to issue their own storm warnings in 1876.[29]

The Army Topographical Engineers

In 1858 the U.S. Army Engineers began a topographical, hydrological, and meteorological survey of the Great Lakes under the direction of Captain George Meade.[30] Meade was eager to share his data with the Smithsonian but needed official permission from the Bureau of Topographical Engineers to do so. The bureau promised to cooperate but was unwilling to hire a clerk to make extra copies of the observations.[31] Henry informed Meade that copies would not be necessary if the original records were sent to the Topographical Bureau in Washington: "We can probably get access to them in the office of the [War] Department. We shall wish to examine them with special reference to the study of winter storms of this continent on which we are at present engaged." Meade persevered, however, and by 1861 was successful in sending the observations of Lake Survey observers directly to the Smithsonian.[32]

Following the model of the Smithsonian meteorological project, the Lake

TABLE 6.1. Observers and stations in the Survey of Northern and Northwestern Lakes, 1860.

Lake Superior
- George R. Stuntz, Superior City, Wisconsin
- H. Selby, Ontonagon, Michigan
- G. H. Blaker, Marquette, Michigan

Lake Michigan
- William Woodbridge, Michigan City, Indiana
- I. A. Lapham, Milwaukee, Wisconsin
- Heber Squier, Grand Haven, Michigan

Lake Erie
- John Lane, Monroe, Michigan
- Benjamin A. Stanard, Cleveland, Ohio
- William S. King, Buffalo, New York

Lake Huron
- J. J. Malden, Thunder Bay, Michigan
- John Oliver, Ottawa Point, Michigan
- Lt. C. N. Turnbull and James Carr, Forestville, Michigan
- E. P. Austin, Sanilac, Michigan
- Lt. C. N. Turnbull and James Carr, Fort Gratiot, Michigan
- Lt. C. N. Turnbull and E. P. Austin, Detroit, Michigan

Lake Ontario
- L. Leffman, Fort Niagara, New York
- Andrew Mulligan, Charlotte, New York
- Henry Metcalf, Sackett's Harbor, New York

Survey adopted Henry's "Instructions to Observers," Guyot's *Meteorological and Physical Tables,* and standard instruments by James Green.[33] Henry praised the Lake Survey system for its military discipline, efficiency, and precision. He saw it as "one of the most perfect which has ever been established," which, if continued for a few years, would "give the local climate of the district, with an accuracy which has never been obtained in any other part of the continent."[34] Data from the Lake Survey were published *in extenso* in the *Annual Reports* of the secretary of war until 1868. Beginning in 1868, summaries of the data appeared in the *Annual Reports* of the chief of engineers.[35] A significant finding of the survey—that the Great Lakes were "rapidly and sensibly affected by winds and storms," with lake levels depressed to windward and raised to leeward—was revealed by the combination of meteorological and hydrographical measurements of rainfall, evaporation, outflow, and lake levels.[36] Lake Survey observations ended at Detroit when the U.S. Signal Office opened a station there in 1870.[37] Other stations continued to report until 1874 (see Table 6.1).[38]

Coffin and the Agricultural Branch of the U.S. Patent Office

In 1839 an appropriation of $1,000 was made by Congress for "the collection of agricultural statistics" and investigations "for promoting agriculture and rural economy."[39] Most of this money was used for distributing free seeds to farmers, but appeals were also heard for a climatological review of the planting, growing, and harvest seasons.[40] Commissioner of Patents Charles Mason considered this information essential for a system of scientific agriculture and promised farmers a new level of support: "The degree of heat, cold, and moisture in various localities, and the usual periods of their occurrence, together with their effects upon different agricultural productions, are of incalculable importance in searching into the laws by which growth of such products is regulated, and will enable the agriculturalist to judge with some degree of certainty whether any given article can be profitably cultivated."[41]

To advance these ends, the Patent Office entered into a cooperative arrangement with the Smithsonian meteorological project. In 1855 twenty-five sets of standard instruments and two hundred simple rain gauges were purchased from James Green for free distribution to the agricultural community.[42] Funds were also provided to collect, transcribe, reduce, and publish the incoming data.[43] This cooperative venture led to increased numbers of agricultural workers joining the corps of observers. Some farmers expected that the new association between meteorology and agriculture would lead to immediate practical benefits, but Henry was quick to deny this hope:

> There are no royal roads to knowledge, and we can only advance to new and important truths along the rugged path of experience, guided by cautious

> induction. We cannot promise to the farmer any great reduction in the time of the growth of his crops, or the means of predicting, with unerring certainty, the approach of storms. But in the course of a number of years the average character of the climate of the different parts of the country may be ascertained, and the data furnished for reducing to certainty, on the principle of insurance, what plants can be most profitably cultivated in a particular place. . . . We make these remarks in order to prevent disappointment and the evils produced by exciting expectations which cannot possibly be realized.[44]

After the Blodget affair, Henry signed a written agreement with Coffin at Lafayette College to tabulate and reduce the incoming and accumulated observations. Coffin provided daily, monthly, and yearly averages of all the meteorological variables, calculations that reduced the observations to sea level and to a temperature of 32° Fahrenheit, tables for computing relative humidity, and special compilations on major storms.[45] During 1856, for example, records of more than five hundred thousand separate observations, each requiring reduction, were received and processed by Coffin. Henry estimated that an average calculation required one minute per observation—enough to keep three full-time computers busy for the year.[46] Coffin employed from twelve to fifteen persons—many of them local schoolteachers—on a part-time basis at a rate of twenty-five cents per hour. This was the equivalent of three to four full-time employees.[47] Between 1858 and 1860, when data were being prepared for a special two-volume report of the Patent Office, Coffin's employees logged an average of 8,276 hours per year at an annual cost of over $2,000—a significant portion of the Smithsonian's total expenditures for meteorology. A summary of Coffin's bills for the period 1857 to 1860 appears in Table 6.2.

Because of the financial contribution and assistance "in kind" of the Patent Office, the Smithsonian was able to compile and publish annual summaries of its observations in the *Agricultural Report* of the U.S. Patent Office. In 1857 a pamphlet entitled *Smithsonian Meteorological Observations for the Year 1855* was distributed to observers "for critical examination." This pamphlet became the prototype for a massive five-year compilation of periodical phenomena, climatic data, and storm observations entitled *Results of Meteorological Observations*

TABLE 6.2. Coffin's expenses for data reduction, 1857–1860, as recorded in bills found in the Henry Papers

Year	Expenses	Hours	Workweeks	Full-Time Equivalents
1857	$1,008.67	4,032	101	2.0
1858	2,190.05	8,760	219	4.4
1859	1,927.15	7,708	193	3.9
1860	2,091.08	8,364	209	4.2

Made under the Direction of the United States Patent Office and the Smithsonian Institution from the Year 1854 to 1859, Inclusive, published in two volumes by the Patent Office in 1861 and 1864 and distributed free to meteorological correspondents.[48] One observer noted the long delay in publication and voiced an opinion on the massive amount of work that went into each volume: "I have this day received the Vol. of Reduced Meteorological Observations for which accept thanks. They were long enough on the road, so long as to have created despair. The arrival however has dissipated all doubts. I think that this Prof Coffin's head was [more] fit for a coffin, than a pillow after having completed such a work."[49] These publications satisfied in part the provision that the Department of the Interior collect more agricultural statistics and blunted the criticism of those who accused the Smithsonian of inaction while coveting its meteorological system. Coffin's "central association" of observers and Loomis's "grand meteorological crusade," both proposed in the 1840s, had finally produced tangible results.[50]

Henry also prepared a series of articles under the general title "Meteorology in Its Connection with Agriculture."[51] The fourth article, on "atmospheric vapor and currents," contained a map of rainfall patterns across the continent illustrating the link between climate, wind patterns, and soil moisture (see Figure 6.1). Henry found a "distressing lack of moisture" in the Far West: "The results of the climatological investigations do not present a very favorable picture of our continent for the sustenance of a dense population. Nearly the whole of the domain of the United States west of the 98th meridian except the narrow strip along the Pacific, is a barren waste. Or in other words one half of the surface of the United States is unfit for agriculture."[52] This result attracted considerable attention, particularly in the agricultural journals and papers.[53]

Cooperative observations by the agricultural division of the Patent Office ended suddenly and unexpectedly with a change in commissioners in 1861 and the introduction of a separate office of superintendent of agriculture.[54] Because of the sudden break in funding, the Smithsonian stopped the process of data reduction but continued to receive the benefit of free postage. Part of the franking privilege ended when a new postage law enacted by Congress in 1863 stipulated that all communications to offices of government after July 1 were free only for "officers and correspondents responsible to the Department."[55] Volunteer meteorological and agricultural correspondents did not qualify under this definition. The correspondence would have been discontinued for lack of funds at this point had not Isaac Newton, the commissioner of the new Department of Agriculture, decided to send a circular letter and the necessary postage stamps to one hundred representative observers.[56] Amid the bureaucratic confusion, Henry also issued a circular letter requesting observers to retain their records until an arrangement to supply postage could be made. Congress then amended the postage law, allowing the Department of Agri-

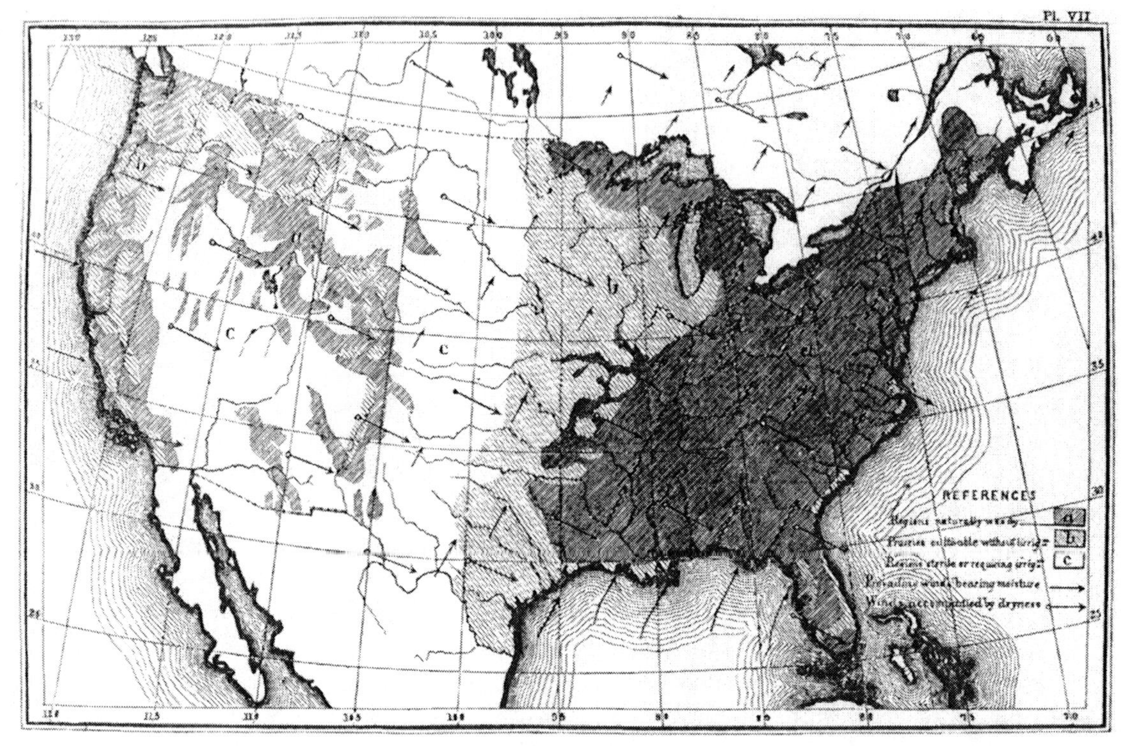

Fig. 6.1. Joseph Henry's map of arable lands. From Henry, "Meteorology in Its Connection with Agriculture," U.S. Patent Office, *Annual Report, Agricultural, 1858,* plate VII; rpt. in *Scientific Writings of Joseph Henry,* Part II continued (1847–78), *Smithson. Misc. Coll.* 30 (1887), 288. Superimposed on the map are arrows depicting surface wind currents.

culture to receive the observations on behalf of the Smithsonian.[57] The Department of Agriculture published tables of temperature, precipitation, and crop conditions in its *Monthly Reports* until 1872.

Bache, Schott, and the Coast Survey

In addition to the systems administered by the War Department, the Smithsonian, and the states of New York and Massachusetts, Henry considered the observers of the U.S. Coast Survey to be a "distinct system" of meteorological observation in the United States.[58] With Bache as the director of the Coast Survey, cooperation with the Smithsonian in meteorology was straightforward. For example, Coast Survey calculators converted Guyot's meteorological tables from French to American scales.[59] The Smithsonian employed Charles A. Schott, an assistant in the Coast Survey well versed in the newly emerging field of statistical data analysis, to interpret the results of three exploring expeditions into Arctic regions.[60]

A meteorological and magnetic observatory was set up in 1854 near the southeast corner of the Smithsonian building at the joint expense of the Coast Survey, which furnished the magnetic instruments, and the Smithsonian, which provided the building. The results were intended to provide a calibration standard for observations made by the western surveys and at more than sixty remote Coast Survey stations, including three permanent ones at Astoria, San Francisco, and San Diego.[61]

In 1860, the Coast Survey provided a meteorological "missionary," W. L. Nicholson, who surveyed tornado damage in Iowa and Illinois at Smithsonian expense. The "Camanche tornado" had destroyed the town of Camanche, Iowa, in June, causing 141 fatalities.[62] Nicholson talked to survivors and, in an attempt to assure a future supply of observers for the Smithsonian corps, "endeavored to diffuse a taste for meteorology among the people."[63] He also collected data on the altitude of the country from records in the archives of the state capitols he visited.

Between 1868 and 1874, Schott prepared two innovative monographs on the rainfall and temperature of the United States using records gathered by the Smithsonian, the Army Medical Department, the Lake Survey, the Coast Survey, the states of New York and Pennsylvania, and other sources extending back into the eighteenth century.[64] He applied statistical techniques borrowed from the astronomical work of Benjamin Peirce and Benjamin Gould, including a discussion of error analysis for meteorological data.[65] In his monographs Schott presented extensive tables, maps, and charts showing the average distribution of rain and temperature across the continent. His maps, shown in Figures 6.2 and 6.3, were more reliable than those of Blodget. They were undoubtedly used to shape government land policy for the arid West. Tables based on Schott's work were prepared for the Census Office in 1873.[66]

Schott also prepared a harmonic analysis of the temperature data to examine secular changes in the climate (see Figure 6.4). His conclusion, which put to rest uninformed speculation about temperature changes caused by settlement of the continent, argued against the notion of a changing climate: "There is nothing in these curves to countenance the idea of any permanent change in the climate having taken place, or being about to take place; in the last 90 years of thermometric records, the mean temperatures showing no indication whatever of a sustained rise or fall. The same conclusion was reached in the discussion of the secular change in the Rain-Fall, which appears also to have remained permanent in amount as well as in annual distribution."[67]

Natural History: A New Theoretical Focus

As the Smithsonian system of meteorological correspondents spread across the nation, natural historians interested in collecting specimens and investigating the "economy of nature" across wide expanses of the continent

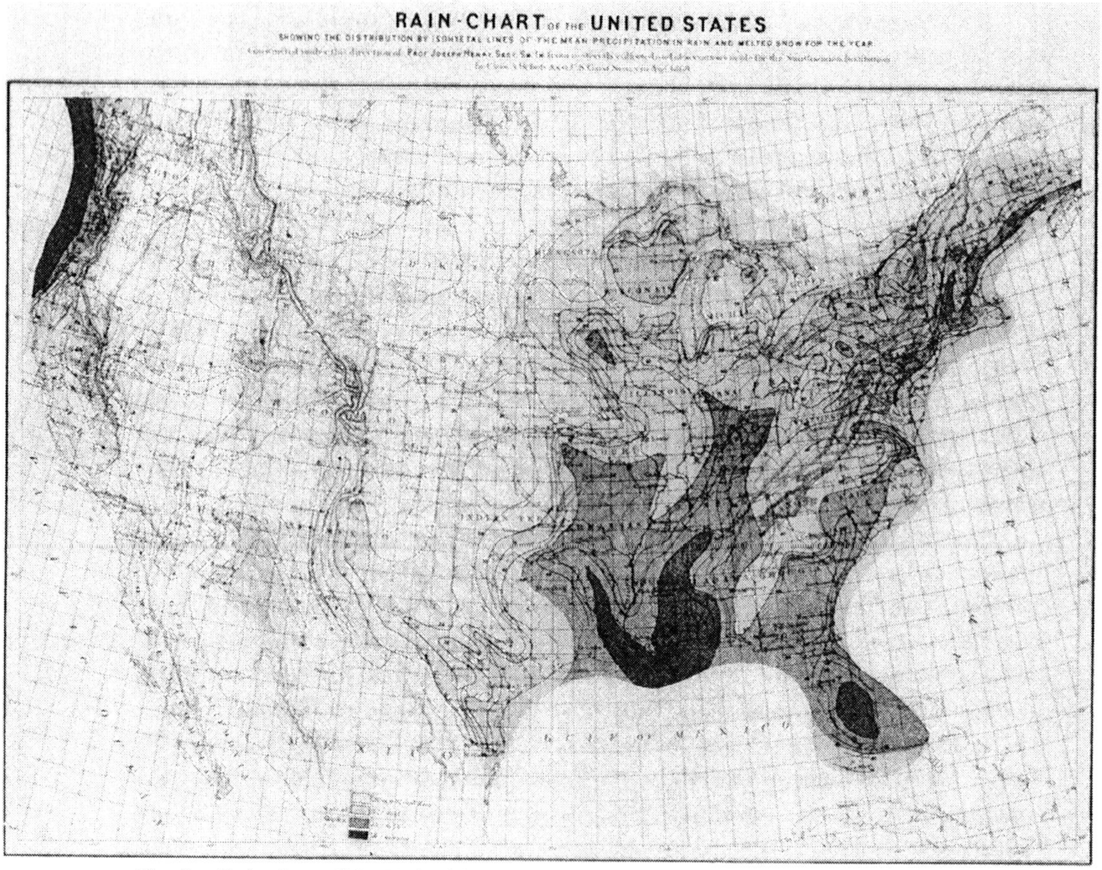

Fig. 6.2. Rain chart of the United States, 1868. From Charles A. Schott, "Tables and Results of the Precipitation, in Rain and Snow, in the United States and at Some Stations in Adjacent Parts of North America and in Central and South America," *Smithson. Contrib.* 18, Article II (1872).

took note of this new scientific resource. Spencer Fullerton Baird provided the direct link between Henry and the naturalists. In 1849, while seeking a permanent position at the new institution, Baird wrote to Henry praising him for his "gigantic plans for the advancement of meteorological science." The compliment may have been sincere, but Baird had little interest in meteorology, although he did make weather observations for the Smithsonian in Carlisle for one year. Rather, he suggested to Henry that the meteorological correspondents could be asked to observe other natural phenomena along with the weather:

> [It is] easy to call upon the trained meteorological correspondents for information upon other subjects, the distribution and local or general appearance of certain forms of animals, vegetables, or minerals; the occurrence of certain

Fig. 6.3. Temperature chart of the United States, 1874, mean temperature of December, January, and February. From Charles A. Schott, "Tables, Distribution, and Variations of the Atmospheric Temperature in the United States and Some Adjacent Parts of North America," *Smithson. Contrib.* 21, Article V (1876).

> diseases over the entire country; the spread and rate of progress of a pestilence as small pox, cholera, or yellow fever through the land; the range of action of various insects, as the Hessian fly, the cotton or tobacco worm and, . . . an infinity of others. I have long dreamed of some central association or influence which might call for such information, digest it, and then publish it in practical form to the world, and I see that my dream is not far from realization.[68]

Baird proposed that a series of skeleton maps be prepared at the Smithsonian's expense on which to mark the distribution of species of animals, plants, and other natural phenomena. This was to be combined with a great system of specimen collection.[69] Duplicates of the maps were to be returned to the institution and preserved for study and generalization.[70] Although Henry saw the Smithsonian as primarily a research institution and had serious reservations about using its funds to maintain large museum collections, he was not

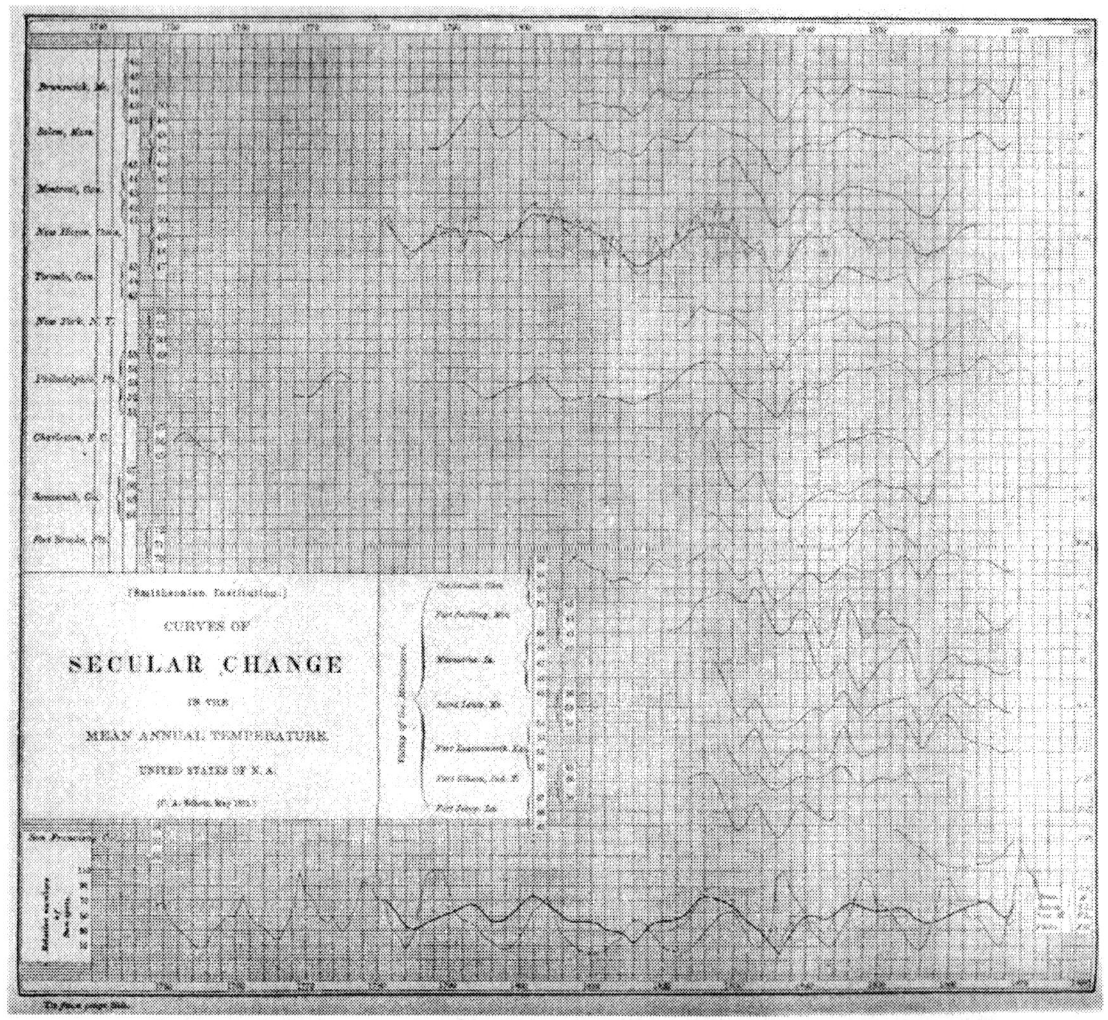

Fig. 6.4. Curves of secular change in the mean annual temperature. From Charles A. Schott, "Tables, Distribution, and Variations of the Atmospheric Temperature in the United States and Some Adjacent Parts of North America," *Smithson. Contrib.* 21, Article V (1876), chart facing p. 310. Curves for Brunswick, Maine; Salem, Mass.; Montreal, Canada; New Haven, Conn.; Toronto, Canada; New York, N.Y.; Philadelphia, Pa.; Charleston, S.C.; Savannah, Ga.; Fort Brooke, Fla.; Cincinnati, Ohio; Fort Snelling, Minn.; Muscatine, Ia.; St. Louis, Mo.; Fort Leavenworth, Kan.; Fort Gibson, Indian Territory; Fort Jessup, La.; and San Francisco, Calif. The time interval examined was 1750 to 1870.

hostile or indifferent to the needs of natural history. During his visit to Europe a decade earlier, an account of Herschel's observations of the stars of the southern skies stimulated the following entry in his diary: "Sir J. cannot fail to make interesting observations for the same reason that a botanist visiting an unexplored region will be sure to discover some new plants."[71] If the southern skies were Herschel's scientific domain, the American continent was Henry's, and the Smithsonian Institution provided him with the resources to explore it.[72] Baird's program presented Henry with the opportunity to include natural history in his program. Within six months, Henry had appointed Baird to the position of assistant secretary of the Smithsonian Institution.[73]

Following up on his dream of a "central association" of collectors, Baird called on the meteorological observers to focus their attention on questions of natural history. For example, at his request, the circulars of John LeConte, soliciting natural history observations and specimens, were distributed to all of the meteorological correspondents.[74] Volunteers were asked to observe "periodical phenomena" such as the blossoming of plants, the migration of birds, and the hatching of insects. Tables of these observations covering the nation were published in the *Report of the Commissioner of Patents for 1864*. This report listed the foliation of eighty-seven species of plants, the blossoming of ninety-two, the ripening of fruit of ten, and the defoliation of eighteen species. It also listed the first appearance of sixteen species of birds, one of reptiles, three of fish, and two of insects. According to Henry, the results of such investigations "have a direct application to meteorological science, by indicating the progress of the seasons in different localities and their relative variability in different years."[75] In the absence of international standards of observation and readily available sources of calibrated instruments, these phenological measurements could be compared with the results from other countries and from earlier periods.[76]

Maps illustrating physical phenomena such as weather and climate or the distribution of species were new and exciting tools for scientific research in the mid-nineteenth century.[77] Several factors contributed to the Smithsonian's initiatives in this field: agricultural and commercial interests were expanding westward into new territories, raising concerns about climatic changes from rapid settlement and the removal of the forests; the institution had a national system of correspondents poised to gather the necessary data; and both Henry and Baird endorsed a physical atlas project for the nation.[78] Although no atlas was produced, the Smithsonian published maps of topography, rainfall, temperature, and other physical and biological factors. For example, Henry used Smithsonian resources to gather data for a report on the distribution of forest trees in the United States. He hired James Cooper, the son of the noted naturalist William Cooper, to arrange, chart, and discuss the data.[79] Cooper's map, compiled from reports received from many sources, including the meteorological observers, is shown in Figure 6.5.

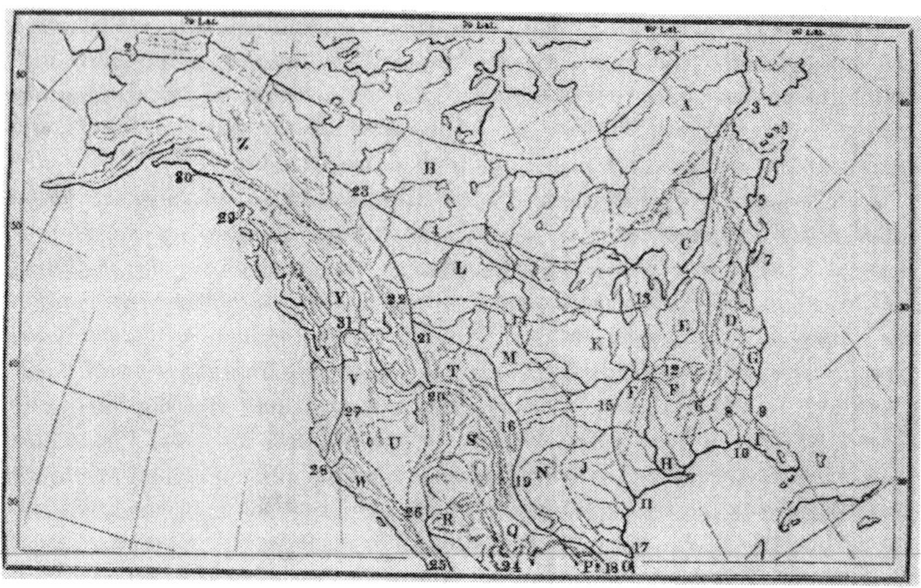

REGIONS		PROVINCES
A	Algonquin	
B	Athabascan	Lacustrian
C	Canadian	
D	Alleghany	
E	Ohio	
F	Tennesseean	Apalachian
G	Carolinian	
H	Mississippian	
I	Floridian	West Indian
J	Texan	
K	Illinois	
L	Saskatchewan	Campestrian
M	Dacotah	
N	Camanche	

REGIONS		PROVINCES
O	Tamaulipan	
P	Choahuilan	Mexican
Q	Chihuahuan	
R	Arizonian	
S	Wasatch	
T	Padoucan	Rocky Mountain
U	Utah	
V	Shoshonee	
W	Californian	
X	Oregonian	
Y	Kootanic	Caurine
Z	Yukon	

Fig. 6.5. James Cooper's map of forest regions of North America. From James Cooper, "On the Distribution of the Forests and Trees of North America, with Notes on Its Physical Geography," *Smithsonian Report, 1858*, 267. Solid lines demarcate natural forest provinces, broken lines indicate forest regions and parts of province boundaries not accurately determined.

Collecting Data and Specimens

Although the correspondents were asked to provide a growing number of observations, meteorological observers differed from specimen collectors in important ways. Weather observers needed regular habits, patience, persistence, and care in handling the instruments and recording the data. They needed to be present at the place of observation at the required hours of the day all year round. Farmers, physicians, lawyers, and clerks were well suited for these duties.[80] Typically, observers would contribute seven to ten years of observations from the same location. Some made voluntary contributions throughout the twenty-five-year duration of the project. In contrast to the sedentary and disciplined life of the weather observer, specimen collectors needed to possess a sense of adventure and sufficient leisure time to roam. Typical skills of the collector—hunting, skinning, and preserving specimens—were quite different from those of a weather observer—reading a barometer, constructing a rain gauge, and keeping a regular journal. Nevertheless, many naturalists contributed meteorological observations to the Smithsonian. Although their measurements were often taken on collecting trips with less-than-reliable portable instruments, they were valuable because they often represented the first information from new and often exotic regions.

An example of a lineage of naturalists, all of whom contributed a year or two of observations to the meteorological project, is Dr. Jared P. Kirtland, Robert Kennicott, and William Healy Dall.[81] Kirtland took weather observations in East Rockport, Ohio, in 1854. Kennicott, who studied natural history with Kirtland, Baird, and others, was the Smithsonian meteorological observer in Hudson's Bay Territory in 1860 and in Alaska in 1861. Dall, who studied under Kennicott and wrote a biography of Baird, contributed weather observations from Alaska in 1866 and 1867.

Other individuals also provided strong links between meteorology and natural history. For example, Dr. George Englemann of St. Louis kept a weather diary for fifty years and published reports on local meteorology and mortality. Most physicians were content to correlate the average temperature with outbreaks of disease, but Englemann kept extensive data on the clearness of the sky, the barometric pressure, and the number of thunderstorms "because it is believed that these different meteorological phenomena have an important bearing on human health perhaps . . . as much as the variations of temperature."[82] Englemann's medical practice prevented him from doing extensive field work in natural history. Nevertheless, he encouraged local collectors to bring him specimens, which he supplied to naturalists in the East.[83] He also coordinated the east-west flow of publications in both natural history and meteorology. For example, he wrote to Baird in 1853 requesting six copies of the latest publications in both fields: "I wish you would send me half a dozen copies of the 'Serpents' to distribute among those that take an interest in these things and assist in collecting; also a similar number of copies of Guyot's tables."[84]

Although the community of naturalists cooperated with the meteorological project, they did so largely in support of their own interests. This resulted in a sense of competition for the observers' limited time and attention. The public's demand for practical information for the settlement of the West and the emerging ecological interests of the naturalists resulted in new theoretical and practical links between meteorology and natural history.[85]

Other Contributions: Coffin and Ferrel

James Coffin's Smithsonian connection—especially the system of international exchanges—provided the resources for him to complete and publish his "Winds of the Northern Hemisphere" in 1853 and "Winds of the Globe" in 1875. Henry assured him, "Of course we will cheerfully do anything in our power to facilitate your new investigation in regard to the Winds, and to publish the result of your labors as they are completed. If you will send us the addresses of the parties with whom you wish us to correspond, we will forward them copies of the 'Meteorological Results.'[86]

In "Winds of the Northern Hemisphere," Coffin presented a series of wind charts using data from 579 stations in the Northern Hemisphere. He demonstrated the existence of three latitudinal surface wind belts and three vertical circulation cells (see Figure 6.6, top and middle). This result proved that George Hadley's hypothesis of a single circulation cell between the equator and the pole was inadequate. "Winds of the Globe," not completed until a year after Coffin's death by his son S. J. Coffin and the Russian climatologist Alexander Woeikof, presented worldwide data from 3,223 stations, including Maury's wind charts for the oceans, all the Smithsonian results from 1856 to 1870, and observations from the Army Medical Department, polar expeditions, and several hundred international stations. These empirical studies provided the materials for Ferrel's theoretical work.

William Ferrel did not scrutinize nature inductively. He used his self-taught mathematical skills, his physical and mechanical intuition, and the observations compiled by others to prepare an innovative theoretical study of the general circulation of the earth's atmosphere.[87] Ferrel used Coffin's wind studies, Espy's *Third Report,* and data newly available from the polar expeditions of Elisha Kent Kane, Francis L. M'Clintock, and Isaac I. Hayes as raw material. He also attacked Maury's poorly reasoned ideas on the general circulation. As one author explained, Ferrel was almost obliged to Maury for presenting such a "convincingly impossible" theory.[88]

Since the time of Hadley (1735), the differential heating of the earth's surface had been considered the *primum mobile* of the winds. In hotter regions the air would tend to rise and expand, causing lower pressure; cooler areas would experience higher pressure. If the earth was stationary, a meridional flow would result, with winds blowing from areas of high to areas of low pressure. Because the earth's rotational speed was greater at the equator, conservation of the angular momentum of the southerly return flow would lead to northeast trade winds in the tropics. But this simple one-cell model of the

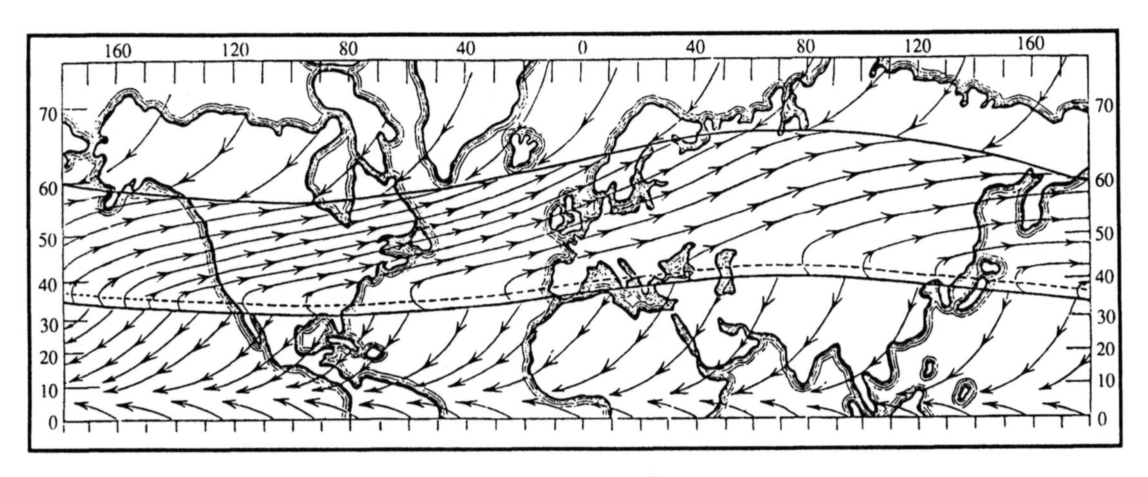

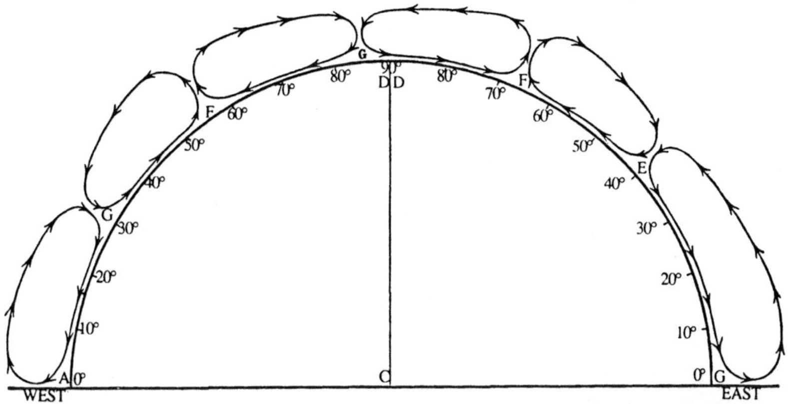

Fig. 6.6. Comparison of Coffin's charts (top and middle) and Ferrel's chart (bottom) of the general circulation of the atmosphere. From James H. Coffin, "On the Currents of the Atmosphere." *AAAS, Proceedings* 12 (1859): 202–203; and William Ferrel, "An Essay on the Winds and Currents of the Ocean," *Nashville Journal of Medicine and Surgery* 11 (1856), 287.

At top, Coffin's chart of the horizontal movements of the atmosphere. Shown are three major surface wind zones: northeasterly trades, prevailing westerlies, and polar easterlies.

Center, Coffin's chart of the vertical movements of the atmosphere in the plane of the 90th meridian. Shown are Hadley cells, reverse Hadley cells (now known as Ferrel cells), and polar cells.

At right, Ferrel's chart of the general circulation of the atmosphere.

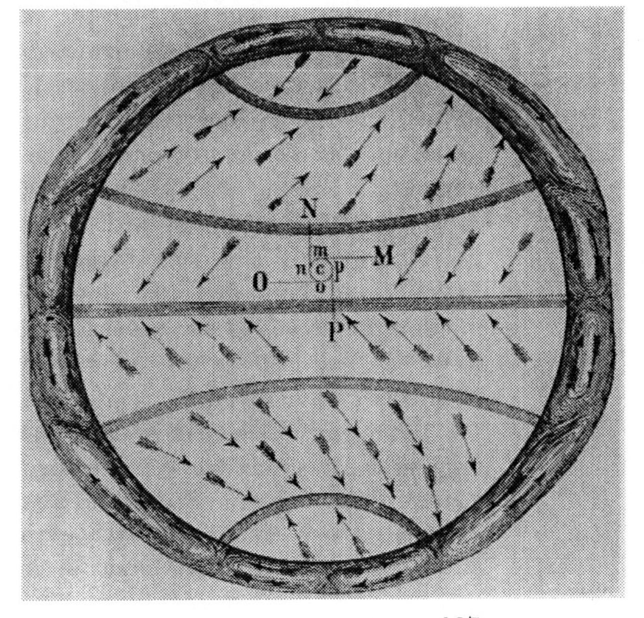

atmosphere's circulation could not explain the wind patterns observed in other regions. Several crucial developments occurred in the first half of the nineteenth century. Laplace, in his *Mécanique céleste* (5 vols., 1799–1825), developed the general form for equations of motion of a fluid at the surface of the earth. In the 1830s Gaspard Coriolis discovered an apparent "force" that appears when a body moves in an accelerated system. Siméon Poisson applied this discovery to explain small but systematic deviations in the trajectories of artillery shells.[89] None of these authors, however, developed a theory of the general circulation of the atmosphere.

In his ground-breaking but obscure "Essay on the Winds and Currents of the Ocean," hidden in the *Nashville Journal of Medicine and Surgery*, and in two papers in the *Astronomical Journal* (1858) and *Mathematical Monthly* (1859–60), Ferrel announced a "new force" that had not been taken into account in previous studies of atmospheric circulation: "If a body is moving in any direction, there is a force, arising from the earth's rotation, which always deflects it to the right in the northern hemisphere, and to the left in the southern."[90] In fact, he had rediscovered the Coriolis acceleration and was applying it to the earth's atmosphere. But Ferrel went far beyond Coriolis, who had only hinted that his discovery might be important to meteorologists.[91] His theory explained both meridional and zonal wind flows on the rotating earth. With this new force taken into account, the earth's atmosphere generated three circulation cells and oblique winds in midlatitudes with high velocities from the west (see Figure 6.6, bottom; the midlatitude circulation is now called the Ferrel cell). Moreover, high pressure at the poles, which would be expected because of low temperatures, was reversed to low pressure by the excessive centrifugal force of the whirling winds. Conversely, high pressure was generated near the tropics. All these theoretical results were in accordance with the latest observations.

Ferrel was "a supporter," with modifications, of Espy's convective theory of storms.[92] He held that gyratory storms were maintained by "Espian thermal processes," but he did not support Espy's centripetal wind patterns: "There cannot be a rushing of air from all sides towards a center, on any part of the earth except at the equator, without producing a gyration." Ferrel explained the calm center of a hurricane as the place where centripetal and centrifugal forces balanced and deduced from his theory that no hurricanes could have their origin directly on the equator. He also established a correlation between convectional (warm-core) cyclones and the general hemispheric (cold-core) circulation.[93]

Ferrel sent his essay to Joseph Henry, who thought the new force was an original concept that could bring together all the accumulated facts into a general circulation theory: "I think there are now sufficient facts accumulated to form the basis of a hypothetical discussion, on strictly mechanical principles, of the motion of the air on the surface of the globe, heated at the equator and cooled at the poles. In this discussion the effect of moisture is properly excluded in obtaining the first approximate solution. The effect, however, of this element should next be considered."[94] Henry immediately referred the essay to Espy for review. Espy expressed disappointment that Ferrel, an outsider to the meteorological community—he was then a schoolteacher in Nashville—had neglected moist-thermal processes as a cause of barometric fluctuations: "[Ferrel does not] agree with Professor Espy as to the cause assigned by him for the belt of high barometer at the outer limits of the trade winds . . . the author completely ignores the fact that the air above being deprived of much of its vapor at the belt of rains near the equator, would be specifically heavier than the air below as soon as it became of the normal temperature by radiation, and would then of course descend . . . *the high barometer in these belts is partly due to the dryness of the air.*" Espy also disagreed with Ferrel over the rotational forces active in storms. Espy's calculations of the forces generated by meridional and zonal flows of air rushing into the low-pressure area of hurricanes and tornadoes led him to conclude that the effect of the earth's rotation would be "vanishingly small" for hurricanes and tornadoes and that "no whirl can take place in tornadoes, from the causes in question."[95]

Although Ferrel did not join the meteorological community until the 1880s, when he worked in the "study room" of the Army Signal Office, his essays earned him a position in Cambridge with the *American Nautical Almanac* in 1857 and with the Coast Survey in Washington in 1867. Shortly after Ferrel's death, his friend and associate Cleveland Abbe wrote that his memoir in the *Mathematical Monthly* was the "starting point" of our knowledge of the mechanics of the atmosphere: "I have often said that that memoir is to meteorology what the 'Principia' was to astronomy . . . as Newton's 'Principia' arrested all further vain speculations and turned the whole trend of thought toward the true celestial mechanics so Ferrel's memoir served to turn all eyes toward the true atmospheric dynamics. . . . Ferrel's memoir will always remain the *principia meteorologica.*"[96]

The Smithsonian's effort to coordinate the meteorological efforts of other groups, although only partially successful, resulted in effective cooperative programs with the states of New York, Massachusetts, and Maine, the Canadian government, the Coast Survey, the Army Topographical Engineers, the Patent Office, and others. Publications from these programs were widely distributed. The Smithsonian's plan, in a milieu in which the contributions of the common people were both valued and actively solicited, was to furnish the published materials in their raw tabular form to all who wished to study them. In Henry's optimistic terms, "It is highly desirable that as many minds as possible should be employed on this subject."[97] Although no flood of theoretical studies resulted, the data compiled by Coffin and Schott provided a basis for Ferrel's theory of the general circulation of the atmosphere. Moreover, the publications were undoubtedly of use to agriculturalists, educators, and settlers on the American frontier. In one noteworthy study, Increase A. Lapham used Smithsonian data in 1870 to show that the path of a storm could be traced across the continent for over seventy-two hours and that a national telegraphic weather service could issue warnings in advance of its path.

CHAPTER 7 Weather Telegraphy

To think of an instantaneous communication between the cities of this great continent!! Well! What next? . . . as a crowning work, *electric circulation through space, is now given to human thought!*
—Redfield to Reid, July 4, 1846, Reid's Correspondence

From the earliest experiments in the late 1840s to the creation of a federal storm-warning service under the U.S. Army Signal Office in 1870, telegraphy was the premier new technology of the meteorological community. At midcentury, a daily weather map displaying current conditions as reported by the telegraph companies hung in the lobby of the Smithsonian building. Within a decade, however, the Civil War and a fire at the Smithsonian halted the telegraphic reports. After the war, as Henry struggled to rebuild the Smithsonian system, Cleveland Abbe began issuing weather reports and forecasts from the Cincinnati Observatory. Abbe's colleague Increase A. Lapham was convinced that the nation needed a federal telegraphic storm-warning system and successfully petitioned Congress to pass the enabling legislation. Although telegraphy was a marvelous new tool that increased the speed of reporting many times over and provided meteorologists with information on current weather conditions over a large area, it also produced an important shift of emphasis in meteorology from weather science to weather service. The efficiency of a network in bringing timely and accurate warnings to the public took priority over more theoretical concerns.[1]

In the wake of Samuel F. B. Morse's federally funded telegraphic experiment between Washington and Baltimore in 1844 and the rapid expansion of lines to Philadelphia, New York, and Boston the following year, American meteorologists' dreams of instantaneous communication of weather data between cities gave way to practical arrangements that would make it a reality. Redfield's opinion, published in 1846, was the first to appear in print: "In the Atlantic ports, the approach of a gale may be made known by means of the electric telegraph, which probably will soon extend from Maine to the Mississippi."[2] Loomis, in his "Report on the Meteorology of the United States," included suggestions for making the telegraph "subservient" to the protection of commerce from the ravages of storms.[3] Henry intended to employ telegraphy in scientific studies at the Smithsonian.

The first daily weather reports were compiled from information sent, not by telegraph but by trains. James Glaisher, secretary of the British Meteorological Society, arranged for observations to be taken each morning at 9:00 A.M. at railway stations, the results to be transported to London for publication the following day. These day-old observations first appeared in tabular form in the *London Daily News* for June 14, 1849.[4] Glaisher also prepared surface weather maps from the data. Two years later, between August 8 and October 11, 1851, the Electric Telegraph Company provided experimental telegraphic weather reports to the Crystal Palace Exhibition. Daily maps, prepared from data, were put on display.[5]

Concurrent with these developments in England, Henry was planning to "establish observations along the line of the telegraph, and to make observations on the origin and progress of storms."[6] By the end of June 1849, Henry had petitioned the telegraph companies for the use of their lines for meteorological purposes. Henry O'Reilly, who along with Amos Kendall and Morse was constructing a telegraph line from Pennsylvania to St. Louis, promised that "every necessary facility" would be afforded by his company free of cost. Similar promises were received from managers of the northern and southern lines.[7] At the Smithsonian, Edward Foreman constructed a large map of the United States, in hopes of displaying the telegraphic reports, and Henry wrote to Glaisher at the Greenwich "Meteorological observatory" for information about procedures and forms used in the British railway system.[8] There is no evidence, however, that a report or map was prepared until 1856. Scattered telegrams were received but not in sufficient numbers to prepare a general picture of the nation's weather. Nevertheless, some early reports were received by telegraph.

Alexander Jones, a reporter for the Associated Press in New York, was a pioneer in organizing cooperative press services and marketing wire reports among American cities. Known for having filed the first news message by telegraph from New York to Washington in 1846, Jones now attempted to begin a commercial system of telegraphic weather reports. On behalf of "Jones & Company," he advertised his services in the *American Journal of Science* in 1848, offering "colleges, universities, and other public institutions" daily meteorological reports from around the country on the "most reasonable terms of 12½ to 25 cents per day per report for each city."[9] Hoping that the government or the Smithsonian might assist his endeavor by supplying telegraph stations with meteorological instruments, he wrote to Henry: "By the way of experiment I obtained the enclosed observations yesterday and the day before, over a large tract up country in this state, which though not as complete as such observations ought to be, yet they are interesting. I am induced to believe that if the Government or the Regents of the Smithsonian Institute were to supply every Telegraph Station in the United States with suitable instruments, . . . correct and extensive observations might be simultaneously obtained over the greater part of the United States, by which the progress of storms could be accurately noted, and their course duly ascertained."[10]

Although Jones was quite aggressive, sending weather information to help make up the circuits each morning was a routine habit, if not a duty, of many operators. Abbe received the following testimony from an early operator, David Brooks: "About the year 1849 I became manager in Philadelphia, and was in the habit of getting information about the condition of the lines westward every morning. If I learned from Cincinnati that the wires to St. Louis were interrupted by rain, I was tolerably sure a 'northeast' storm was approaching (from the west or southwest). For cold waves we looked in Chicago."[11] Jeptha Homer Wade, one of the founders and later president of Western Union, confirmed these practices in a letter to Abbe:

> I commenced operation in a telegraph office in 1846. With the small amount of commercial business then on the lines, the employees had less to do than they have now, and it was quite common for the operators in different parts of the country to enquire of each other about the weather, such as the direction and force of the wind as nearly as we could guess it, together with the temperature and its changes from time to time at different points. . . . I would frequently write upon the bulletin board in my office, what and when weather changes were coming. Frequently this was with such accuracy as to create considerable comment and wonder. Later on, when I was promoted to the presidency of the company, I suggested to, and urged upon the Smithsonian Institution, the advantageous use that could thus be made of the telegraph. I succeeded in getting them to take an interest, and test reports as we were able to give them. These reports were continued free for a long time, hoping and expecting it would lead to an appropriation by the Government, resulting in an important addition to our business.[12]

Captain Joseph Brooks, manager of a line of steamers between Portland and Boston, recalled that in 1849 he received three telegraphic reports daily, morning, noon, and evening, from New York, Albany, and Plattsburgh: "'If the weather looked bad in the morning, the agent in New York was to send a dispatch at 8 o'clock; if a storm came up, to send another about noon; and then at 3 o'clock to give a full statement of its condition.' With these data Captain Brooks could tell when the storm would reach New Haven, Springfield, Boston, Portland, and other points. If the storm had been raging with severity in New York for some hours, he knew it would not be safe for the Portland boat to leave."[13] Figure 7.1 shows telegraph lines in the United States in 1846. By 1848 there were 2,100 wire miles; by 1850, 12,000 wire miles; and by 1852, 23,000 wire miles in service.

In 1856 the Smithsonian began to display the telegraphic weather messages it received on a large map of the United States which hung in the great hall of the institution. A piece of iron wire was driven into the map at each point of observation, and colored circular cards of about an inch in diameter were hung on the wires to indicate the weather and wind direction. White represented fair weather, blue indicated snow, black meant rain, and brown, cloudiness. Each card had an arrow painted on it and eight holes punched around

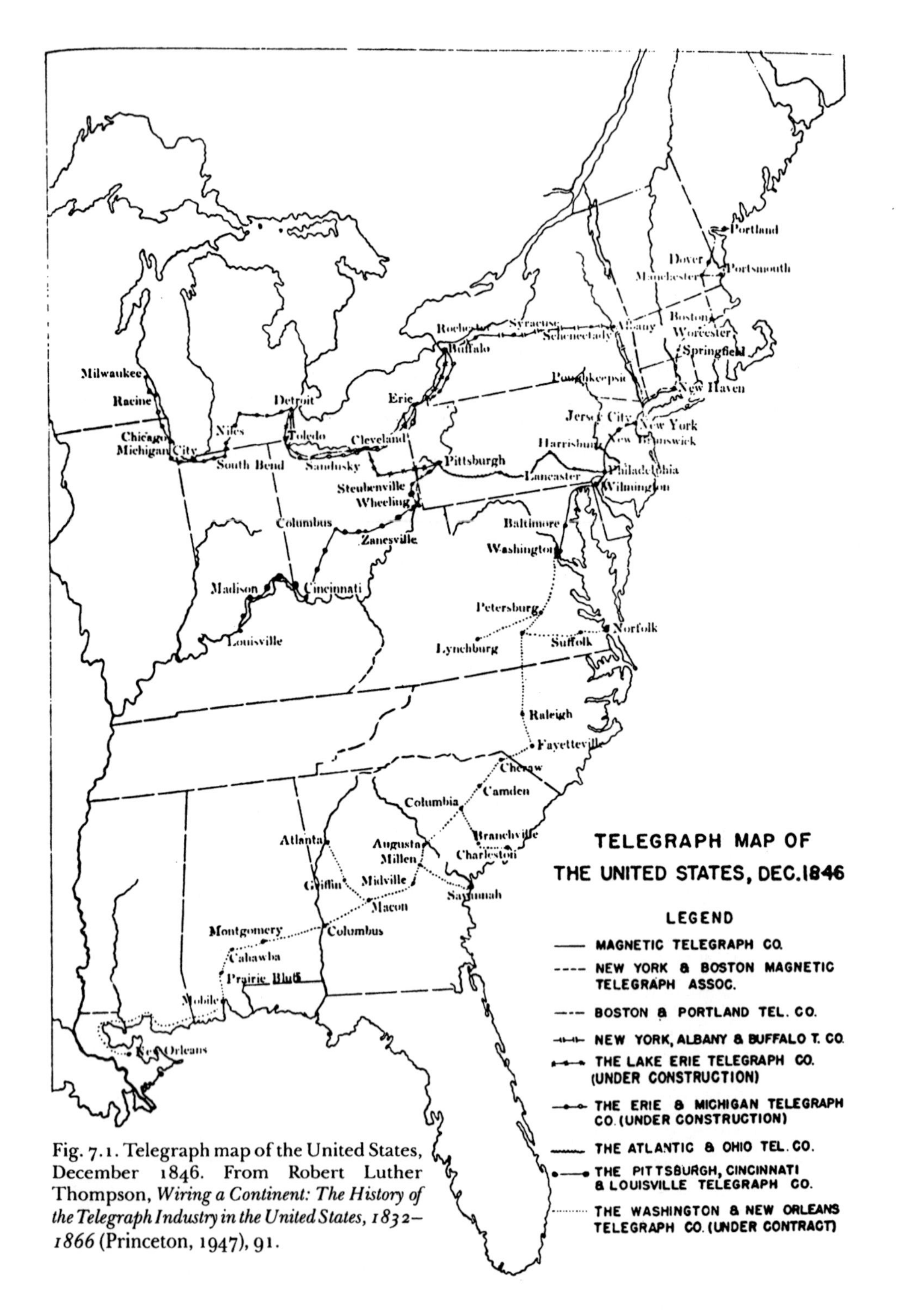

Fig. 7.1. Telegraph map of the United States, December 1846. From Robert Luther Thompson, *Wiring a Continent: The History of the Telegraph Industry in the United States, 1832–1866* (Princeton, 1947), 91.

Fig. 7.2. Smithsonian telegraph network, 1860.

the circumference so that the card could be hung to indicate the direction of the prevailing winds.[14] Reports were received about ten o'clock each morning and were changed as often as new information was received. Special weather flags were also flown from the high tower of the Smithsonian building to indicate forthcoming weather conditions, especially on nights when Smithsonian lectures were scheduled.[15] On May 1, 1857, the *Washington Evening Star* first published a report of the current weather at nineteen telegraph stations, all near the eastern seaboard on the line from New York to New Orleans. A week later, on May 7, the *Evening Star* published the first weather forecast in America, probably prepared by Henry and Espy: "Yesterday there was a severe storm south of Macon, Ga; but from the fact that it is still clear this morning at that place and at Wheeling, it is *probable* that the storm was of a local character."[16] These telegraphic investigations served the dual purpose of exciting public interest in meteorology and confirming Espy's earlier inferences on the eastward propagation of storms, "that as a general rule the storms of our latitude pursue a definite course."[17] The Smithsonian telegraph network in 1860 is illustrated in Figure 7.2.

The Civil War and Other Shocks to the Smithsonian System

> Our system of Meteorology has been sadly broken in upon by the war. All our southern and many of our western observers have discontinued sending.
> —Henry to Coffin, December 10, 1861, Henry Papers

The Civil War and a fire at the Smithsonian dealt severe blows to the meteorological project. Although Henry gradually rebuilt the system to its prewar size, the project never regained its antebellum significance. Theoretical interest in storms had been all but buried with the original controversialists; cooperation with the Department of Agriculture brought more farmers and fewer scientists into the system; voluntary arrangements with telegraph companies foundered. Even Henry, now the aging patriarch of American science, had lost a measure of his earlier zeal.

The Smithsonian system of meteorology was in "prosperous condition" in 1860. A year later, however, after the initial devastation of the war, Henry reported that the system had "suffered more from the disturbed condition of the country than any other part of the operations of the Smithsonian establishment."[18] A letter from Henry to Colonel Edward Sabine in London, begun in March 1861 and completed in July, illustrates both the progress in telegraphic meteorology and the sad impact of the war: "When our reports are from parties nearly from points directly west of this city, I scarcely ever fail to be able to foretell for nearly a day in advance the state of the weather." This was written in March. The conclusion of the letter contains a grim report of the battle of Manassas and the turbulent political situation in Washington.[19]

Because of the war, telegraph lines to the South were cut, observations ceased, and only a few meteorological records were preserved. Henry Ravenel of South Carolina urged Smithsonian observers in the region to continue their work and to hold their monthly returns until the Confederate government was ready to receive them, but very few records survived.[20] One observer in Key West, Florida, far from the war zone, kept a continuous record, but the only other reports from the South during the war years were two fragmentary reports from South Carolina and three from Texas. The Smithsonian income also suffered a $4,000 shortfall because Virginia and Tennessee defaulted on interest payments due on investments.[21]

The northern telegraph lines, preoccupied with war messages, stopped carrying weather reports, and the ranks of monthly correspondents were depleted when observers reported for military duty. W. G. Fuller, an observer in Ohio, wrote to Henry that he had to close his observations abruptly because he was called "without an hour's notice to leave my home and enter the field to aid our country's cause" and had no time to find a substitute. Another observer,

Lucius E. Ricksecker of Pennsylvania, reported that he had received a package of the wrong forms, but he had just been drafted so all that "is of no consequence now."[22]

The confusion over the status of the cooperative agricultural and meteorological correspondents of the Smithsonian, caused by numerous changes in the office of the commissioners of patents and agriculture, was mentioned in Chapter 6. Understandably, scientific cooperation from the Army Medical Department was also interrupted by the exigencies of war and by the short tenure of Surgeon General Clement A. Finley (1861–62). His successors, William A. Hammond (1862–64) and Joseph K. Barnes (1864–82), although preoccupied with battle casualties, were supportive of cooperative meteorological ventures.[23]

As the Civil War was drawing to a close and plans were being formulated with the North American Telegraph Association to revive the system of weather reporting, another unexpected calamity struck the Smithsonian.[24] On the afternoon of January 24, 1865, a fire began in the woodwork near the roof of the lecture room and spread to the secretary's office, destroying Henry's official correspondence—fifty thousand pages of letters received and thirty-five thousand pages of copies of letters sent, a large collection of scientific instruments including the donation of Robert Hare, many of the personal effects of James Smithson, and some of the meteorological records.[25] The desk of William Q. Force, in charge of the meteorological records since 1858, was also lost in the fire.

The fire disrupted the work of the meteorological project, but two lists compiled by Force after the fire provide detailed documentation of the contents of the meteorological office and indicate some of the activities that went on there.[26] According to Force's memorandum, material lost in the fire included twelve bound volumes of instrumental observations for 1860, all the wind records of the self-registering apparatus at the Smithsonian, all the newspaper meteorological clippings, the observations made at lighthouses of the United States during the year 1860, a set of the "International Bulletin" published by the Imperial Observatory in Paris containing daily telegraphic notices of the weather for places in Europe, records of exploring expeditions, a list of corrections for the volumes printed by the Patent Office, and a case of pigeonholes containing summaries of meteorological observations, notices of amounts of rain, earthquakes, auroras, meteors, storms, diagrams of temperature, descriptions of instruments, notices of balloon voyages, and other materials. A number of meteorological instruments in use at the time were destroyed, including a Newman standard barometer, a Green psychrometer, a Kendall thermometer, a common London thermometer, the self-recording wind apparatus, and a box containing extra instruments. The fire also destroyed a bundle containing the telegraphic weather dispatches received at the Smithsonian.

Four main categories of material were saved from the fire: the administrative records of the meteorological project, manuscripts in preparation for publication, manuscript copies of original observations, and published reference material. The first category (fortunately for the historian of meteorology) included thirty volumes of letters received from the observers; record books containing lists of correspondents, registers received, and instruments distributed. The other categories included nine bundles of manuscripts by Coffin used for the Patent Office publications, thirty-two bundles of manuscript meteorological data from 1785 to 1864, all the registers of meteorological observations in all parts of the United States made three times daily on Smithsonian forms since 1849 (except for a few gaps), six bound volumes of original observations made at the Smithsonian Institution from 1858 to 1864, and the office collection of published reference works.

The fire, though a "sad calamity," was not as disastrous as it might have been.[27] The loss of the secretary's correspondence and the physical apparatus was most deeply felt. Although there was time to save many of the important meteorological papers, the project was thrown into considerable confusion. Moreover, repairs and fireproofing of the building cost $125,000. Research funds were diverted for this purpose, further slowing the restoration of the war-damaged system of meteorology.[28]

Reconstruction, 1865–1870

By 1865 the Smithsonian meteorological project was in its fifteenth year of operation. Henry, wiser and wearier for the experience, had moderated his initial optimism about meteorological research and the prospects of the system: "There is, perhaps, no branch of science relative to which so many observations have been made and so many records accumulated, and yet from which so few general principles have been deduced. This has arisen, first, from the real complexity of the phenomena, or, in other words, from the number of separate causes influencing the production of the ordinary results; second, from the improper methods which have been pursued in the investigation of the subject, and the amount of labor required in the reduction and discussion of the observations." He called on the federal government to relieve the Smithsonian of its burden by funding a national weather service with an annual appropriation of $50,000, "to advance the material interests of the country and increase its reputation."[29]

Congressional approval was not forthcoming, and the Smithsonian—as in antebellum times—prepared a circular letter again requesting voluntary cooperation from the telegraph companies. The circular announced the resumption of the system of receiving and publishing daily reports of the weather in all parts of the United States:

> Will you please make the following observations every morning (~~Except Sunday~~) and forward them immediately to Washington addressed to this Institution:
> 1) Whether it is *clear, cloudy, raining,* or *snowing.* . . .
> 2) The direction from which the *wind* is blowing, and whether it is moderate, strong or very strong. . . .
> 3) The height of the *thermometer.*
> 4) If there is a *barometer* in the office, and the mercury is steadily rising or falling, state when it began to rise or fall.
> 5) When any remarkable meteorological phenomena occur.[30]

The telegraphic reports were to be displayed as before with the addition of a second map in the rotunda of the Capitol. Forty-two major stations were proposed, reaching as far west as Salt Lake City and Santa Fe and from Halifax in the North to New Orleans in the South. Because the telegraphic reports were destroyed in the Smithsonian fire, it is impossible to say how well the system functioned. Probably only scattered reports were received. One surviving telegram, although dated 1864, a year earlier than the fire, illustrates the potential Henry saw in the telegraphic reporting system. The American Telegraph Company operator in Omaha announced, "Heavy fall Snow West Yesterday—Commenced Snowing here at noon today—no Wind." Two days later, Henry scribbled at the bottom of the telegram, "Began to snow at Washington 9 am January 4."[31]

Henry repeatedly had to ask telegraph companies to cooperate voluntarily. Eventually he had to promise remuneration while asking for less information than before:

> We do not wish to tax the time of the operators unnecessarily and nearly all we ask is what may be gathered by a casual inspection of the sky and a moment's reference to the thermometer or barometer if either or both be at hand. . . . It was not our intention to promise that dispatches should be sent free of telegraphic dues but rather of any charges by the Institution for the information conveyed. If the Mercantile Associations or Boards of Trade of the different cities do not consider the warnings that it may be in our power to give of impending changes of weather, worth paying for, we do not consider the Western Union be called upon to furnish them gratuitously.[32]

Restoration of the monthly climatological reports from the South was also proceeding slowly. In 1866 there were only thirty observers in the eleven southern states contributing a total of 159 monthly reports. Louisiana had no observers, South Carolina had one (who reported for only two months with no instruments), and Georgia had one (equipped with only a thermometer and a rain gauge). Three sets of standard instruments, including a total of six barometers, covered the entire region. By comparison, the entire Smithsonian system had 352 observers contributing 3,038 monthly reports. The states of Illinois, New

York, and Ohio each had more observers, more complete sets of instruments, and contributed about 100 more monthly observations than the entire South.[33] One would-be observer in Arkansas reported pathetically that he had no instruments with which to observe because "the rebels cleaned me out, as they did all other northern men."[34] Moreover, the Smithsonian had no money for new instruments because, in addition to the expense of repairs to the building, postwar inflation had cut purchasing power by 50 percent.[35]

Henry hoped that a "new epoch" of climatological observations could begin by cooperating with the Light House Board and restoring relations with the Army Medical Department.[36] For reasons of finance and convenience, however, the Smithsonian established stronger ties with the Department of Agriculture to gain the needed numbers of correspondents. In return for the raw data, Henry provided a summary of the weather for the Agriculture Department's *Monthly Report.*[37] Henry again rehearsed the reasons why meteorology was of practical benefit to agriculture. Some farmers responded with imaginative proposals. One thought that telegraphic reports to population centers could be combined with the ringing of church bells and the firing of cannon to warn farm workers of approaching storms.[38]

By 1868 cooperation had largely ceased; the Smithsonian was forced to receive its meteorological returns only after the Department of Agriculture was finished with them.[39] When in 1872 Commissioner Frederick Watts discontinued the *Monthly Report,* the rationale for providing a free frank to the Smithsonian ended as well. As a result, the Smithsonian actively sought its own franking privilege so its projects would be freed from the whims of other agencies.[40] Developments elsewhere, however, were leading to increased public support for telegraphic meteorology.

Cincinnati's Role

> We shall soon have a Meteorological Observatory in this city furnished with NY recording instruments and shall publish a daily Weather Bulletin compiled from telegraphic dispatches & covering the greater part of the Ohio Valley. If it would not interfere with any plans of the Smithsonian I should be glad to see this Bulletin extend its field to cover the whole of the United States E[ast] of the Rocky Mts.
> —Abbe to Henry, October 29, 1868, OSI-SIA

When Cleveland Abbe issued his inaugural report to the Board of Control of the Cincinnati Observatory on June 30, 1868, he proposed to make the observatory more "useful" by extending its activities beyond scientific astronomy to include basic investigations in meteorology and magnetism and the application of these sciences to geography and geodesy, to storm prediction and surveying, and to other needs of the general population.[41] Abbe, who had been trained at Pulkova under Otto Struve, argued that astronomers needed widespread meteorological observations to investigate the effects of atmospheric refraction. Most likely, however, Abbe was trying to stimulate public interest and raise additional funds by arguing for the utility of an observatory that combined "the usefulness of Greenwich and the science of Poulkova."[42]

John A. Gano, president of the Cincinnati Chamber of Commerce and publisher of the *Commercial* newspaper, chaired the local committee that approved Abbe's meteorological plans and promised to underwrite the expenses of the first trial. In 1868 Abbe proposed a cooperative effort between the Cincinnati Observatory, the Associated Press, the Smithsonian meteorological observers, and the Western Union Telegraph Company to establish telegraphic reports from "at least 100 stations." The observatory promised to receive the reports, "submit them to a careful discussion," and "within a few hours, return them systematically arranged and condensed, to the Associated Press."[43] The complexity and expense of this system, however, were deemed too great for an experimental program and nothing was done for a year.

On May 7, 1869, Abbe submitted a more modest proposal to Gano, omitting reference to the Associated Press, which was happy with the weather reports it already printed:

> I propose to inaugurate such a system, by publishing in the daily papers, a weather bulletin, which shall give the probable state of the weather and river for Cincinnati and vicinity one or two days in advance. . . . The accuracy of the prediction will increase from month to month and year to year, in proportion as we are able to increase the number of our stations. As Cincinnati is very favorably situated with respect to the proposed outlying stations, it is most probable that 90% of our weather predictions will be verified. It is thus evident that we do not propose to guess at the weather (leaving that to the almanac makers), but we shall be able to assert with confidence the nature of the weather for one, two or four days in advance, as well as the stand of the water in the river.[44]

Expenses were not to exceed $1,000 per year. A sheet of instructions for observers was printed in which details were "to conform to the instructions issued by the Smithsonian Institute, or by the Patent Office, or by the Surgeon General of the Army, or by the Western Meteorological Association, or by this Observatory, to the observers corresponding with these respective institutions."[45] Reports from sixteen locations and a daily chart were to be prepared for the exclusive use of the Cincinnati Chamber of Commerce. Abbe asked Henry for his opinion on how large a chart should be to display weather changes and requested that the Smithsonian observers be allowed to report to him as well. He promised Henry several copies daily of his weather bulletin. Abbe's responsibilities included finding and training observers, testing and standardizing instruments, preparing report blanks and charts, devising telegraphic code for brevity and accuracy, and keeping within his meager budget. Western Union offered him free reports like those given to the Smithsonian Institution and the newspapers.

Initiating such a complex system, however, was not an easy task. In July 1869 Abbe wrote letters to prospective observers and received positive responses from individuals in Indianapolis, Chicago, St. Louis, Springdale, Milwaukee, Leavenworth, Omaha, Memphis, Lookout Mountain (Chattanooga), Louisville, and Pittsburgh. The observer in Springdale wrote that he lived several miles from the telegraph but would send observations with a neighbor who commuted daily to Louisville—with one crucial exception: "on days when inclement weather or sickness . . . prevent[ed] the trip."[46] Edward Goldsmith in Memphis received Abbe's cryptic telegraphed instructions: "Begin weather report on Wednesday, September First. . . . Fuller Instructions by Mail." Goldsmith indicated his willingness to comply but needed "fuller instructions" before he could do so: "Please, therefore, direct me at what hours of the day you desire the Observations to be made, of what instruments do you wish to obtain the indications, whether you wish (in addition to instrumental observations) the appearance of Sky, Kind and motion of Clouds, Rain fall, Casual or Incidental Phenomena &c. Also whether you wish the indications of therm. and wet & dry bulbs in full degrees or tenths. . . . Please inform me on all these points."[47]

With only two cities reporting, Abbe issued the first *Weather Bulletin of the Cincinnati Observatory* on September 1, 1869, and announced "probabilities" for the next day. The second *Bulletin* carried reports from three cities—Chicago, St. Louis, and Leavenworth—along with a trial weather forecast.[48] On the third day Abbe offered a daily telegram to Urbain Le Verrier in Paris.[49] The experiment ran for three months. Bulletins were prepared by Abbe, copied on manifold paper by clerks in the Western Union office, and delivered to subscribers by messengers. Following the principles of Espy's *Philosophy of Storms,* Abbe placed his forecasting formula at the base of the bulletin: "A falling barometer, increased temperature and humidity, and nimbus or nimbus-stratus clouds, precede rainstorms. The winds blow towards the center of a storm, or toward the point where the barometer is below its average height."[50] At the expense of astronomical research at the observatory, Abbe continued to prepare daily weather bulletins with the assistance of the Western Union Company. As he wrote to his father, George Abbe, at the inception of the project, "I have started that which the country will not willingly let die."[51]

Congressional Action

In October 1869, Increase Lapham, the Milwaukee correspondent to the Smithsonian Institution, Lake Survey, and Cincinnati *Bulletin,* prepared a memorial urging the Wisconsin Academy of Sciences and the Chicago Academy of

Sciences to follow Abbe's example and prepare telegraphic weather reports for the Lake Michigan region. Lapham also sent a copy to the National Board of Trade meeting in Richmond, Virginia, which read as follows: "Resolved, that it is expedient to inaugurate in the United States a system of meteorological observations by communicating telegraphic information of the occurrence of destructive storms and winds, thus preventing much of the present loss of life and property upon the ocean and lakes."[52] The memorial was supported by the representative from Milwaukee, C. B. Holton, and two delegates from Cincinnati, William Hooper and Gano.[53] Simultaneously, Lapham sent a petition to Wisconsin Congressman Halbert E. Paine and published a series of articles in the *Chicago Bureau* calling for a storm warning system. Citing recent successes in England, France, and Italy and the experiments he had conducted eight years earlier to confirm Espy's theory of storm movement, Lapham called for "the services of a competent meteorologist at some suitable point on the lakes, with the aid of a sufficient corps of observers, with compared instruments, at stations located every two or three hundred miles toward the west, and the cooperation of the telegraph companies." Government patronage was requested so that destructive storms could be discovered "in time to give warning of their probable effects upon the lakes."[54] Alternatively, Lapham asked, "Have we not a Smithson, Cooper or Peabody in Chicago?" To bolster his case, Lapham provided a list of recent marine disasters caused by storms on the northern lakes. In 1868 there were 1,164 casualties, 321 lives lost, and property damage valued at $3.1 million; in 1869 there were 1,914 casualties, 209 lives lost, and property damage valued at $4.1 million. Lapham argued that an annual appropriation of $100,000 for storm warnings represented only 5 percent of the estimated $2 million in losses during only two storms, November 16 and 19, 1869.[55]

As an example of what he hoped a telegraphic weather service could accomplish, Lapham mapped the "origin and progress over the country" of the storm of March 13–17, 1859, tracing its path for over seventy-two hours. Information on pressure, temperature, cloudiness, wind, and rain was taken from tables prepared by Coffin and published by the Patent Office in 1864. Lapham also included topographical features, annual isotherms, and arrows showing the mean wind direction: "The map brings out, very prominently, the fact that the storm first struck our coast in Western Texas, about 2 P.M. of the 13th; from thence it moved to the northward and eastward, touching Lake Michigan twenty-four hours, and the Atlantic coast forty-eight hours later, thus allowing ample opportunity, with the aid of the telegraph, to prepare for its dangers, both upon the lakes and upon the sea. This storm occupied one more day in reaching Nova Scotia, and another before it finally left the continent, at St. John, Newfoundland." Lapham concluded that this case was "another very complete confirmation of the general deductions of Prof. Espy" (see Figure 7.3).[56]

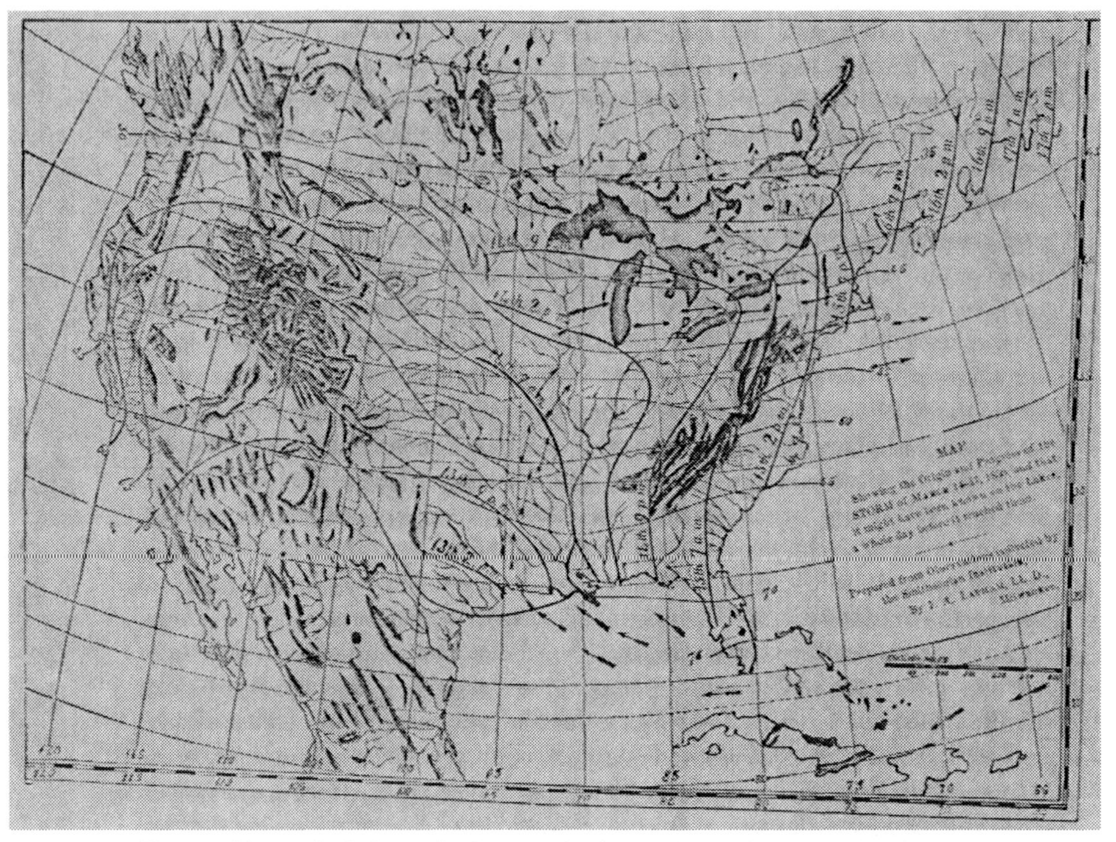

Fig. 7.3. Meteorological map by Increase Lapham using Smithsonian data. From *Bureau,* January 1870, 93; copy in the Cleveland Abbe Papers, MS 60, Special Collections, Milton S. Eisenhower Library, Johns Hopkins University.

Congressman Paine was impressed by Lapham's memorial and solicited recommendations from scientists on how to proceed. Paine's former teacher at Western Reserve, Elias Loomis, suggested that a select number of the Smithsonian observers combined with observations taken at military posts would provide a "tolerably complete" network of "great value to commerce."[57] Henry had no doubts that a "properly devised and intelligently conducted" system of weather telegrams, like those already in operation in England, France, Holland, and Italy, should be supported at government expense. Indeed, with storms typically moving eastward, the United States had a geographical advantage over the nations of Europe: "The Atlantic seaboard of the North American continent is much more favorably situated for receiving intelligence of approaching storms than the western coast of Europe, since, as a general rule, the storms which visit the latter coast are generated on the ocean, from which no telegraphic signals can be sent; while a large majority of those which prove disastrous to the shipping of our eastern coast, have their origin on the land, and moving eastward may consequently be telegraphed in advance to the principal commercial cities of the east."[58] Surgeon General Barnes also suggested that medical officers of the army stationed throughout the country could provide notices of passing storms if the proper telegraphic connections could be provided.[59]

Although the original draft of the bill named the Smithsonian Institution administrator of the new system, Paine's bill, H.R. 602, introduced on December 16, 1869, and reported back from the Committee of Commerce, named the secretary of war as the responsible party.[60] Colonel Albert J. Myer, the chief signal officer of the army, immediately called on Paine to stake his claim.[61] Paine recalled later that Myer "was greatly excited and expressed a most intense desire that the execution of the law might be intrusted to him."[62] The weather service bill was ready-made for an aggressive administrator looking for a mission.

Myer started his military career in 1854 as an assistant surgeon in the Army Medical Department, where his duties included filing weather reports to the surgeon general from a post in Texas. Based on his innovative work in military signaling, he was appointed army signal officer with the rank of major in 1860 to assist in the Navajo Indian campaigns.[63] With the outbreak of the Civil War, he was called to organize a Signal "Corps" for the war effort. Although relieved of his duty in 1863, largely because of a conflict with the U.S. Military Telegraph, Myer was befriended by General U.S. Grant and left the war with the rank of colonel. In 1867, again thanks to Grant, now the secretary of war, Myer resumed his position as chief signal officer and was breveted brigadier general, retroactive to March 15, 1865, "for distinguished services in organizing, instructing, and commanding the Signal Corps of the Army and for special service on Oct. 5, 1864."[64] Still, the Signal Corps had been mustered out with the volunteer army in 1865, and Myer was a bureau chief without a bureau, in charge of one lieutenant and two clerks. The remnants of his once proud command practiced their signals with salvaged war materials in the abandoned Forts Greble and Whipple. His request for a budget of $10,000 in 1869 had been cut in half.[65] Myer was "grasping for something that would keep the Signal Corps alive" when the new weather bill in Congress suggested to him that the motions of storms would provide a peacetime enemy worthy of his signals. He argued convincingly that his job would be the gathering, analysis, and transmission of storm intelligence. Myer believed that the telegraph could announce "meteorological observations, statistics, and reports giving the presence, the course, and the extent of storms . . . and their probable approach, as it would, in time of war, those of an enemy."[66] Military posts, scattered from the Pacific coast, throughout the interior, to the Atlantic, were increasingly linked to the branching commercial telegraph lines rooted in the business centers of the East. Myer prepared elaborate maps showing proposed locations for coast telegraphic and signal stations for the War Department and their relationship to existing lines. Paine was persuaded by Myer's zeal and the promise of military discipline in the system. Arrangements were made between them for Congress to assign the weather service to the secretary of war.

The Joint Resolution establishing a telegraphic weather service under the secretary of war was introduced by Congressman Paine and Senator Henry Wilson on February 2, 1870. It passed Congress without debate two days later and was signed into law by Myer's old friend President Grant on February 9, 1870. The law "authorized and required" the secretary of war to take meteorological observations at military stations and other points in the interior of the continent and to give notice "on the northern lakes and on the seacoast by magnetic telegraph and marine signals, of the approach and force of storms." Soon after, Congress passed a deficiency appropriation of $15,000 "for the benefit of the commerce of the northern lakes and seaboard."[67] Myer rejoiced.

As the weather service bill moved from the desk of Congressman Paine to the desk of President Grant, the "Western Union Telegraph Company Weather Report Compiled at the Cincinnati Observatory," an outgrowth of Abbe's experiment, began a ten-month run.[68] A daily map was prepared by the manager of the telegraph office, and reports were printed in the papers. After Abbe's marriage in May 1870, the observatory relinquished its role in preparing the report. With the Signal Service planning to establish a station in Cincinnati that would "dwarf" his efforts, and with strained relationships between him and the observatory's board of control, Abbe resigned his position in October, ostensibly to become the astronomer with the Darien expedition to Panama.[69] However, when he heard that the leader of the expedition was a harsh taskmaster and that he could not take his wife along, Abbe decided not to accept the position. In the meantime, he advised Myer on opening the signal station in Cincinnati and contributed sample charts and telegraphic ciphers. Impressed with Abbe's efforts, Myer invited him to Washington for an interview at government expense.[70]

Telegrams and Reports for the Benefit of Commerce

> The undertaking, upon a scale of such magnitude, . . . had not been generally contemplated even in this country. The benefits to be had, if fair success could be obtained, were vast and lasting. It was not a subject for trivial contemplation.
> —Myer, *Signal Office Report, 1870*

When the weather service bill became law, Myer submitted his plan and budget estimate to the secretary of war. Within a month the chief signal officer was formally assigned the duties of storm signaling.[71] The organizational structure of the army, however, provided no regular staff for the Signal Office. Thus it was viewed with suspicion by the regular army. Scientists had their doubts as well. Abbe wrote to Lapham that civilian telegraph operators would probably do a better job because the army's meteorological observations had been generally "very unreliable. . . It would I think have been wiser if the bill had recommended that Congress appoint a committee of three (Henry, Coffin and a navy or army officer) to report some plan of action. And I am specially of opinion that the money expended would do more towards effecting good results if it goes through the hands of meteorologists rather than through the hands of army officers. . . . It would be a pity to see the country swindled with an inefficient meteorological office as [the country] has already enough to do to carry on the naval observatory with the present objectionable system of management."[72] Lapham responded that the aid of army officers "may do some good" but only if the right director were chosen: "Should the matter fall into the hands of such Army officers as Gen. A. A. Humphreys, Chief of Engineers, Maj. R. S. Williamson, Gen. W. S. Raymond, Sup. of our Lake Survey, or Col. Wheeler in charge of our Harbors, I should have no fears of the result."[73]

The Signal Service used Fort Whipple, Virginia, as a school of instruction and practice for its officers. Basic training—a six-week course—began in the summer of 1870. To qualify as "observer-sergeants," enlisted men were required to pass examinations in meteorology, including the operation and repair of common instruments, and in signaling, including the maintenance of the telegraphic equipment. Candidates learned the outlines of their science from Loomis's *Treatise on Meteorology* and Alexander Buchan's *Handy Book of Meteorology*.[74] The curriculum rapidly expanded to include practical study of observational guidelines published by the Signal Office and the Smithsonian and studies of the daily weather maps and their variation from day to day. Candidates also read Espy's writings on storms and were introduced to Redfield's theory through books by Piddington and F. P. B. Martin on hurricanes.[75]

On October 10, 1870, twenty-five observer-sergeants—the first graduating class—received orders to report to their posts (compare Figure 7.4 with Figure 7.2). Each station was equipped with a barometer, a thermometer, a hygrometer, an anemometer, an "anemoscope" or wind vane, and a "pluviameter" or rain gauge. The instruments, when necessary, were compared with a common standard at Washington. In addition to the reference books the observer-sergeants had studied during training, each station received a manual of signals, a copy of Guyot's *Tables*, Ferrel's papers on fluid motion from the *Mathematical Monthly*, Coffin's "Winds of the Northern Hemisphere," Blodget's *Climatology*, and extracts from Redfield's papers and the *Barometer Manual* of the London Meteorological Office.[76]

Three daily synchronous observations, by Washington time, were ordered, at "about" 8:00 A.M., "about" 6:00 P.M., and midnight. These times were chosen not because they represented the average meteorological conditions for a given day but because they avoided the "press of business at other times upon the telegraphic lines." Initially, reports were limited to a simple statement of prevailing conditions. Abbe's code of cipher and his maps, forms, bulletins, and circulars were adopted for use by the Signal Service when it began operations on November 1, 1870.[77]

The wires of a score of different telegraph companies all converged on the Telegraph Room in Myer's office at 1719–21 G Street, N.W., placing him at the center of an electric network. Initially, the Signal Service was dependent on the commercial lines, notably Western Union, and observer-sergeants were located only in areas where such access was available. Soon, however, Myer received authorization to construct military telegraph lines to lighthouses and mountain stations and into the southwestern and northwestern frontiers.[78]

Myer sought and found civilian scientists to help as assistants. Lapham received $167 per month to supervise the service on the Great Lakes but soon retired to private life in Milwaukee.[79] Two months later, on January 3, 1871, Abbe began work as a civilian assistant in the Signal Office in Washington.[80] His salary began at $3,000 a year and reached $4,500 in his second year on the job. He was one of the nation's highest paid scientists.[81] He was also one of the most

Fig. 7.4. Initial stations of the Signal Service, 1870; compare with Smithsonian Telegraphic Stations, 1860 (Fig. 7.2). Stations included Washington, New York City, Boston, Chicago, St. Louis, Cincinnati, New Orleans, Nashville, Mobile, Montgomery, Augusta (Georgia), Buffalo, Rochester, Oswego, Cleveland, Toledo, Detroit, Milwaukee, St. Paul, Duluth, Omaha, Cheyenne, Pittsburgh, Key West, and Lake City.

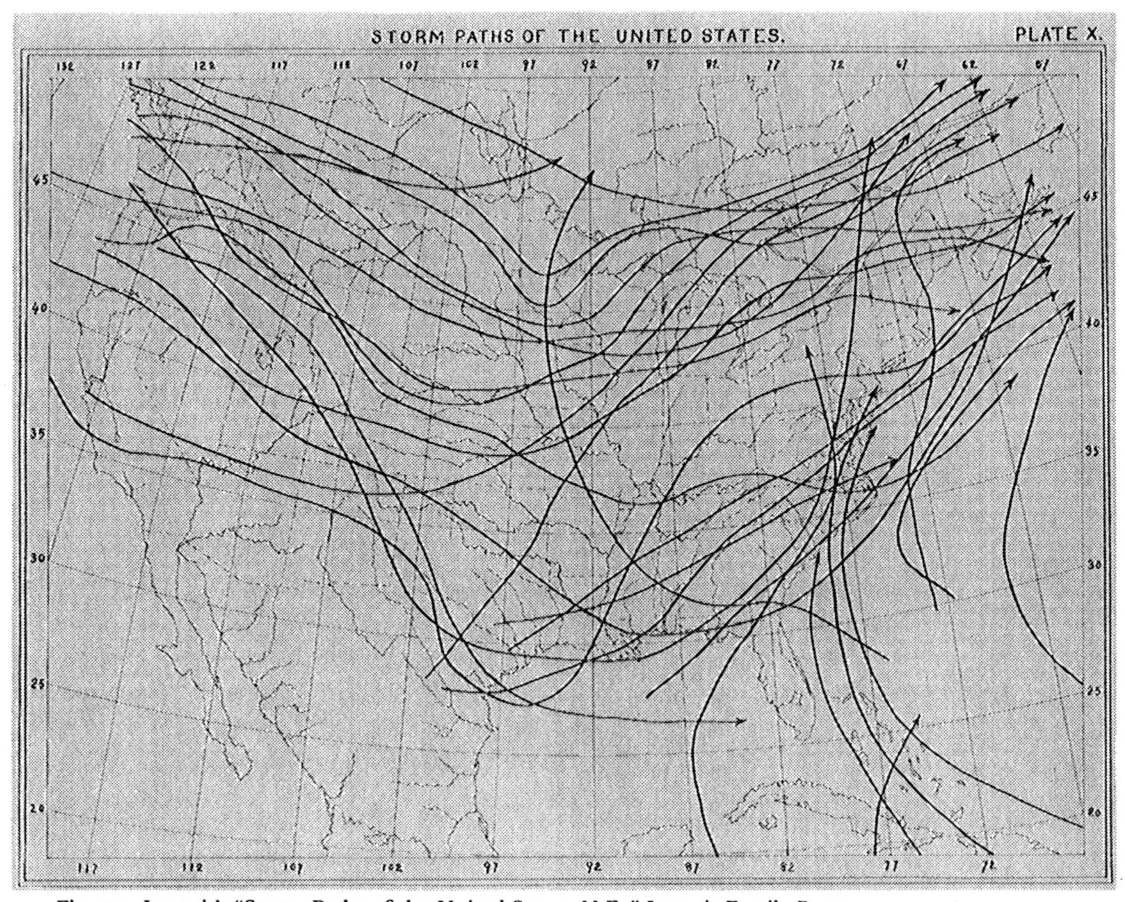

Fig. 7.5. Loomis's "Storm Paths of the United States, N.E.," Loomis Family Papers.

insecure because civilians were not regular employees of the Signal Office and their salaries depended on temporary annual appropriations.

Loomis, using Signal Office data, plotted the tracks of storms across the continent in an attempt to find a coherent pattern. Figure 7.5, one of a series of charts constructed by Loomis, illustrates storm paths toward the Northeast. Others showed tracks toward the Southeast and the West. These diagrams, from Loomis's unpublished papers, revealed no obvious regularities, but they inspired the Signal Office to publish similar charts of the "mean track of low barometer" (Figure 7.6) in an attempt to improve storm warnings. Beginning in 1874 and until his death in 1889, Loomis read papers before the National Academy of Sciences and published twenty-three articles in the *American Journal of Science* analyzing the effects of pressure, moisture, and winds on the paths of weather systems.[82]

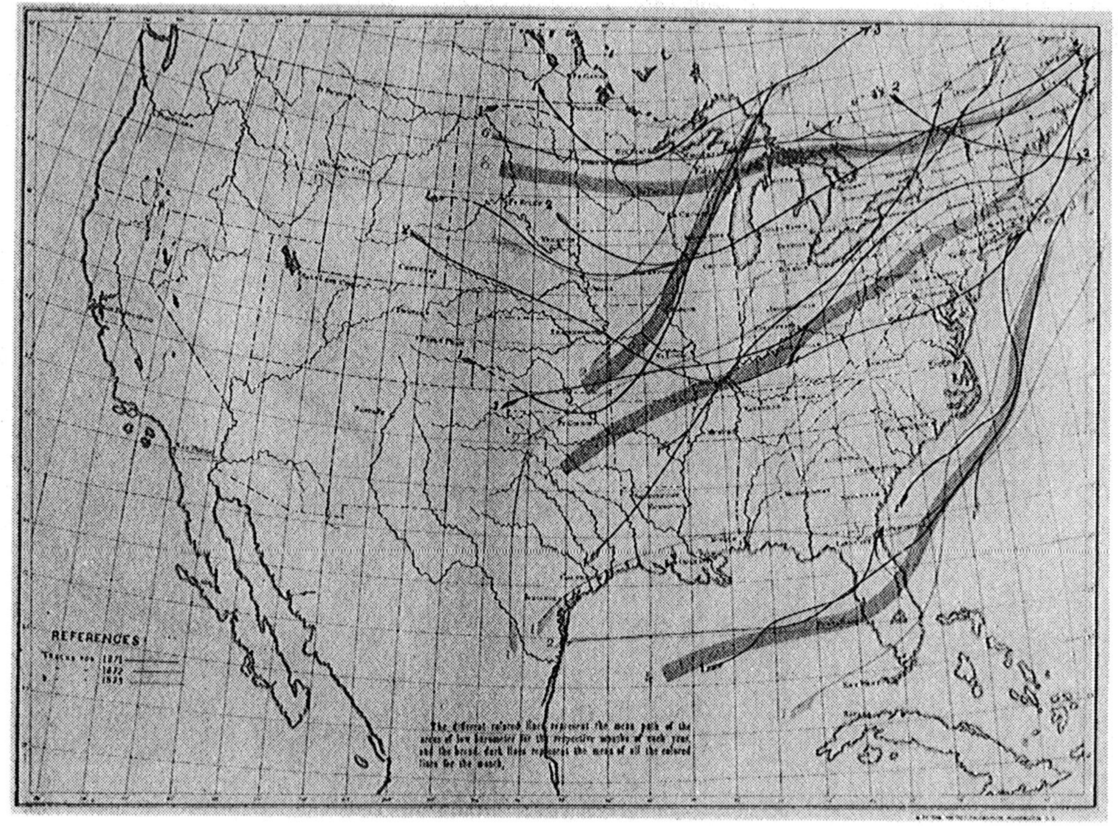

Fig. 7.6. Signal Office map, "Mean Tracks of Low Barometer for Dec. 1871, '72 and '73." From *Signal Office Report, 1874*, map 53.

The Consolidation of Other Systems

> Long interview with General Myer relative to the telegraphed signals—Had a free talk with him—was not pleased that he did not confer with me as director of the Institution—It was coming on our ground without our consent.
>
> —Henry Desk Diary, May 16, 1870, Henry Papers

The inauguration of a federal storm warning system did not mean that all meteorological activity was suddenly concentrated in the Signal Office. Smaller projects like the Survey of the Northern and Northwestern Lakes under the Army Topographical Engineers ceased their observations almost immediately.[83] Surgeon General Barnes, however, maintained his jurisdiction over the army post observations. Beginning in 1874 he offered the Signal Office access to the reports that had accumulated, unpublished, since 1860. At that time reports from 123 post surgeons were being received, but the number decreased rapidly because of the declining fortunes of medical climatology among physicians and because Myer gave little encouragement to the post surgeons and their work.[84]

The Smithsonian had consistently advocated the propriety and necessity of congressional support for a public weather service. Although Henry was the sentimental and popular favorite of many for heading the system, he did not want the institution's research efforts burdened with operational tasks which others could do as well or better.[85] Moreover, in 1870 he was seventy-three years old.

As early as 1870, Henry indicated that the Smithsonian would "willingly relinquish" the field of meteorology to the Signal Office because weather prediction was clearly in the realm of "immediate practical utility" and not part of the Smithsonian mandate to support only new research projects. He cautioned, however, that an ongoing meteorological and climatological research program was still needed. The needs of the scientist, the physician, and the agriculturist demanded more extensive data collection, as well as a small army of calculators and facilities to publish the results. He reminded the Signal Office of its debt to "abstract science," citing the research of Espy and Redfield as examples, and urged Congress to appropriate additional funds for research.[86]

With the loss of funds in 1872 from the Department of Agriculture for printing, postage, and publication and the loss of income from the Panic of 1873, the Smithsonian could no longer support its meteorological project. Henry begged Myer to take over the administration of the volunteer observers and provide a small appropriation to preserve the system.[87]

> It has been from the first a part of the policy of this institution to devote its energies to no field of research which can be as well cultivated by other means; and the United States Government having established a system of meteorological observations, and having made appropriation for its support, we have thought it for the best interest of the science to transfer the system of meteorological observations which has been so long continued by the Institution to that of the War Department, under the Chief Signal-Officer, General Myer.
>
> The propriety of this transfer will be evident from the fact that the Institution has not the means of paying for printing blanks, postage, and the calculations and monthly publication of the results, and that the assistance which has heretofore been rendered in this way, by the Department of Agriculture, is now discontinued; furthermore, General Myer can combine these observations with those made with standard instruments now under his charge, and out of the whole form a more extended and harmonious system than any at present in existence.
>
> We shall retain, for the present, all the records of observations which have been accumulating at the Institution during the last twenty-five years, and

> continue the work of their reduction and discussion up to the end of the year 1873.[88]

Myer reprinted Henry's letter to the Smithsonian observers and invited them "to continue their labors" as voluntary observers of the Signal Office.[89] Although 383 Smithsonian observers complied, the Signal Office placed top priority on the work of the observer-sergeants and devoted few resources to the volunteers.[90]

Although experiments with the telegraph in the 1850s produced interesting results, a series of calamities—war, fire, financial panic, and changing cooperative arrangements—dealt severe blows to the Smithsonian meteorological project and opened the way for other projects to take the lead. France and Britain had founded state meteorological offices in the 1860s. As the era of Reconstruction advanced, the United States followed suit.[91]

CHAPTER 8 The Worldwide Horizon

No other part of the world has offered such facilities for the collection of meteorological data, the system extending over so large a portion of the earth's surface; the observers, with a few exceptions all speaking the same language, and many of them being furnished with full sets of compared standard instruments.
—Henry, *Smithsonian Report, 1864*

During his second tour abroad, Joseph Henry received a warm welcome from his European colleagues. In part this was because he was the elder statesman of American science, but it was also because some Europeans truly admired the recent gains made by scientists in America. Throughout the nineteenth century, American observational systems grew roughly apace of their counterparts in Europe and Russia. By the 1870s, the system of telegraphic storm warnings under the U.S. Army Signal Office was the largest and best funded in the world. In scale if not in theoretical insights, American contributions to meteorology merited international attention.

Henry's Second Tour of Europe

On May 17, 1870, Henry received an invitation to attend the meeting of the BAAS in Liverpool. It was, he noted, "just 33 years after my previous attendance"; the paper he chose to read was entitled "On the Rainfall of the United States."[1] During his first tour of Europe in 1837, Henry was confronted by those skeptical of the quality of American science. His diary records his envy of the scientific infrastructure of Europe and his early attempts to bring Americans into the Atlantic community of science. On his second tour between May and October 1870—really a vacation—the situation was dramatically different. Henry was honored as the chief representative and living patriarch of American science: "I have been constantly occupied, night and day, since my arrival in London and received more invitations than I have been able to attend to. I have attended the meetings of the Royal, the Linnean, the Statistical, the Geological and the Anthropological societies and am engaged to attend others. The Institution is well known and its operations highly prized."[2] The Athenaeum Club provided Henry with lodgings and an honorary membership; while in London, he gave a talk on the Smithsonian and the advance of science to members of Parliament.[3] Henry was also welcomed in Ireland, Scotland, Belgium, Prussia,

Switzerland, and France, though the tour on the Continent was curtailed by the outbreak of the Franco-Prussian War.[4]

The meteorological dimension of Henry's journey, as on his earlier tour, was prominent. In addition to presenting his paper on rainfall, Henry visited with scientists involved in calibrating instruments and collecting meteorological data. He toured the Kew Observatory and took copious notes on its photographic registry of meteorological instruments. For amusement on the return voyage, he plotted the course of an Atlantic storm with his pocket aneroid barometer.[5] Beyond Henry's personal interest in the weather, his second tour reveals the progress of American science in general and meteorology in particular in the three decades since his first tour. A letter Henry received in London from Baird illustrates these gains. Baird wrote that the congressional budget hearings for 1870 had granted $10,000 to the museum, $10,000 for the Smithsonian building, $25,000 for Ferdinand Hayden's western geological survey, $50,000 for Arctic expedition, $29,000 for solar eclipse measurements by the Coast Survey, $50,000 (likely) for a new telescope at the Naval Observatory, and $12,000 for John Wesley Powell's expedition to Colorado—"So you see that Science has been much better recognized this session than usual."[6] Baird could well have added to his list $25,000 appropriated for storm warnings under the Signal Office. In contrast to their usual disdain of "colonial" science, European savants now expressed admiration for the amount of government funding available for science in America. U.S. scientific education in meteorology was also praised by Alexander Buchan, Secretary of the Scottish Meteorological Society:

> In the schools of the United States of America, meteorological observations and the keeping of meteorological registers, form a part of the common education of the people. . . . But in this country [Scotland] few even of the liberally educated classes are able to read from vernier; they are ignorant of the moveable cistern of a barometer; they have not the elementary knowledge to give an intelligible interpretation of the fluctuations of the barometer as indicative of the coming changes of the weather; and when required to send their barometers to a distance for repair; so ignorant are they of their construction, that they forward them by rail as ordinary parcels, thus almost to a certainty securing their destruction. This state of things is the necessary consequence of the general neglect which meteorology receives in our educational system . . . the objects of meteorology can never hold that place in the public mind to which they are entitled, till the science becomes, as in America, a recognised branch of education.[7]

The International Meteorological Community

A study of meteorology in nineteenth-century America would not be complete without an attempt to locate national developments within the context of

TABLE 8.1. Dates of founding of weather services and directors, according to Kutzbach, *Thermal Theory*, pp. 12–13.

1826	Belgium	Observatoire Royal: A. Quetelet
1847	Germany	Meteorologische Institut Berlin: H. W. Dove
1849	Russia	Central Physical Observatory: A. T. Kupffer
1854	Great Britain	Meteorological Department of the Board of Trade: R. Fitzroy
1855	France	Observatoire: U. J. Le Verrier
1870	United States	Weather Bureau (Signal Service): A. J. Myer

the larger community of scientific nations. One author used the dates of the founding of national weather services, loosely defined, to compare developments in different nations. Simple compilations of the various dates of the founding of national weather services in the nineteenth century, however, do not reveal the complex texture of events in this predisciplinary era of science. The result was a rather inaccurate and misleading chronology, summarized in Table 8.1, which indicates that developments in the United States trailed those in Europe and Russia by several decades.[8] The result is vastly different when meteorological systems, rather than weather services are made the basis for comparison (see Table 8.2).

Regular observations of pressure, temperature, and humidity were begun at the Brussels Observatory on January 1, 1833, not in 1826 as Table 8.1 indicates. The earlier date is when Quetelet's plan was approved for an astronomical observatory, including "observations having to do with the physical structure of the atmosphere and earth."[9] Construction of the observatory began in 1827, but as a result of the revolution of 1830, the observatory established by the Dutch regime was deprived of its astronomical instruments. Quetelet therefore emphasized meteorological research. Special observations were made on solar radiation, the soil temperature at different depths, and the periodic phenomena of plants.[10] Because of the small size of the country and the relative uniformity of its climate, the establishment of a meteorological network was not of primary importance. A weather service was eventually established in 1878 by Quetelet's successor, Jean Houzeau, and consisted of only three weather stations and thirty climatological stations.

Until the 1870s, systems of meteorological observation in the German states suffered from the political fragmentation of the region. Nevertheless, inspired by the early example of Mannheim, numerous Germanic unions were attempted, including the Anstalten für Witterungskunde in Sachsen-Weimar-Eisenach, begun in 1821, and the Süddeutsche meteorologische Verein begun in 1841, which included Baden, Württemberg, Norddeutschland, Austro-Hungary, and parts of France, Italy, Belgium, and Holland.[11]

The Prussian Meteorological Institute was established in October 1847 in Berlin with Carl H. W. Mahlman, not Dove, as director. In 1848 its scope was

TABLE 8.2. Comparisons of observational systems in Europe, Russia, and the United States in the nineteenth century

Belgium	
1833	Climatic observations begun at the Brussels Observatory
1878	Meteorological service with three weather stations and thirty climatological stations established by J. C. Houzeau
Germany	
1821	"Witterungs-Anstalten" in Sachsen-Weimar-Eisenach
1841	Süddeutsche meteorologische Verein
1847	Prussian Meteorological Institute, Berlin; thirty-five stations (1848), budget of 9,000 marks ($3,000), Carl H. W. Mahlman, director
1868	Norddeutsche Seewarte
1872	Deutsche Seewarte
1875	Telegraphic storm warnings
Great Britain	
1823	Meteorological Society of London
1850	British Meteorological Society
1854	Meteorological Department of the Board of Trade, R. Fitzroy, director
1867	Meteorological Office
1874	Budget for meteorology, £ 4,500 ($21,000)
France	
1849	*Annuaire Météorologique de la France*
1852	Société Météorologique
1854	Le Verrier replaced Arago as director of Paris Observatory
1864	Paris Observatory became a central superintending station, observing stations established at normal schools
1878	Central Meteorological Office under the Ministry of Public Education, E. Mascart, director
Russia	
1820–35	Thirty private observers, no unity of plan
1835–41	Institute of Mining Engineers, seven to ten stations, A. Ya. Kuppfer, director
1849	Central Physical Observatory, St. Petersburg; ten principal stations (hourly observations), fifty private observers (1850) declining to twenty-four (1864), no standard instruments, methods, or times of observation, budget of 9,000 rubles ($7,000)
1872	Hydrographic Department issued telegraphic bulletins from thirty-six Russian stations
United States	
1819	U.S. Army Medical Deparment began observations
1825	New York State Board of Regents began observations at academies
1836–39	Joint Committee in Philadelphia investigated storms
1849–74	Smithsonian meteorological project, up to six hundred observers, annual budget of $4,000, telegraphic experiments
1870	U.S. Army Signal Office telegraphic weather reports from twenty-five stations. Budget of $25,000 (1870), $400,000 (1874); *International Bulletin of Simultaneous Observations* (1875).

roughly comparable to that of the early Smithsonian meteorological project: there were thirty-five stations, a staff consisting of two persons, and a budget of 9,000 marks (approximately $3,000). Mahlman's *Manual for Observers,* was used throughout Dove's tenure as director (1849–79). Bavaria organized a meteorological network centered in Munich in 1878. Small networks were also established in Baden, Württemberg, and Saxony.[12]

The Norddeutsche Seewarte (1868), although primarily a maritime institution, was also the central bureau that issued telegraphic weather announcements from data provided by the diverse state land services. It became the Deutsche Seewarte in 1872 and began issuing storm warnings in 1875.[13]

Until 1823 there was no meteorological society or association in England. Long series of individual records were kept, but there was no uniformity or combined effort in observation. The Meteorological Society of London, founded in 1823, set no higher standards. Members needed no qualification other "than a desire to promote the science of Meteorology."[14] The group soon sank into quiescence "from a want of zeal."[15]

The society was revived in 1836 and produced the first English rainfall map in 1840 with results from fifty-two stations.[16] In the first volume of its *Transactions* (1839), John Ruskin of Oxford set out the ambitious goals of the society: "The Meteorological Society . . . has been formed, not for a city, nor for a kingdom, but for the world. It wishes to be the central point, the moving power, of a vast machine, and it feels that unless it can be this, it must be powerless; if it cannot do all, it can do nothing."[17] But the Meteorological Society of London held meetings for only three more years and ceased its publications soon thereafter.[18] Results from this group of enthusiasts were far from trustworthy. As George Symons noted in his history of English meteorological societies, "I have seen results published as air temperatures obtained from thermometers inside a hen house, I have seen a rain gauge under the eaves of a cottage, and another under a tree."

In 1850 the British Meteorological Society was founded to establish a "general system of observation, uniformity of registry, systematic communication, and other measures for insuring precision to the advancement of the aerostatical branch of physics."[19] James Glaisher was elected secretary and organized the first current daily weather reports from reports sent by train along lines running to London. [20] A royal charter was granted to the Meteorological Scoiety in 1866.

In 1854 the British Board of Trade established a Meteorological Department with Robert Fitzroy, former captain of the *Beagle,* as director; he served until 1865. Fitzroy coordinated observations, reports, and the compilation of data but issued no forecasts. It was not until 1859, when 457 people died in the wreck of the luxury ship *Royal Charter* during a storm off the coast of Wales, that Fitzroy set up a coastal warning system in Britain. The network included fifteen stations in Britain and additional reports from Paris. Fitzroy's budget was £218 in 1860. Three years later the budget stood at £2,989. The Meteorological Office was established in 1867, but the British Meteorological Council did not provide a daily weather map until 1872. The budget for meteorology was £4,500 ($21,000) in 1874.[21]

The meteorological records of the Paris Observatory from 1798 to 1822 were published *in extenso* in the *Journal de Physique,* from 1816 in the *Annales de Chimie et de Physique,* and from 1835 in the *Comptes rendus de l'Académie des Sciences,* but the only extended observations in France in the 1830s were the rain measurements made by the Service des Ponts et Chaussées begun in 1835.[22] In 1849 the *Annuaire météorologique de France* (1849–52) was published. This led to the founding of the Société Météorologique de France in 1852. The society published the *Annuaire de la Société Météorologique de France* beginning in 1852. It also published reports and monographs in the *Annales.*

In 1854 Le Verrier replaced Arago as director of the Paris Observatory. The same year a destructive gale in the Crimean Sea near the port of Balaclava wrecked Anglo-French transport ships during the Crimean War. In response, the observatory began an experiment in telegraphing weather facts. It did not, however, issue warnings. It was not until 1863 that Le Verrier telegraphed forecasts of impending weather. That was largely because the Paris Observatory was limited to its own observations. Although there were many observers in France, there was no central organization. The first notable government organization began in 1864, when Le Verrier, with the approval of the minister of public instruction, invited the councils-general to establish observing stations at normal schools in France. Fifty-eight schools responded. In addition, various departmental commissions collected observations made under their control and forwarded them to the Paris Observatory. In addition to the observers, clergy, medical men, teachers, and others maintained a meteorological correspondence. Most, however, observed without instruments. One of the products of this joint venture was the *Atlas météorologique de la France* (1865–76), which presented, among other things, a study of thunderstorms. Another series, *Nouvelles météorologiques* (1868–76), contained detailed observations from approximately sixty stations, but the stations were not inspected or standardized until 1873. It was not until after the death of Le Verrier in 1877 that the Bureau Central Météorologique de France was founded. Its director was Eleuthère Mascart, and its budget in 1878 was about $12,000.[23]

Before 1835, Russia had only a few widely scattered private meteorological observers.[24] Because of the Magnetische Verein, eight magnetic observatories were set up in Russia. By 1837 A. Ya. Kupffer had established meteorological stations at seven locations run by the Institute of Mining Engineers and at four other observatories. The results were published in the *Annuaire magnetique et météorologique* (St. Petersburg, 1837–48). Coffin used the published results for his "Winds of the Northern Hemisphere."[25]

The Central Physical Observatory published its observations separately in the *Annales de l'observatoire physique central* (1849–64). By 1849, the observatory was sponsoring a meteorological project roughly comparable to that of the Smithsonian. The staff of six had a budget of 9,000 rubles (about $7,000), and received data from eight principal stations (hourly observations) and forty-eight private observers (monthly journals). The number of private observers declined to twenty-four by 1864. Moreover, they had no standard instruments, methods, or times of observation. Results were published quarterly in *Correspondence Météorologique* (1850–64). In 1857, Constantine Wesselovski, the permanent secretary of the Russian Academy of Sciences, published *O Klimate Rossia* in two volumes with data taken from his collection of old journals. The work is similar to Blodget's *Climatology of the United States*, published in the same year. The director of the observatory, Kupffer, was also in charge of the department of standard weights and measures for the Russian Empire. He established the first Russian telegraphic weather reports in 1864, but his network was limited to nine inland and two foreign stations, and no particular use was made of the reports. Kupffer died in 1865. His successor, Ludwig Kämtz, served only two years until his death and was replaced by Heinrich Wild.[26]

Wild, director from 1868 to 1895, was trained in Switzerland, where he had been in charge of the Swiss meteorological stations and had established standardized instruments and procedures. When he came to Russia, he found numerous errors in the observations and was surprised that none of the meteorological stations had been inspected for the past twenty years. Wild brought a new standard of observation to the Russian Empire.[27] Perhaps Khrgian in his "History of Meteorology in Russia" can be excused for his patriotism when he said that Kupffer's systems "were models for that time and were adopted since, with minor modifications, as an international standard."[28] In 1872, with the cooperation of the Hydrographic Department, Wild began a lithographed meteorological bulletin which contained telegraphic reports from fifty-five stations (thirty-six in Russia and nineteen in Europe and Asia) and provided synoptic charts for Russia and parts of Asia. Telegraphic storm warnings commenced in 1874.

As Table 8.2 and the foregoing discussion show, the development of meteorological systems in the United States did not lag behind Europe and Russia and was indeed quite similar. The nineteenth-century era of expanding horizons was in fact an international phenomenon. Climatic surveys of various nations were widespread in the first half of the century, telegraphic experiments began in the 1850s, and storm warning services were established in the 1860s and 1870s. International cooperation soon followed.

Because of its well-funded weather service (over $400,000 in 1874), America soon took the lead in international cooperation in meteorology. General Myer, representing the United States at the Vienna conference of directors of weather services in 1873, proposed that the nations of the world prepare an international series of simultaneous observations to aid the study of world climatology and weather patterns. Myer's suggestion led to the *Bulletin of International Simultaneous Observations,* published by the Signal Office beginning in 1875. The *Bulletin* contained worldwide synoptic charts and summaries of observations recorded simultaneously at numerous locations around the world.[29] The "metrological" standards established by the work of these international congresses initiated a new era of worldwide observation and more homogeneous data inscription as a practical result.[30]

Meteorology in American Science

Perhaps more strongly than in most other sciences, natural history excepted, meteorological research was (and is) shaped by national resources, both cultural and geographical. Indeed, the expansion of meteorological systems into the frontier mirrored the development of the nation as a whole. The nation was young, it lacked scientific societies of high reputation, and its educational system was underdeveloped. Yet these limitations were not always detrimental to the advance of meteorology. Compared to other scientific fields, meteorology had greater potential for rapid growth because it was highly unorganized and, until midcentury at least, relatively lacking in theoretical development. American meteorologists did not need formal schooling, special mathematical abilities, large libraries, or expensive technical equipment to make important contributions to the science. Meteorological and climatological research demanded patience, persistence, and large-scale organization. The invention of the electric telegraph made timely weather warnings possible; the expansion of telegraph networks allowed wide distribution of these warnings, making them useful to the general public.

Meteorology was a science in which amateurs were able to make significant contributions; indeed groups of volunteers working in conjunction with centrally located administrators, clerks, and theorists were indispensable for its progress. The meteorological project at the Smithsonian Institution epitomized the antebellum epoch of voluntarism during which science, specifically meteorology, was "organized" but not yet professional. David Allen, in *The Naturalist in Britain,* makes a similar case for nineteenth-century British naturalists. For example, the Botanical Society of Edinburgh (1836), "from the start, resolved to make one of its primary functions the organized exchange of specimens on a national scale." This was a "new feature" for an amateur society.[31] Naturalists in America, such as John Torry, Asa Gray, Spencer Baird, and Louis Agassiz, were the recipients of a flood of data and collections from exploring expeditions and from responses to the thousands of circular letters they sent out. Indeed, Baird used the Smithsonian meteorological correspondents for these ends. Like his colleagues, he was "happily drunk on data."[32] The British, working in a much smaller geographic region than the Americans, formed amateur societies and interest groups that grew into larger networks for collecting specimens. The Americans conceptualized large projects covering the continent and then sought the resources to accomplish them.

The expansion of volunteer observational systems in America was aided by a common language, polity, and sense of democratic values. Although the American West was largely unsettled, many public-minded citizens—physicians, educators, and agriculturists—interested in building local institutions in remote areas corresponded with the Smithsonian Institution; military physicians at frontier posts corresponded with the surgeon general. Combined with general prosperity, a rising sense of national power, and westward expansion, these strengths far outweighed the temporary disruption caused by the Civil War.

The republican ideology valuing the participation and contributions of the "common man," although widely assumed to be a source of tension between scientific elites and the general public and to exert a leveling influence on the quality of basic research, was a major advantage in American meteorology. Tensions between public values and research ideals were not pronounced. Many people became convinced of the ultimate utility of meteorological science for commerce, agriculture, and medicine. Commercial interests, especially the telegraph companies and shipping lines, and Congressional funding aided the development of a national storm warning system.

When civil strife came to America, the national meteorological system at the Smithsonian suffered. As Peter Hall noted, "The war changed everything." It transformed and revitalized the elite "culture of organization," introduced new models of organization, and, through national networks—railroads, telegraph lines, and the banking system—infused new technologies into American society.[33] More than preserving political union, the war, according to Allan Nevins, "measurably transformed an inchoate nation, individualistic in temper and wedded to improvisation, into a shaped and disciplined nation, increasingly aware of the importance of plan and control."[34] In the more managerial era that followed the war, a large-scale storm warning service under the federal government meant the end of the leadership of voluntary organizations. Having quelled the rebellion to the south, the army and Congress found in storms new and unusually violent threats to national tranquillity. Paid observer-sergeants and multi-million-dollar budgets tried to accomplish what volunteer observers and several thousand dollars had attempted to do earlier.

In biological studies, the advent of Darwinism and the increasing use of laboratory analysis marked a parting of the ways between laymen and professionals.[35] Meteorology had no such abrupt break. Systems of volunteer weather observers, though no longer predominant after the Civil War, did not disappear entirely. The Signal Service used "voluntary observers," many of them recruited from the ranks of the Smithsonian meteorological project, to supply climatological records from areas not served by the military stations. These amateur observers fueled a growing movement for state "climatological services" in the 1880s. A similar situation existed in astronomy. According to John Lankford's study of the American Association of Variable Star Observers, professionals representing exact knowledge, esoteric research, and costly programs, never had a complete monopoly over amatuer astronomers who stood for "careful and persistent observing of a routine nature."[36]

American meteorology also developed with some distinct geographic advantages. By definition, the physical geography of a region determines its climate and helps to shape the impact of its major weather systems. American storms are more numerous and more violent than their European counterparts and so attracted the attention of the storm controversialists. Moreover, the weather for Washington and the populous eastern seaboard typically came from the western states and possessions of the United States, which made weather systems easier to investigate and tempting to try to predict. As an illustration, consider the offer from the Anglo-American Telegraph Company in London to the Smithsonian to transmit weather messages from Europe for an annual fee of about £200, and Henry's quick reply: "Telegrams relative to the weather from Europe this way of no special value but in the opposite direction may be of use."[37] Henry further noted in his Smithsonian report: "We are much more favorably situated for predicting the coming weather than the meteorologists of Europe. The storms in our latitude generally move from west to east, and, since our seaboard is on the eastern side of a great continent, we can have information of the approaching storm while it is still hundreds of miles to the west of us. Not so with the meteorologists of Europe, since they reside on the western side of a continent, and can have no telegraphic dispatches from the ocean."[38]

Because of the small size of the nations of Europe, weather services had to cooperate across national boundaries if meaningful investigations and timely forecasts were to be made. Because major storm systems were typically larger than the states of Europe, such scale considerations were very important. The weather for the Russian Empire typically came from Europe. Moreover, because of the sponsorship of the Department of Mines, the Russians were wedded to magnetic, not synoptic, observations. The British, because of their location and naval interests, focused on maritime wind patterns and storms. France and Germany emphasized agricultural meteorology. Political, cultural, and linguistic boundaries confused the already complex task of gathering meteorological data.

It is typical of meteorological science today that differences in national cultural and geographic resources seem stronger than differences in the national content of particular scientific theories. This is in part because of widespread international exchange and training programs and the existence of organizations such as the World Meteorological Organization and international projects such as the Global Atmospheric Research Programme that promote common standards and practices. In the nineteenth century, however, international science had no formal infrastructure. By necessity, theoretical programs closely followed national boundaries. The Americans—the storm controversialists and the Smithsonian—emphasized the "problem of American storms." Espy, Loomis, and Ferrel all considered thermally induced convection and the release of latent heat in moist air of primary importance. Although their thermodynamic theories were not particularly innovative or mathematically sophisticated, their meteorological contributions were sound. In the eyes of Harvard's geographer and meteorologist William Morris Davis, they represented the leading school of meteorological thought of the last half of the nineteenth century. If it is "inconceivable that science in America was a unique case because of American conditions," it is surely now within the realm of possibility that, at least in the formative years, meteorological science took its shape largely from national cultural and geographic conditions.[39]

American meteorologists joined the Atlantic community of scientists in the fourth decade of the nineteenth century. Thereafter they made distinctive theoretical, observational, and institutional contributions. American investigators attacked physical problems on the forefront of natural philosophy and pursued related goals of constructing medical, topographical, and climatological maps of the continent. The storm controversy of the mid-nineteenth century led to the founding of new observational programs and ultimately to new theoretical and practical approaches in meteorology. Systems sponsored by the Joint Committee and the Smithsonian Institution also had a significant impact on meteorological theory, providing mechanisms for professional entry and advancement in science and better control over the quality and amount of data. By the 1850s, no responsible meteorological theorizing could be conducted in America without access to the results of these systematic projects. Loomis, Schott, Abbe, and Ferrel represented the first generation of theorists to respond to these changes.

After the death of the three storm controversy protagonists, the Smithsonian's research program, driven by the expanding needs of agriculture, the exploration of the American West, and the research interests of the natural history community, emphasized climatological mapping and specimen collecting. With the introduction of statistical analysis, especially under the Signal Service, hopes for an exact storm law were abandoned in favor of a program of issuing daily weather probabilities for the nation using the telegraph networks. Thus the horizons of meteorological theory and the canons of acceptable theoretical discourse were conditioned by the growing observational systems as the systems themselves grew out of an attempt to resolve an earlier theoretical controversy. By the 1870s meteorology's observational horizons had reached the worldwide level.

Between 1800 and 1870, meteorology emerged as both a legitimate science and a government service in America. Meteorologists were driven by fundamental questions about the nature of storms and climatic change, aided by new technologies such as the telegraph, and funded by government sources expecting practical results. The observational systems they established and managed grew from small bands of isloated observers in 1800 to national and international networks seven decades later. Systems of observation—the army post surgeons, the states of New York and Pennsylvania, the Smithsonian meteorological project, the Signal Service, and others—served as scientific instruments used by meteorologists to examine the weather far beyond their local horizon. More than the contributions of a particular theorist, it is the linkage of institutional and theoretical developments—the crusades and the controversies—which broadens our historical horizon. This approach provides new and clearer perspectives on nineteenth-century American meteorology and the history of science in general.

APPENDIX Collective Biography of a Sample of Smithsonian Observers Active in 1851, 1860, and 1870

The first line of each entry gives the name of each observer, followed by dates of birth and death (if known) and the location of the observations. The second line lists occupation, the total number of years the person supplied observations (fractions of years are not available), the dates of observation, and the instruments used. Brief biographic details (if known), begin on the third line, including indication of membership in the AAAS (when appropriate) and the number of publications listed in the *Royal Society Catalogue of Scientific Papers.* Extensive use was made of biographical dictionaries, the correspondence of the meteorological project (on thirty-one reels of microfilm), and genealogical and census records. Often only the observer's occupation and age were available.

Key:

Name of observer (life span if known), location of observations
OCCUPATION, number of years as an observer (dates), instruments (B. = barometer, T. = thermometer, R. = rain gauge, P. = psychrometer, A. = all, N. = none)
Biographical information (if available), including source of published biography
AAAS = Member of the AAAS
RSC (number of articles) = Number of articles by the author listed in the *Royal Society Catalogue of Scientific Papers*

Biographical Information

1851

R. F. Astrop, Diamond Grove, Virginia
MILITARY, 13 (1849–61).
Traveled in Mexico, 1848, and reported on the climate. AAAS.

George A. Atkinson, Oregon City, Oregon
MINISTER OF THE GOSPEL, 2 (1851–52).

Jacob Batchelder, Jr., Lynn, Massachusetts
NATURALIST, LYNN NATURAL HISTORY SOCIETY, 4 (1849–52).
Published a pamphlet on cholera.

Levi A. Beardsley, South Edmeston, New York
FARMER, 3 (1849–51).
P. P. Brown, Upper Alton, Illinois
PROFESSOR, SHURTLEFF COLLEGE, 3 (1849, 1851–52).
Rufus Buck, Bucksport, Maine
CUSTOMHOUSE OFFICER, 4 (1849–52).
Egg collector.
James Henry Coffin (1806–73), Easton, Pennsylvania
METEOROLOGIST, MATHEMATICIAN, EDUCATOR, 1 (1851).
Graduate of Amherst College (1828), teacher at Fellenberg Academy (1829–37), principal of Ogdensburg Academy (1837–40), tutor at Williams College (1840–43), principal of Norwalk Academy (1843–46), professor of mathematics and natural philosophy at Lafayette College, author of *Solar and Lunar Eclipses* (1843), *Analytical Geometry* (1849), *Conic Sections* (1852), "Winds of the Northern Hemisphere"(1853), and "Winds of the Globe"(1875). In charge of data reduction for the Smithsonian meteorological project. *DAB, Natl. Acad. Sci. Biog. Mem.* 1 (1877): 257–64; J. C. Clyde, *Life of James Henry Coffin* (1881), AAAS, *RSC* (11 articles).
John P. Fairchild (1804–52), Seneca Falls, New York
JEWELER, 4 (1849–52).
C. M. Freeman, Williamstown, Massachusetts
UNKNOWN, 2 (1851–52).
Charles Christopher Frost (1806–80), Brattleboro, Vermont
BOTANIST, SHOEMAKER, 3 (1849–51).
Author and member of scientific societies. *ACAB*, AAAS.
John Gridley, Kenosha, Wisconsin
MINISTER, 10 (1849, 1851–52, 1857–63).
Albert Hosmer, Buffalo, New York
YOUNG MEN'S ASSOCIATION OF BUFFALO, 4 (1849–52).
Jos. W. Holt, Fond du Lac, Minnesota Territory
MINISTER, 3 (1849–51), T. (received from the Smithsonian).
Gardiner Jones, South Bend, Indiana
PROFESSOR, UNIVERSITY OF NOTRE DAME DU LAC, 1 (1851).
Josiah Jones, Walkersville, Maryland
FARMER, 3 (1849–51).
Stephen P. Lathrop, Beloit, Wisconsin
PROFESSOR, BELOIT COLLEGE, 6 (1849–54). AAAS.
Levi Washburn Leonard (1790–1864), Dublin, New Hampshire
MINISTER, TEACHER, AUTHOR, 3 (1849, 1851–52).
Unitarian clergyman and teacher, graduated Harvard College (1815) and Harvard Divinity School (1818), teacher, Bridgewater Academy (1818–20), organized lyceum, library, schools, and town government of Dublin,

author of *The Literary and Scientific Classbook* (1826), *The North American Spelling Book* (1835), and *The History of Dublin, New Hampshire* (1855), and editor of the *Exeter Newsletter* (1854–62). *DAB*, AAAS.

S. Y. McMasters, Drennon Spring, Kentucky

PROFESSOR, WESTERN MILITARY INSTITUTE, 2 (1849, 1851).

Cleric in charge of the Protestant Episcopal church, geologist.

R. M. Manley (b. 1824), Randolph, Vermont

PRINCIPAL OF ACADEMY, 3 (1849–51).

William Williams Mather (1804–59), Athens, Ohio

PROFESSOR, GEOLOGIST, ACTING PRESIDENT, OHIO UNIVERSITY, 3 (1849–51).

Graduate of West Point (1828), professor of chemistry and mineralogy West Point (1829–35), author of textbooks, topographical engineer, Espy volunteer observer, director of geological survey of Ohio (1837), state geologist of Kentucky (1838), professor of natural science, Ohio University (1842–45), vice-president and acting president (1847–59), author of *Elements of Geology for the Use of Schools* (1833), editor of the *Western Agriculturist* (1851–52). *DSB*, AAAS, *RSC* (19 articles).

C. F. Maurice, Sing Sing, New York

PRINCIPAL, MOUNT PLEASANT ACADEMY, 4 (1849–52).

John T. Mettaner, Prince Edward, Virginia

PHYSICIAN, 4 (1849–52).

Oran Wilkinson Morris, New York, New York

PROFESSOR, INSTITUTE FOR THE DEAF AND DUMB, 25 (1849–73), B.P.T. (received from the Smithsonian).

Graduate of Albany Academy, Espy volunteer observer. Torrey Botanical Club, *Bulletin* 6 (1877): 166–68, AAAS, *RSC* (6 articles).

John Newton (b. 1814), Chestnut Hill, Florida

SCHOOLTEACHER, 4 (1851–54). AAAS.

Thomas Oakley, Jackson, Mississippi

TEACHER, OAKLAND INSTITUTE, 4 (1849–52).

David Dale Owen (1807–60), New Harmony, Indiana

GEOLOGIST, CHEMIST, ILLUSTRATOR, 3 (1849–51).

Graduate of Ohio Medical College (1836), state geologist of Indiana, Kentucky, and Arkansas, U.S. geologist, author of numerous reports. Walter B. Hendrickson, *David Dale Owen: Pioneer Geologist* (1943), *DSB*, AAAS, *RSC* (15 articles).

Charles F. Percivall, Benton, Alabama

PHYSICIAN, 3 (1849–51).

Josiah Little Pickard (1824–1914), Platteville, Wisconsin

EDUCATOR, PRINCIPAL OF PLATTEVILLE ACADEMY, 9 (1851–59).

Graduated Bowdoin College (1844), A.M. (1847), superintendent of public instruction of Wisconsin (1860–64), superintendent of public schools of

Chicago (1864–77), president, University of Iowa (1878–87), president of the National Educational Association (1871), and president of Iowa Historical Society, author of *School Supervision* (1887) and *History of Political Parties of the United States* (1889). *NCAB*.

Ovid Plumb, Salisbury, Connecticut

PHYSICIAN, 5 (1849–51, 1853–54). AAAS.

Christopher Prince, Thomaston, Maine

CUSTOMS COLLECTOR, INSURANCE AGENT, 4 (1849–52).

A. Spear, ed., *Letters to Christopher Prince, 1855–1865* (1969).

Martin N. Root, Byfield, Massachusetts

PHYSICIAN, 1 (1851).

Andrew Roulston, Freeport, Pennsylvania

MECHANIC, OPTICAL INSTRUMENT MAKER, 3 (1849–51). AAAS.

William Skeen (b. 1818), Huntersville, Virginia

ATTORNEY AT LAW, 6 (1851–56).

Lewis Henry Steiner (1827–92), Frederick, Maryland

PHYSICIAN, PROFESSOR, LIBRARIAN, 1 (1851), B.T.

A.M. Marshall College (1846), M.D. University of Pennsylvania (1849), professor at Columbian College and National Medical College (1853–56), reorganized Maryland College of Pharmacy, professor of chemistry (1856–61), assistant editor, *American Medical Monthly* (1859–61), chief of sanitation, Army of the Potomac (1863–64), president of Frederick County School Board (1865), member of Maryland Senate (1871–83), founder of American Academy of Medicine and president (1878), librarian of Pratt Free Library, Baltimore (1886–92). *DAB*, AAAS.

William M. Stewart, Clarksville, Tennessee

PLANTER, 23 (1851–73)

Kept a "plantation journal." AAAS.

Zadock Thompson (1796–1856), Burlington, Vermont

PROFESSOR OF CHEMISTRY AND NATURAL HISTORY, UNIVERSITY OF VERMONT, STATE NATURALIST, GEOLOGIST, HISTORIAN, 6 (1849–54).

Graduated University of Vermont (1823), wrote almanacs and gazetteers, schoolbooks, editor of *Iris and Burlington Literary Gazette* and *Green Mountain Repository*, author of histories, especially *History of Vermont, Natural, Civil, and Statistical* (1841–53), state geologist of Vermont (1845–48), professor, University of Vermont (1851–56). W. H. Crocket, ed., *Vermonters: A Book of Biographies* (1931), 217–20, *DAB*, AAAS, *RSC* (3 articles).

1860

H. L. Alison, Carlowville, Alabama

PHYSICIAN, 12 (1856–60, 1867–73), T.R.

Melvin H. Allis (1836–92), Gonzales, Texas

PROFESSOR OF MATHEMATICS, GONZALES COLLEGE, 3 (1859–61), N.

Honorary A.M., University of Rochester (1860). S. W. Geiser, *Men of Science in Texas* (1958).

B. F. Anthonioz, Grand Coteau, Louisiana
UNKNOWN, 1 (1860), T.

Melvin C. Armstrong, Chicago, Illinois
MOLDER, 2 (1860–61), B.T.

Sanford Armstrong (b. 1834), Caldwell Prairie, Wisconsin
FARMER, 3 (1859–61), T.

Charles R. Barney, University Place, Tennessee
MILITARY ENGINEER, UNIVERSITY OF THE SOUTH, 3 (1859–61), T.R.

Edward L. Berthoud, Leavenworth, Kansas
PHYSICIAN, NATURALIST, 3 (1858–60), R. AAAS.

George W. Bowlsby (b. 1826), Monroe, Michigan
BUSINESS, JEWEL CARRIER, 3 (1859–61), B.T.R.
Private journalist for twenty years.

N. A. Chapman, Twinsburg, Ohio
BUSINESS, 1 (1860), T.

W. W. Curtis, Rocky Run, Wisconsin
FARMER, 15 (1859–73), T.R.

Charles Davis, Cannonsburg, Pennsylvania
UNKNOWN, 1 (1860), A.

Chester Dewey (1784–1867), Rochester, New York
CLERGYMAN, EDUCATOR, SCIENTIST, PROFESSOR, UNIVERSITY OF ROCHESTER, 13 (1855–67), B.T.R.
Graduate of Williams College (1806), professor of mathematics and natural philosophy, Williams College (1810–27), principal of the Berkshire Gymnasium (1827–36), principal and professor at the high school, collegiate institute, and University of Rochester (1836–61), contributor to fifty-three volumes of the *American Journal of Science*. *DAB*, AAAS, *RSC* (26 articles).

George C. Dickinson, Cobham Depot, Virginia
PLANTER, FORMER CIVIL ENGINEER, 3 (1859–61), T.R.

W. B. Flippin, Yellville, Arkansas
ACADEMY TEACHER, 2 (1859–60), T.

Robert Hallowell Gardiner (1782–1864), Gardiner, Maine
AGRICULTURALIST, PUBLIC BENEFACTOR, 10 (1855–64), A.
Graduated Harvard College (1801), founder of the Gardiner Lyceum (1821, a forerunner of American agricultural and technical schools), member of the Maine House of Representatives (1822), president of Maine Historical Society (1846–55). *DAB*, *RSC* (1 article).

T. Gibbs, Huntsville, Texas
BUSINESSMAN, 3 (1858–60), R.

David J. Heaston, (b. 1833) Bethany, Missouri
LAWYER, 2 (1859–60), N.
John C. House (b. 1831), Waterford, New York
FLOUR MERCHANT, 8 (1856–63), A.
Daniel Hunt, Pomfret, Connecticut
MINISTER, 17 (1853–69), A.
Gustavus A. Hyde, Cleveland, Ohio
CIVIL ENGINEER, 10 (1849, 1851, 1855–61, 1868), B.T.R.
Cleveland Acad. Nat. Sci., author of *The Weather at Cleveland, Ohio: What It Has Been for Forty Years* (1896), Espy volunteer observer. Bruce Sinclair, "Gustavus Hyde, Professor Espy's Volunteers, and the Development of Systematic Weather Observation," *Amer. Metl. Soc., Bulletin* 46 (1965): 779–87.
Samuel K. Jennings, Jr., Orrville, Alabama
PHYSICIAN, 8 (1849, 1858–60, 1868–71), T.P.R.
Oliver Hudson Kelley (1826–1913), Itasca, Minnesota
FOUNDER OF THE PATRONS OF HUSBANDRY, 3 (1860–61, 1863), T.
Reporter for the *Chicago Tribune,* telegraph operator, clerk in the Department of Agriculture (1864), founder and first secretary of the Patrons of Husbandry, the Grange (1868–78), founder of the town of Carabelle, Florida, author of *The Origins and Progress of the Order of the Patrons of Husbandry* (1875), *DAB.*
T. R. Kibble, Downieville, California
PHYSICIAN, 1 (1860), T.R.
James A. Kirkpatrick, Philadelphia, Pennsylvania
TEACHER, CENTRAL HIGH SCHOOL, 21 (1852–60, 1862–73), A.
AAAS, *RSC* (1 article).
John H. Luneman, St. Louis, Missouri
PROFESSOR, PRIEST, ST. LOUIS UNIVERSITY, 4 (1860–62, 1864), A.
E. M. Murch, Russelville, Kentucky
TEACHER, RUSSELVILLE ACADEMY, 1 (1860), N.
Professor of natural science, University of Public Schools, Louisville, Kentucky (1865).
Freeman Norvell, Greenwood, Dakota Territory
U.S. INDIAN AGENT, 3 (1859–61), T.R.
William Parry, Cinnaminson, New Jersey
NURSERYMAN, POMONA GARDEN AND NURSERY, 2 (1859–60), N.
William Wines Phelps (1792–1872), Salt Lake City, Utah
MORMON ELDER, 12 (1859–61, 1863–71), A.
Self-educated, editor of *Ontario Phoenix* (New York, 1820), established newspaper in Independence, Missouri (1832), served in Utah legislature

(1850–57), Speaker of the House, justice of the peace, "astronomer, astrologer and almanac maker" for the Mormons, author of the "Deseret Alphabet." *ACAB*.

R. C. Phillips, Cincinnati, Ohio
ENGINEER AND SURVEYOR, 14 (1859–72), T.R.

Zina Pitcher (1797–1872), Detroit, Michigan
PHYSICIAN AND NATURALIST, 5 (1858–62), A.
M.D., Middlebury College (1822), assistant surgeon, U.S. Army (1822–36), collaborator with botanists Torry and Gray, private medical practice in Detroit (1836), mayor of Detroit (1840–44), member of Michigan Board of Regents (1837–52), founder of the medical department, University of Michigan (1850), president of the Territorial Medical Society (1838–51), president of the Michigan State Medical Society (1855–56), and president of the American Medical Association (1856). *DAB*, AAAS.

Henry C. Prentiss, Worcester, Massachusetts
PHYSICIAN, STATE LUNATIC HOSPITAL, 5 (1859–64), A.

E. S. Robinson, Prairie Line, Mississippi
MINISTER, ACADEMY TEACHER, GEOLOGIST, 3 (1859–61), A.

Samuel Rodman, New Bedford, Massachusetts
BUSINESSMAN, 21 (1853–73), A.
Private journalist, 1823–53.

James Stephenson, St. Ingoes, Maryland
MINISTER, 13 (1859–71), A.
Botanical observer since 1857.

E. Ware Sylvester, Lyons, New York
PHYSICIAN, 4 (1859–62), B.T.

Henry A. Titze (b. 1839), West Salem, Illinois
FARMER, 6 (1855–60), T.R.

W. J. Vankirk (b. 1835), Bolivar, Missouri
SURVEYOR, 9 (1859–61, 1866–71), N.

O. G. Wagner, Santa Fe, New Mexico
MILITARY, TOPOGRAPHICAL ENGINEERS, 1 (1860), A.

Charles C. Wakeley, New York, New York
UNKNOWN, 4 (1860–63), A.

Helen I. Whelpley (b. 1839), Monroe, Michigan
SURVEYOR'S DAUGHTER, 7 (1855–61), T.R.
One of three in the family who took observations for twenty years (1852, 1855–73); the others were Thomas (b. 1794) and Florence (b. 1844).

Henry Willis, Portland, Maine
COUNSELOR-AT-LAW, 6 (1855–60), A.
Helped secure Maine meteorological appropriation.

1870

E. W. Adams, Goldsboro, North Carolina
PROFESSOR, 4 (1860–61, 1867, 1872), T.R.

J. A. Applegate and daughter, Mount Carmel, Indiana
AGRICULTURE, 5 (1869–73), T.R.

William Bacon, Richmond, Massachusetts
FARMER, 23 (1849–63, 1865–72), T.R.

Taylor J. Bingman (b. 1850), Quaker Ridge, Ohio
FARM WORKER, 3 (1870–72), T.R.

C. B. Blackburn, Louisville, Kentucky
PHYSICIAN, 1 (1870), A.

George A. Bowman (b. 1858) and Joshua B. Bowman (b. 1827), Vienna, Virginia
FARMERS, 4 (1870–73), T.R.

Frederick Brendel, Peoria, Illinois
PHYSICIAN, 19 (1855–73), A.
Botanical collector.

D. H. Chase, Louisville, Illinois
PHYSICIAN, 5 (1869–73), T.R.
RSC (1 article).

William Cheyney, Minneapolis, Minnesota
UNKNOWN, 10 (1864–73), A.

A. J. Compton, Watsonville, California
MINISTER, PHYSICIAN, 3 (1869–72), T.R.
Interested in scientific agriculture.

William H. Cook, Carlisle, Pennsylvania
PHYSICIAN, 6 (1868–73), B.T.R.

J. C. Covell, Staunton, Virginia
PRINCIPAL, DEAF, DUMB, AND BLIND ASYLUM, 5 (1868–72), A.

Nicholas DeWyl, Jefferson City, Missouri
PHYSICIAN, 5 (1868–72), B.T.

A. Failor, Newton, Iowa
FARMER, 2 (1869–70), T.R.

Robert Hallowell Gardiner (1855–1924), Gardiner, Maine
TRUSTEE, LAWYER, 9 (1865–73), A.
Graduated Harvard College (1876), Harvard Law School (1880). *WWW1*.

Edward Goldsmith (b. 1836), Memphis, Tennessee
BANK CASHIER, 4 (1867–70), A.

John Grant, Manchester, Illinois
FARMER, 20 (1854–73), A.
Assisted by three of his children.

J. St. Julien Guerard, Bluffton, South Carolina
PHYSICIAN, 1 (1870), T.R.
Henry Haas, Depauville, New York
FARMER, 9 (1865–73), T.R.
Ebenezer Hance, Fallsington, Pennsylvania
FARMER, 25 (1849–64, 1865–73), B.T.R.
Advocate of weather modification. AAAS.
Henry E. Hanshew, Frederick, Maryland
BUSINESSMAN, 9 (1852–54, 1856, 1863, 1869–72), B.T.R.
Asa Horr (1817–96), Dubuque, Iowa
PHYSICIAN, SCIENTIST, 11 (1851–55, 1857–58, 1868–71), A. (received Smithsonian instruments in 1851).
Graduate of Cleveland Medical College (1846), president and founder of the Iowa Institute of Science and Arts, experimented with Increase Lapham with telegraphic techniques for forecasting the weather. *ACAB*.
William H. Hunt (b. 1821), Biscayne, Florida
STATE SENATOR AND FARMER, 1 (1870), T.R.
Gustavus A. Hyde, Cleveland, Ohio
(See entry for 1860).
Samuel K. Jennings, Jr., Havana, Alabama
(See entry for 1860).
L. D. Kidder (b. 1835), Whitefield, New Hampshire
FARM WORKER, 5 (1869–73), T.R.
Assisted by wife.
Charles McCall, Cathlamet, Washington Territory
FARMER, 4 (1870–73), T.
E. Melton McConnell (b. 1825), Newcastle, Pennsylvania
GARDENER, 8 (1866–73), T.R.
Longtime private journalist.
Samuel D. Martin, Pine Grove, Kentucky
PHYSICIAN, 9 (1865–73), A.
Elias Nason (1811–87), North Billerica, Massachusetts
CLERGYMAN, SCHOOLMASTER, WRITER, LECTURER, 8 (1866–73), B.T.
Papermaker, graduate of Brown University (1835), editor and teacher in Waynesboro, Georgia (1835–40), teacher, high school and Latin school principal, Newburyport, Massachusetts (1840–52), Congregational minister (1852), editor of the *New-England Historical and Genealogical Register* (1866–67), author of some thirty-nine books and pamphlets on religion and history. *DAB*.
J. D. Parker, Mount Desert, Maine
SURVEYOR, 25 (1849–73), T.R.
Helped secure Maine meteorological appropriation.

J. M. Patrick (b. 1834), North Volney, New York
BOATMAN, 6 (1868–73), T.
J. P. Reinhard (b. 1840), Missoula, Montana
HARDWARE DEALER, 1 (1870), T.
Alexander P. Rodger, Gallipolis, Ohio
FARMER, 12 (1857–58, 1864–73), T.R.
James M. Sherman (b. 1815), Hampton, Virginia
HORTICULTURIST, 5 (1869–73), T.R.
William Soule, Cazenovia, New York
PROFESSOR OF NATURAL SCIENCE, ONEIDA CONFERENCE SEMINARY, 8 (1865, 1867–73), B.T.
Stillman Spooner, Oneida, New York
PHYSICIAN, 10 (1864–73), T.R.
D. Thompson, Milnersville, Ohio
MINISTER, 12 (1862–73), T.R.
W. J. Vankirk (b. 1835), Fish River, Alabama
SURVEYOR, 9 (1859–61, 1866–67, 1868–71), T.R.
(See entry for 1860).
J. Walters, Holton, Kansas
PHYSICIAN, 7 (1867–73), T.
J. Gilbert Williams, Utica, New York
SON OF FARMER, 1 (1870), T.R.
Milo G. Williams, Urbana, Ohio
PROFESSOR, UNIVERSITY OF URBANA, 19 (1855–73), B.T.R.
Espy volunteer observer.
P. Zahner, Nebraska City, Nebraska
FARMER, 3 (1868–71), T.

ABBREVIATIONS

AAAS: American Association for the Advancement of Science

Abbe Papers, JHU: Cleveland Abbe Papers, MS 60, Special Collections, Milton S. Eisenhower Library, JHU

Abbe Papers, LC: Cleveland Abbe Papers, Manuscript Division, LC

ACAB: *Appleton's Cyclopedia of American Biography*, edited by J. G. Wilson and J. Fiske. 6 vols. New York, 1888–89.

AES: Atmospheric Environment Service

Amer. Acad.: American Academy of Arts and Sciences

Amer. Journ. Sci.: *American Journal of Science and Arts* (Silliman's *Journal*)

Amer. Metl. Journ.: *American Meteorological Journal*

Amer. Metl. Soc.: American Meteorological Society

Amer. Phil. Soc.: American Philosophical Society

Amer. Phil. Soc., Early Proceedings: *Early Proceedings of the Amer. Phil. Soc., Manuscript Minutes of Its Meetings, 1744–1838* (1884).

BAAS: British Association for the Advancement of Science

Bache Papers: Alexander Dallas Bache Papers, RU 7053, SIA

Baird Papers: Spencer Fullerton Baird Papers, RU 7002, SIA

BDAS: *Biographical Dictionary of American Science: The Seventeenth through the Nineteenth Centuries*, edited by Clark A. Elliott. Westport, Conn., 1979.

Biog. Mem.: Biographical Memoir

Brown Collection: Samuel Gilman Brown Collection, Houghton Library, Harvard University

DAB: *Dictionary of American Biography*. 20 vols. New York, 1928–36.

DNB: *Dictionary of National Biography from the Earliest Times to 1900*. 22 vols. Oxford, 1885–1900.

DSB: *Dictionary of Scientific Biography*, editor-in-chief, Charles C. Gillispie. 16 vols. New York, 1970–80.

Frankl. Inst.: Franklin Institute

Franklin Papers: *The Papers of Benjamin Franklin*, edited by Leonard W. Labaree et al. Vols. 1–10. New Haven, 1959–66.

Guyot Letters: Arnold Guyot Letters (1849–50), Smithsonian Meteorological Project Records, Records of the Weather Bureau, RG 27, NA.

Guyot Papers: Arnold Guyot Papers, Historical Society of Princeton, N.J.

Hare Papers: Papers of Robert Hare, Amer. Phil. Soc. Library, Philadelphia, Pa.

Henry Papers: Joseph Henry Collection, RU 7001, SIA

Herschel Papers: J. F. W. Herschel Papers, Royal Society Library, London

Jefferson Papers: *The Papers of Thomas Jefferson*, vols. 1–20 edited by Julian P. Boyd, vols. 21–22 edited by Charles T. Cullen. Princeton, N.J., 1950–86.

JHP: *The Papers of Joseph Henry*, edited by Nathan Reingold et al. 5 vols. Washington, D.C., 1971–85.

JHU: Johns Hopkins University

Journ.: Journal

Journ. Smithsonian Regents: *Smithsonian Institution: Journals of the Board of Regents, Reports of Committees, Statistics, Etc.*, edited by William J. Rhees. *Smithson. Misc. Coll.* 8 (1880).

Lake Survey Records: Letters Sent and Letters Received by the U.S. Lake Survey, 1845–1913, Records of the Office of the Chief of Engineers, RG 77, NA
Lapham Papers: Increase A. Lapham Papers, State Historical Society of Wisconsin, Madison
LC: Library of Congress
Loomis Family Papers: Elias Loomis Family Papers, MSS 331, Manuscripts and Archives, Yale University Library
Loomis Papers: Elias Loomis Papers, Beinecke Rare Book and Manuscript Library, Yale University
LR: Letters Received
LS: Letters Sent
Merchants' Mag.: Hunts's Merchants' Magazine and Commercial Review
Metl. Mag.: Meteorological Magazine
Myer Papers: Albert James Myer Papers, Manuscript Division, LC. Includes four reels of microfilm from originals in the U.S. Army Signal Corps Museum, Fort Monmouth, N.J. (originals now at the U.S. Army Center for Military History, Carlisle Barracks, Pa.)
NA: National Archives and Records Administration
Natl. Acad. Sci.: National Academy of Sciences
Naval RC: Naval Records Collection of the Office of Naval Records and Library, RG 45, NA
NCAB: The National Cyclopaedia of American Biography, 57 vols. New York, 1892–1977.
NOAA: National Oceanic and Atmospheric Administration
NOR-LC: Naval Observatory Records, Manuscript Division, LC
NOR-NA: Records of the Naval Observatory, RG 78, NA
OSI-SIA: Office of the Secretary, Incoming Correspondence, RU 26, SIA
OSO-SIA: Office of the Secretary, Outgoing Correspondence, 1865–91, RU 33, SIA
Phil. Mag.: London, Edinburgh, and Dublin Philosophical Magazine and Journal of Science
Phil. Trans.: Philosophical Transactions of the Royal Society
Quetelet Correspondence: Selected correspondence of Adolphe Quetelet, 2 reels of microfilm, H.S. Films 11, Amer. Phil. Soc. Library, Philadelphia, Pa.
Redfield Family Papers: William Cox Redfield Papers, Manuscript Division, LC
Redfield Letterbooks: William C. Redfield Letterbooks, 3 vols., Beinecke Rare Book and Manuscript Library, Yale University
Redfield Papers: William C. Redfield Papers, Beinecke Rare Book and Manuscript Library, Yale University
Reid's Correspondence: William Reid's Correspondence with W. C. Redfield, 3 vols., Beinecke Rare Book and Manuscript Library, Yale University
RG: Record Group
Rhees Collection: William Jones Rhees Collection, Henry E. Huntington Library and Art Gallery, San Marino, Calif.
RU: Record Unit
Sabine Papers: Edward Sabine Papers, Royal Society Library, London
SIA: Smithsonian Institution Archives
Signal Office Records: Letters Sent and Letters Received, U.S. Army Signal Office, Records of the Weather Bureau, RG 27, NA
Signal Office Report: Annual Report of the Chief Signal Officer, in *Annual Report of the Secretary of War*
SIMC-NA: Smithsonian Institution Meteorological Correspondence, Letters Received, Records of the Weather Bureau, RG 27, NA
SIMC-SIA: Smithsonian Institution Meteorological Correspondence, Letters Received, RU 60, SIA

Smithson. Contrib.: *Smithsonian Contributions to Knowledge*
Smithson. Misc. Coll.: *Smithsonian Miscellaneous Collections*
Smithsonian Report: Annual Report of the Smithsonian Institution
WCC Correspondence: Correspondence, Wind and Current Chart Agent, Superintendent's Office, Records of the U.S. Naval Observatory, Manuscript Division, LC
WWW 1: Who Was Who in America, Vol. 1, 1897–1942. Chicago, 1943.
WWW Historical: Who Was Who in America, Historical Volume, 1607–1896. Chicago, 1963.

NOTES

Introduction

1. See Ralph Waldo Emerson, "The American Scholar," in Alfred R. Ferguson, ed., *The Collected Works of Ralph Waldo Emerson*, vol. 1, *Nature, Addresses, and Lectures* (Cambridge, Mass., 1971), 52–70.

2. Richard H. Shryock, "American Indifference to Basic Research during the Nineteenth Century," *Archives internationales d'histoire des sciences* 28 (1948): 50–65.

3. Nathan Reingold, "American Indifference to Basic Research: A Reappraisal," in George Daniels, ed., *Nineteenth-Century American Science: A Reappraisal* (Evanston, Ill., 1972), 54–55; Robert V. Bruce, *The Launching of Modern American Science, 1846–1876* (New York, 1987), 1.

4. See, for example, Samuel P. Langley, "The Meteorological Work of the Smithsonian Institution," *U.S. Weather Bureau, Bulletin* 11 (1893): 216–20; and Marcus Benjamin, "Meteorology," in G. B. Goode, ed., *The Smithsonian Institution, 1846–1896: The History of Its First Half Century* (Washington, D.C., 1897), 647–78. Notable exceptions include the works of W. E. Knowles Middleton and David M. Ludlum.

5. James R. Fleming, *Guide to Historical Resources in the Atmospheric Sciences: Archives, Manuscripts, and Special Collections in the Washington, D.C. Area*, NCAR Technical Note 327+IA (Boulder, Colo., 1989). The National Archives, the Smithsonian Institution, and the Library of Congress all have rich archival and printed collections relevant to nineteenth-century meteorology. The single best research collection is the Joseph Henry Papers at the Smithsonian, including part of Henry's personal library. Other abundant sources are the papers of William C. Redfield and Elias Loomis at Yale University, the Robert Hare Papers in the Library of the American Philosophical Society, and the Cleveland Abbe Papers at Johns Hopkins University. The papers of James P. Espy are found in many other collections. Small collections for William Ferrel, James Henry Coffin, and Arnold Guyot are also valuable.

6. I. Bernard Cohen, *Revolutions in Science* (Cambridge, Mass., 1985), 150.

7. August Schmauss (1929), quoted in Karl Schneider-Carius, *Weather Science, Weather Research: History of Their Problems and Findings from Documents during Three Thousand Years* (German ed. 1955; trans. New Delhi, 1975), 409.

8. Bruno Latour, "Visualization and Cognition: Thinking with Eyes and Hands," *Knowledge and Society: Studies in the Sociology of Culture Past and Present* 6 (1986): 22–23.

9. *The Oxford English Dictionary* (Oxford, 1888–1928), s.v. "system: a unitary whole composed of parts in orderly arrangement according to a certain scheme or plan."

10. James R. Fleming, "Storms, Strikes, Indian Uprisings, and Other Threats to Domestic Tranquility: The U.S. Army Weather Service and the Telegraph, 1870–1891," paper presented at the History of Science Society Meeting, Cincinnati, 1988.

11. *Oxford English Dictionary*, s.v. "network: an interconnected chain or system of immaterial things"; Walter F. Cannon, "Scientists and Broad Churchmen: An Early Victorian Intellectual Network," *Journal of British Studies* 4 (1964): 66–72; Derek J. de Solla Price, *Little Science, Big Science* (New York, 1963); M. J. Mulkay, G. N. Gilbert, and S. Woolgar, "Problem Areas and Research Networks in Science," *Sociology* 9 (1975): 188. The term *net-work*, used by Elias Loomis and others in the 1870s in reference to telegraphic weather reporting, is a later connotation.

Chapter 1. Early Issues and Systems of Observation

1. See Alfred J. Henry, "Early Individual Observers in the United States," *U.S. Weather Bureau, Bulletin* 11 (1893): 291–302; James M. Havens, ed., *An Annotated Bibliography of Meteorological Observations in the United States, 1731–1818,* Florida State University Department of Meteorology, Technical Report No. 5 (Tallahassee, Fla., 1956); Harriette M. Forbes, comp., *New England Diaries, 1602–1800* (Topsfield, Mass., 1923); William Matthews, *American Diaries: An Annotated Bibliography of American Diaries prior to . . . 1861* (Berkeley, 1945); and Laura Arksey, Nancy Pries, and Marcia Reed, *American Diaries: An Annotated Bibliography of Published American Diaries and Journals,* vol. 1, *1492–1844* (Detroit, 1983).

2. Gustav Hellmann lists 156 commentaries on Aristotle's meteorology published before 1650 and only 18 after that date ("Entwicklungsgeschichte des meteorologischen Lehrbuches," *Beiträge zur Geschichte der Meteorologie,* vol. 2 [Berlin, 1917], 9–10). On meteorological instruments see W. E. Knowles Middleton, *The History of the Barometer* (Baltimore, 1964); *A History of the Thermometer and Its Use in Meteorology* (Baltimore, 1966); *Invention of the Meteorological Instruments* (Baltimore, 1969).

3. "Dichiarazione d'alcuni Strumenti per Conoscer l'Alterazioni dell'Aria," *Saggi di naturali esperienze fatte nell'Accademia del Cimento* (Florence, 1666); *Archivo Meteorologico Centrale Italiano* (Florence, 1858), reproduced in Gustav Hellmann, *Neudrucke von Schriften und Karten über Meteorologie und Erdmagnetismus* no. 7 (Berlin, 1897), 9–17. Brief mention of this system also appears in H. H. Frisinger, *The History of Meteorology to 1800* (New York, 1977).

4. Gustav Hellmann, "Die ältesten instrumentellen meteorologischen Beobachtungen in Deutchland," *Beiträge zur Geschichte der Meteorologie,* vol. 1 (Berlin, 1914), 103–7; Edme Mariotte, *Oeuvres de Mariotte,* 2 vols. in 1 (Leiden, 1717).

5. Or *Sammlung von Natur- und Medicin-, wie auch hiezu gehörigen Kunst- und Literatur-Geschichten,* published with the assistance of Johann Christian Kundmann, and J. G. Brunschweig. Gustav Hellmann, "Die Vorläuffer der Societas Meteorologica Palatina," *Beiträge zur Geschichte der Meteorologie,* vol. 1, (Berlin, 1914), 139–47; Hellmann, "Umriss einer Geschichte der meteorologischen Beobachtungen in Deutchland," *Repertorium der Deutchen Meteorologie* (Leipzig, 1883), 884–86; Abraham Wolf, *A History of Science, Technology, and Philosophy in the Eighteenth Century* (New York, 1939), 284; and Emil J. Walter, "Technische Bedingungen in der historichen Entwicklung der Meteorologie," *Gesnerus* 9 (1952): 55–66.

6. James Jurin, "Invitatio ad Observationes Meteorologicas communi consilio instituendas," *Phil. Trans.* 32 (1723): 422–27; Roger Pickering, "A Scheme of a Diary of the Weather, Together with Draughts and Descriptions of Machines Subservient Thereunto," *Phil. Trans.* 43, no. 473 (1744–45): 1–18.

7. Partial results appear in Louis Cotte's *Traité de météorologie* (Paris, 1774). See also E. I. Tichomirov, "Instructions for Russian Meteorological Stations of the 18th Century" (in Russian, English summary), *Central Geophysical Observatory Proceedings* (1932): 3–12.

8. Charles C. Gillispie, *Science and Polity in France at the End of the Old Regime* (Princeton, 1980), 226–29; Cotte, *Traité de météorologie,* and Cotte, *Mémoires sur la météorologie pour servir et de supplément au Traité de Météorologie publie en 1774,* 2 vols. (Paris, 1788). On Cotte see J. A. Kington, "A Late Eighteenth-Century Source of Meteorological Data," *Weather* 25 (1970): 169–75; Theodore S. Feldman, "The History of Meteorology, 1750–1800: A Study in the Quantification of Experimental Physics" (Ph.D. dissertation, University of California, Berkeley, 1983; Ann Arbor: University Microfilms, 84-13376), 214; M. G. Pueyo, "Un contintuateur des travaux concernant la météorologie agricole . . . L. Cotte," *Comptes rendus. Academie d'Agriculture de France* 68 (1982): 604–9; and Pueyo, "Les observations météorologiques des correspondants de L. Cotte," *Comptes rendus. Academie d'Agriculture de France* 68 (1982): 658–

63, 1429–35. On Lavoisier see *Oeuvres de Lavoisier: Correspondance*, ed. René Fric, vol. 3 (Paris, 1964), 658; also Lavoisier, "Règles pour prédire le changement de temps d'après les variations du baromètre," in Etiene Chiron (ed.), *Extraits des memoires de Lavoisier concernant la métérologie et l'aéronautique* (Paris, 1926), 139–45. The *DSB* article by Henry Guerlac does not mention Lavoisier's meteorological work.

9. David C. Cassidy, "Meteorology in Mannheim: The Palatine Meteorological Society, 1780–1795," *Sudhoffs Archiv für Geschichte der Medizin und der Naturwissenschaften* 69 (1985): 8–25; Friedrich Traumüller, *Die Mannheimer meteorologische Gesellschaft (1780–1795): Ein Beitrag zur Geschichte der Meteorologie* (Leipzig, 1885); A. Kh. Khrgian, *Meteorology: A Historical Survey*, 2d ed., vol. 1 (Leningrad, 1959), trans. Ron Hardin (Jerusalem, 1970), chap. 6; Helmut E. Landsberg, "A Bicentenary of International Meteorological Observations," *World Meteorological Organization, Bulletin* 29 (1980): 235–38; Albert Cappel, "Societas Meteorologica Palatina (1780–1795)," *Annalen der Meteorologie* n.s., 16 (1980): 10–27, 255–61; J. J. Hemmer, "Historia Societas Meteorologicae Palatinae," in Societatis Meteorologicae Palatinae, *Ephemerides, 1781* (1783): 1–54; Malcolm Rigby, "Ephemerides of the Meteorological Society of the Palatinate," *Environmental Data Service*, Feb. 1973, pp. 10–16.

10. Quoted in Karen Ordahl Kupperman, "The Puzzle of the American Climate in the Early Colonial Period," *American Historical Review* 87 (1982): 1270.

11. William Strachey, *The Historie of Travell into Virginia Britania* (1612), ed. Louis B. Wright and Virginia Freund (London, 1953), 37–38.

12. Campanius (1601–83), diary quoted in Oliver L. Fassig, "A Sketch of the Progress of Meteorology in Maryland and Delaware," *Maryland Weather Service* 1 (1899): 333–34; see also Thomas Campanius Holm, *Kort Beskrifning om Provincieu Nya Sverige uti America* . . . (Stockholm, 1702); Nicholas Collin, "Observations Made at an Early Period on the Climate of the Country Along the River Delaware, Collected from the Records of the Swedish Colony," *Amer. Phil. Soc., Transactions* n.s. 1 (1818): 340–52.

13. MacSparran (d. 1757), *DNB;* James MacSparran, *America Dissected, Being a Full and True Account of All the American Colonies, Shewing the Intemperance of the Climates, Excessive Heat and Cold, and Sudden Violent Changes of Weather, Terrible and Mischievous Thunder and Lightning, Bad and Unwholesome Air, Destructive to Human Bodies, etc.* (Dublin, 1753).

14. Kenneth Thompson, "Forests and Climate Change in America: Some Early Views," *Climatic Change* 3 (1980): 47–64; Ferdinand Columbus, *The Life of the Admiral Christopher Columbus by His Son Ferdinand*, trans. Benjamin Keen (New Brunswick, N.J., 1959), 142–43.

15. William Wood (fl. 1629–35), *DAB;* Wood, *New England's Prospect* (London, 1634), 7–8.

16. Cotton Mather (1662/63–1727/28), *DAB;* Mather, "Essay XIX. Of Cold," in *The Christian Philosopher: A Collection of the Best Discoveries in Nature with Religious Improvements* (London, 1721; Charlestown, 1815), 81.

17. Benjamin Franklin (1706–90), Ezra Stiles (1727–75), both *DAB;* Franklin to Stiles, May 29, 1763, *Franklin Papers*, 10:264–67.

18. Hugh Williamson (1735–1819), *DAB;* Williamson, "An Attempt to Account for the Change of Climate, Which Has Been Observed in the Middle Colonies in North-America," *Amer. Phil. Soc., Transactions* 1 (1771): 272–78; Williamson, *Observations on the Climate of Different Parts of America* (New York, 1811).

19. Thomas Jefferson to Jean Baptiste Le Roy, Nov. 13, 1786, *Jefferson Papers*, 10:524–28.

20. Ibid., 529–30.

21. Jefferson (1743–1826), *DAB;* Thomas Jefferson, *Notes on the State of Virginia* (1785; rpt. Gloucester, Mass., 1976), 79.

22. Dunbar (1749–1810), *DAB;* Dunbar to Jefferson (1801), quoted in David M. Ludlum, *Early American Winters, 1604–1820,* vol. 1 (Boston, 1966), 163; see also Ludlum, "Thomas Jefferson and the American Climate," *Amer. Metl. Soc., Bulletin* 47 (1966): 974–75; William Dunbar, "Meteorological Observations," *Amer. Phil. Soc., Transactions* 4 (1809): 48.

23. Ramsay (1749–1815), *DAB;* Ramsay, *History of South Carolina* (1809), quoted in Ludlum, *Early American Winters,* 1:162.

24. David M. Ludlum, *Early American Hurricanes, 1492–1870* (Boston, 1963), 77–81; John Farrar, "An Account of the Violent and Destructive Storm of the 23rd of September 1815," *Amer. Acad., Memoirs* 4 (1821): 92–97. Observations were published in *North American Review* 6 and 7 (1816 and 1817).

25. Job Wilson, "A Meteorological Synopsis, in Connection with the Prevailing Diseases for Sixteen Years, as They Occurred at Salisbury, Massachusetts," *Medical Repository* 22 (1822): 409–13.

26. Hippocrates, "On Airs, Waters, and Places," *Hippocratic Writings,* trans. Francis Adams (Chicago, 1952), 9–19. See also "Of the Epidemics," ibid., 44–49; and Frederick Sargent II, *Hippocratic Heritage: A History of Ideas about Weather and Human Health* (New York, 1982), 49–59.

27. See, for example, John Arbuthnot, *An Essay Concerning the Effects of Air on Human Bodies* (London, 1733); and William Falconer, *Remarks on the Influence of Climate, Situation, Nature of Country, Population, Nature of Food, and Way of Life, on the Disposition and Temper, Manners and Behavior, Intellects, Laws and Customs, Form of Government, and Religion, of Mankind* (London, 1781).

28. T. R. Forbes, *Chronicles from Aldgate: Life and Death in Shakespeare's London* (New Haven, 1971); John Grant, comp., *Natural and Political Observations . . . upon the Bills of Mortality* (London, 1662); C. Singer and E. A. Underwood, *A Short History of Medicine,* 2d ed. (London, 1962), 179–80; and Sargent, *Hippocratic Heritage,* 165–67.

29. John Fothergill, "Some Remarks on the Bills of Mortality in London with an Account of a Late Attempt to Establish an Annual Bill for This Nation," in John C. Lettsom, ed., *The Works of John Fothergill, M.D.* (London, 1783), 2:107–13.

30. Lining (1708–60), *DAB;* Lining to Secretary of the Royal Society, Jan. 22, 1741, *Phil. Trans.* 42 (1742–43): 491–509; (1744–45): 318–30; and Lining, "A Description of American Yellow Fever," *Essays and Observations* (Edinburgh, [1771]), 2:404–32. See also Robert C. Aldredge, "Weather Observers and Observations at Charleston, South Carolina from 1670–1871," in *Year Book of the City of Charleston,* Historical Appendix (Charleston, 1940), 190–257; Frederick P. Bowes, *The Culture of Early Charleston* (Chapel Hill, 1942), 81–82; Henry, "Early Individual Observers," 295; Helmut E. Landsberg, "Early Weather Observations in America," *EDIS: Environmental Data and Information Service* 10, no. 4 (1979): 21–23; and J. H. Cassedy, "Meteorology and Medicine in Colonial America: Beginnings of the Experimental Approach," *Journal of the History of Medicine and Allied Sciences* 24 (1969): 193–204.

31. Chalmers (1715–77), *BDAS;* Holyoke (1728–1829), *DAB.* Edward A. Holyoke, "Observations on Weather and Diseases at Salem, Massachusetts, for the Year 1786, Extracted from a Communication by Edward Augustus Holyoke, M.D. to the Massachusetts Medical Society," in William Currie, comp., *An Historical Account of the Climates and Diseases of the United States of America* (Philadelphia, 1792), 13–28.

32. *The Charter, Constitution and Bye Laws of the College of Physicians of Philadelphia* (1790), 3, quoted in Brooke Hindle, *The Pursuit of Science in Revolutionary America, 1735–1789* (Chapel Hill, 1956), 296. A prominent contribution to this genre was by Benjamin Rush, "Account of the Climate of Pennsylvania, and Its Influence upon the Human Body," *American Museum* 4 (1789): 26.

33. Webster (1758–1843), *DAB;* Noah Webster, *A Brief History of Epidemic and Pestilential Diseases with the Principal Phenomena of the Physical World, Which Precede and Accompany Them, and Observations Deduced from the Facts Stated,* 2 vols. (Hartford, Conn., 1799).

34. On Franklin's geophysical interests see Martin A. Pomerantz, "Benjamin Franklin—The Compleat Geophysicist," *Eos: American Geophysical Union, Transactions* 57 (1976): 492–505; William M. Davis, "Some American Contributions to Meteorology," *Frankl. Inst. Journ.* 127 (1889): 104–15, 176–91; and Cleveland Abbe, "Benjamin Franklin as Meteorologist," *Amer. Phil. Soc., Proceedings* 45 (1906): 117–28.

35. Eliot (1685–1763), *DAB;* see Herbert Thoms, *The Doctors Jared of Connecticut* (Hamden, Conn., 1958), 3–31; Franklin to Eliot, July 16, 1747, *Franklin Papers,* 3:149.

36. A. D. Bache, Franklin's great-grandson, concluded that the eclipse in question occurred in 1743 (Bache, "Attempt to Fix the Date of Dr. Franklin's Observation, in Relation to the North-east Storms of the Atlantic States," *Frankl. Inst. Journ.* n.s., 12 [1833]: 300).

37. Franklin's reasoning on northeast storms is given in Franklin to Eliot, Feb. 13, 1750, *Franklin Papers,* 3:463–65; and Franklin to Alexander Small, May 12, 1760, ibid., 9:110–12. Franklin's opinion on thundergusts is expressed in Franklin to John Mitchell, Apr. 29, 1749, ibid., 3:365–76.

38. E.g., Increase Mather, *The Voice of God in Stormy Winds* (Boston, 1704); Eliphalet Adams, *God Sometimes Answers His People by Terrible Things in Righteousness. A Discourse Occasioned by that Awful Thunderclap which Struck the Meeting-house in N. London, Aug. 31st, 1735* (New London, Conn., 1735); and Thomas Prince, *The Natural and Moral Government and Agency of God in Causing Droughts and Rains* (Boston, 1751).

39. Andrew Dickson White, *A History of the Warfare of Science with Theology in Christendom* (New York, 1896), 1:323–72. Weather inscriptions from bells of the twelfth to the sixteenth centuries are given in ibid. and in Schneider-Carius, *Weather Science, Weather Research,* 56.

40. William Prout, *Chemistry, Meteorology, and the Function of Digestion, Considered with Reference to Natural Theology,* Bridgewater Treatise VIII, 2d ed. (London, 1834); John Lauris Blake, *Wonders of the Earth* (Cazenovia, N.Y., 1845); Henry to Henry James, Aug. 22, 1843, *JHP* 5:387–88; Marc Rothenberg, "The Educational and Intellectual Background of American Astronomers, 1825–1875" (Ph.D. dissertation, Bryn Mawr College, 1974).

41. Mary Ellen Bowden, "The Weather Observations of Johannes Kepler and David Fabricius: Astrologically Motivated Research in the Early Seventeenth Century," paper presented at the History of Science Society, Raleigh, N.C., 1987; also Schneider-Carius, *Weather Science, Weather Research,* 59–63. One of Kepler's popularizers was John Goad, *Astro-meteorologica, or Aphorisms and Discourses of the Bodies Coelestial, Their Nature and Influences . . . Collected from the Observations at Leisure Times, of Above Thirty Years* (London, 1686).

42. Fassig, "Sketch of the Progress," 347. The first edition of the *Bauern-Praktik* is reproduced in Gustav Hellmann, ed., *Neudrucke von Schriften und Karten über Meteorologie und Erdmagnetismus,* No. 5, Appendix (Berlin, 1896).

43. Madison (1749–1812), *DAB;* Jefferson, *Notes on the State of Virginia,* 74–78. Cf. C. E. Kemper, "Letters of Rev. James Madison . . . to Thomas Jefferson," *William and Mary College Quarterly Historical Magazine* 2d ser., 5 (July 1925): 145–46; Kemper doubts that Jefferson or Madison could have taken these observations; cf. Alexander McAdie, "Simultaneous Meteorological Observations in the United States during the Eighteenth Century," *U.S. Weather Bureau, Bulletin* 11 (1893): 303–4; Frederick J. Randolph and Frederick L. Francis, "Thomas Jefferson as a Meteorologist," *Monthly Weather Review* 23 (1895): 456–58.

44. Jefferson to James Madison, Feb. 20, Mar. 16, 1784, *Jefferson Papers,* 6:545 and 7:31–32.

45. James Madison, Meteorological journals kept at his plantation, with notes on sowing

and harvesting, migration of birds, etc., 1784–93, 1798–1802, 2 vols. A volume for 1793–96 is in the Presbyterian Historical Society, Philadelphia; cited in Stephen J. Catlett, ed., *A New Guide to the Collections in the Library of the American Philosophical Society* (Philadelphia, 1987), entry 719.

46. David Rittenhouse, Meteorological observations (Pennsylvania), 1784–1805, 2 vols., ibid., entry 1016; Peter Legaux, Observations météorologique faites à Springmill, 1787–1800, with an English translation, ibid., entry 663; William Adair, Meteorological diary, Warrenton, N.C., 1789, ibid., entry 758; William Bartram, Meteorological diary (Philadelphia), Jan. 1790–Sept. 1791, 1 vol., and Diary (record of weather, appearances of birds, flowers, insects, and so on), 1802–22, ibid., entries 78 and 80; George Hunter, Journals, 1796–1809, 4 vols., ibid., entry 544; John Gottlieb Ernestus Heckewelder, Meteorological observations (Pennsylvania), 1802–14, 1 vol., ibid., entry 494; John Breck Treat, Meteorological observations made at Arkansas, 1805–8, 1 vol., ibid., entry 1226; and Benjamin Vaughan, Meteorological notes and memoranda, ibid., entry 1247. See also Jefferson to Giovanni Fabbroni, June 8, 1778, *Jefferson Papers,* 2:195–96.

47. Jefferson to Volney, Jan. 8, 1797, quoted in Ralph Brown, "The First Century of Meteorological Data in America," *Monthly Weather Review* 68 (1940): 130–33.

48. E.g., John C. Greene, *American Science in the Age of Jefferson* (Ames, Iowa, 1984).

49. Adams (1773–1864), *DAB; Medical and Agricultural Register* 1 (1806): 30.

50. Correspondents of the *Medical and Agricultural Register* included Rev. Ebenezer Hill, Mason, N.H.; Abijah Bigelow, Leominster, Mass., Dr. Isaac Hurd, Concord, Mass., Charles Peirce, Portsmouth, N.H.; Epaphras Hoyt, Deerfield, Mass., W. Cobb, Warwick, R.I.; Rev. Abel Flint, Hartford, Conn.; and B. Fowler, Windsor, Vt.

51. "Williams College 1816. Meteorological Observations to be made, by agreement with Prof. Hall of Mid. Col.," MS reproduced in Willis I. Milham, *The History of Meteorology in Williams College* (Williamstown, Mass., 1936). The table correlating rainfall with phases of the moon is in *Literary and Philosophical Repertory* 2 (1817): 486.

52. Dewey (1784–1867), Day (1773–1867), both *DAB;* See *Literary and Philosophical Repertory* 1 (1812–14): 165, [2]38, 393, 473; and 2 (1815–17): 78, 316, 393, 484.

53. John Griscomb, "Hints Relative to the Most Eligible Method of Conducting Meteorological Observations," *Transactions of the Literary and Philosophical Society of New York* 1 (1815): 343.

54. Farrar (1779–1853), Cleaveland (1780–1858), both *DAB;* Eights (1773–1848), *JHP* 1:68. For the observations see *North American Review* (1815–18): 1:122–26, 285–90, 437–39; 2:131–34, 277–80; 3:132–36, 284, 426; 4:129, 283, 420–22; 5:149–51, 299–300, 439–40; 6:149–50, 292, 438; 7:148.

55. Probably Robert Hamilton Bishop (1777–1855), *DAB.*

56. Professor [Chester] Dewey, "An Attempt to Ascertain at Which Three Hours of the Day the Thermometer Will Give Nearly the Mean Temperature," *North American Review* 6 (1818): 436–3[7].

57. Bigelow (1787–1879), *DAB;* Jacob Bigelow, "Facts Serving to Show the Comparative Forwardness of the Spring Season in Different Parts of the United States," *Amer. Acad., Memoirs* 4 (1818): 77–85; Silliman's report on Bigelow is "On the Comparative Forwardness of the Spring, in Different Parts of the United States, in 1817," *Amer. Journ. Sci.* 1 (1818): 76–77. See also *Niles' Weekly Register* 12 (May 10, 1817): 167.

58. Tilton (1745–1822), *DAB;* Edgar Erskine Hume, "The Foundation of American Meteorology by the United States Army Medical Department," *Bulletin of the History of Medicine* 8 (1940): 202–38; and J. H. Hagarty, "Dr. James Tilton, 1745–1822," *Weatherwise* 15 (1962): 124–25.

59. Lovell (1788–1836), *DAB;* quoted in Harvey E. Brown, *The Medical Department of the United States Army from 1775 to 1873* (Washington, D.C., 1873), 88.

60. Army physician Hanson Catlett to Major McClelland, Jan. 17, 1815, RG 94, M566, roll 67, NA, quoted in Mary C. Gillett, *The Army Medical Department, 1775–1818* (Washington, D.C., 1981), 194.

61. This was the opinion of Surgeon Henry Huntt, quoted in James Mann, *Medical Sketches of the Campaigns of 1812, 13, 14* (Dedham, Mass., 1816), 136; quote from p. vii.

62. U.S. Army Medical Department, "Regulations for the Medical Department," in *Military Laws and Rules and Regulations for the Army of the United States* (Washington, D.C., 1814), 227–28.

63. Gillett, *Army Medical Department,* 152–54. Tilton was surgeon general of the army from 1813 to 1815.

64. Benjamin Waterhouse, "Topographical-medical remarks, together with meteorological tables, or diary of the weather, made near the headquarters of the 2d military department, by Benj. Waterhouse, Hosp. Surg. and Director," MS (1816) in the Army Medical Library, Bethesda, Md. The first observations, from March 1, 1816, are reproduced in Helmut E. Landsberg, "Early Stages of Climatology in the United States," *Amer. Metl. Soc., Bulletin* 45 (1964): 268–74. Waterhouse noted the barometer, thermometer, face of the sky, wind, rain, and snow for the hours of 7:00 A.M., 2:00 P.M., and 9:00 P.M.

65. Gillett, *Army Medical Department,* 193.

66. Lovell, "Remarks on the Sick Report of the Northern Division for the Year Ending June 30, 1817," quoted in Brown, *Medical Department,* 102–7.

67. U.S. Army Medical Department, *Regulations of the Medical Department,* September 1818, 31, quoted, ibid., 111.

68. *Medical Repository* 21 (1821): 107, 383, 491; and 22 (1822): 115, 250, 283, 495. The first issue of this journal lists meteorological observations and statistics on medical patients and diseases for New York City (*Medical Repository* 1 [1797–98]: 99, 245, 373, 557).

69. *Medical Repository* 21 (1821): 107. Lovell also published the observations in the *National Intelligencer,* July 25, 1820.

70. U.S. Army Medical Department, *Meteorological Register for the Years 1822, 1823, 1824, and 1825 from Observations made by the Surgeons of the Army at the Military Posts of the United States* (Washington, D.C., 1826), 1–2.

71. "Notice of a Meteorological Register for the Years 1822, 1823, 1824 and 1825; from Observations Made by the Surgeons of the Army, at the Military Posts of the United States," *Amer. Journ. Sci.* 12 (1827): 149–54.

72. Alexander von Humboldt, "Uber die Haupt-Ursachen der Temperatur-Verschiedenheit auf dem Erdköper" (read before the Akad. Wissenschaften, Berlin, July 3, 1827), quoted and translated in Landsberg, "Early Stages of Climatology," 270.

73. Josiah Meigs (1757–1822), *DAB.* See also William M. Meigs, *Life of Josiah Meigs* (Philadelphia, 1887), and Horace S. Carter, "Josiah Meigs, Pioneer Weatherman," *Weatherwise* 13 (1960): 166–67, 181. Josiah Meigs, "Circular to the Registers of the Land Offices of the United States," Apr. 29, 1817, in *Niles' Weekly Register* 12 (May 10, 1817): 167–68.

74. Daniel Drake (1785–1852), *DAB;* Meigs to Drake, Feb. 1, 1817, quoted in Meigs, *Life of Josiah Meigs,* 82. On Drake's interest in medical topography see Sargent, *Hippocratic Heritage,* 258–68.

75. Meigs, "Circular."

76. *Niles' Weekly Register for 1818,* 152; ibid., *1819,* 234, 320, 362, 400; and *Supplement to vol. 15,* 16–23.

77. *Niles' Weekly Register for 1819: Supplement to vol. 15,* 16–23, quoted in Meigs, *Life of Josiah Meigs,* 84.

78. Josiah Meigs, "Geometric Exemplification of Temperature, Wind and Weather for 1820 at Washington," *Amer. Phil. Soc., Early Proceedings,* 505.

79. Henry, "Early Individual Observers," 300–301.

80. Josiah Meigs, "Weather Record for 1821," MS in Records of the Smithsonian Meteorological Project, RG 27, NA.

81. DeWitt (1756–1834), *DAB;* T. Romeyn Beck, *Eulogium on the Life and Services of Simeon DeWitt, Surveyor General of the State of New York, Chancellor of the University, &c.* (Albany, 1835); Simeon DeWitt, "Respecting a Plan of a Meteorological Chart, for Exhibiting a Comparative View of Climates of North America, and the Progress of Vegetation," and DeWitt, "Result of Thermometrical Observations for the Years 1795 and 1796 Made at the City of Albany," *Society for the Promotion of Agriculture, Arts and Manufactures, Transactions* 1, 2d ed., rev. (Albany, 1801): 88–92 and 287–88.

82. An account of the regents' interest in meteorology is given in Franklin B. Hough, comp., *Historical and Statistical Record of the University of the State of New York during the Century from 1784 to 1884* (Albany, 1885), 766–74.

83. Greig (1779–1858), *JHP* 1:106; Lansing (unidentified). Resolution dated Mar. 1, 1825, Trustees' Minutes 1 (Sept. 9, 1825), 251–52, Albany Academy Archives; see *JHP* 1:xxvi. A printed circular letter dated Nov. 24, 1825, and a four-page copy of instructions to observers, addressed to John Vaughn, are in MS Communications to the American Philosophical Society, Natural Philosophy, 2:77, Amer. Phil. Soc. Library, Philadelphia, Pa.

84. Resolution dated Apr. 12, 1825, Trustees' Minutes 1 (Sept. 9, 1825), 251–52, Albany Academy Archives; see *JHP* 1:106–7.

85. *Amer. Journ. Sci.* 11 (1826): 59–66.

86. Hawley (1785–1870) and Beck (1791–1855), both *DAB*. T. Romeyn Beck, "Address Delivered before the Lyceum of Natural History (Now the Second Department of the Institute) at Its First Anniversary, March 1, 1824," *Albany Institute, Transactions* 1 (1830): 145.

87. Henry (1797–1878), *DSB;* Simon Newcomb, "Memoir of Joseph Henry," *Natl. Acad. Sci., Biog. Mem.* 5 (1905): 1–45; Beck to Martin Van Buren, Mar. 10, 1826, *JHP* 1:118, 65.

88. Henry to Stephen Alexander, Dec. 5, 1829, *JHP* 1:252.

89. *JHP* 1:xxv–xxvi and passim; T. R. Beck and Joseph Henry, "Abstract of the Returns of Meteorological Observations Made to the Regents of the University by Sundry Academies in the State," *Albany Institute, Transactions* 3 (1855): 234.

90. Alexander (1806–83), *DAB*. Joseph Henry, "Topographical Sketch of the State of New-York, Designed Chiefly to Show the General Elevations and Depressions of Its Surface," *Albany Institute, Transactions* 1 (1830): 87–112; Alexander to Henry, Mar. 13–15?, 1830, *JHP* 1:264.

91. Joseph Henry, "On a Disturbance of the Earth's Magnetism in Connexion with the Appearance of an Aurora Borealis, as Observed at Albany, April 19th, 1831," Regents of the State of New York, *Annual Report, 1832,* 107–19; *Amer. Journ. Sci.* 22 (1832): 143–55; *JHP* 1:213, 213n, 310. See also Middleton, *Invention of the Meteorological Instruments,* 81–132.

92. Henry to John Maclean, June 28, 1832, *JHP* 1:435–36.

93. Bache (1806–67), *DSB;* Merle M. Odgers, *Alexander Dallas Bache: Scientist and Educator, 1806–1867* (Philadelphia, 1947); Joseph Henry, "Memoir of A. D. Bache, 1806–1867," *Natl. Acad. Sci., Biog. Mem.* 1 (1877): 181–212d; American Philosophical Society, "Commemoration of the Life and Work of Alexander Dallas Bache," *Amer. Phil. Soc., Proceedings* 84 (1941): 124–86; Henry to Bache, Mar. [19–27], 1839, *JHP* 4:186.

94. Henry to Loomis, Apr. 22, 1847, Loomis Papers; W. B. Taylor, "The Scientific Work of Joseph Henry," in *A Memorial of Joseph Henry* (Washington, D.C., 1880), 205–425. For Henry's investigations in meteorology, primarily in atmospheric electricity, ibid., 258–63 and

320–21; for his activities as an administrator of the Smithsonian meteorological project, ibid., 286–90.

95. Nathan Reingold, "Cleveland Abbe at Pulkowa: Theory and Practice in the Nineteenth Century Physical Sciences," *Archives internationales d'histoire des sciences* 17 (1964): 144–45.

96. Nathan Reingold, "The New York State Roots of Joseph Henry's National Career," *New York History* 54 (1973): 133–44.

97. Hindle, *Science in Revolutionary America,* 184.

Chapter 2. The American Storm Controversy, 1834–1843

1. Redfield (1789–1857), *DSB,* is treated in his son's autobiography, John Henry Redfield, *Recollections of John Howard Redfield* (Philadelphia, 1900), and in Denison Olmsted, "Biographical Memoir of William C. Redfield," *Amer. Journ. Sci.* 2d ser, 24 (1857): 355–73, which contains a bibliography. On his transportation business, see Redfield Family Papers.

Espy (1785–1860), *DSB;* short autobiographical sketch, MS in the Joseph LeConte Papers, Amer. Phil. Soc. Library; L. M. Morehead, *A Few Incidents in the Life of Professor James P. Espy* (Cincinnati, 1888); Bache's necrology of Espy is in *Smithsonian Report, 1859,* 108–11; another necrology is in *Dial* 1 (Cincinnati, Feb. 1860): iii–vi, 102.

Hare (1781–1858), *Amer. Journ. Sci.* 11 (1826): 59–66; Edgar F. Smith, *The Life of Robert Hare, an American Chemist, 1781–1858* (Philadelphia, 1917).

On the storm controversy see Charles H. Davis, "Redfield, Reid, Espy and Loomis on the Theory of Storms," *North American Review* 58 (1844): 335–71; William Morris Davis, "The Redfield and Espy Period," *U.S. Weather Bureau, Bulletin* 11 (1893): 305–16; David M. Ludlum, "The Espy-Redfield Dispute," *Weatherwise* 22 (1969): 224–29, 245, 261; and Ludlum, *Early American Tornadoes, 1586–1870* (Boston, 1970), 192–94.

2. Martin Rudwick, *The Great Devonian Controversy* (Chicago, 1985), xxi.

3. Walker (1805–53), Olmsted (1791–1859), and Davis (1807–77), all *BDAS;* Peirce (1809–80), Herschel (1792–1871), Arago (1786–1853), Babinet (1794–1872), and Pouillet (1790–1868), all *DSB;* Reid (1791–1858), *DNB.*

4. George H. Daniels, *American Science in the Age of Jackson* (New York, 1968), 92ff.

5. Gregg De Young, "The Storm Controversy (1830–1860) and Its Impact on American Science," *Eos: American Geophysical Union, Transactions* 66 (1985): 657–60.

6. Richard Yeo, "An Idol of the Marketplace: Baconianism in Nineteenth Century Britain," *History of Science* 23 (1985): 282; *JHP* 3:475n. Brewster's anti-Baconianism is spelled out in his *Memoirs on the Life, Writings, and Discoveries of Sir Isaac Newton* (1855; rpt. New York, 1965), 2:400–406.

7. Henry, "Visit to Melrose Abbey," *JHP* 3:475–77.

8. E.g., Elias Loomis, William Ferrel, and Cleveland Abbe (Gisela Kutzbach, *The Thermal Theory of Cyclones: A History of Meteorological Thought in the Nineteenth Century* [Boston, 1979]).

9. Joseph Henry, "Syllabus of a Course of Lectures on Physics," *Smithsonian Report, 1856,* 187–91.

10. See Reid's Correspondence, and O. M. Blouet, "Sir William Reid, F.R.S., 1791–1858: Governor of Bermuda, Barbados, and Malta," *Notes and Records of the Royal Society of London* 40 (1985–86): 169–91.

11. William Redfield, "Summary Statements of Some of the Leading Facts in Meteorology," *Amer. Journ. Sci.* 25 (1834): 122.

12. On Espy's caloric theory, see James E. MacDonald, "James P. Espy and the Beginnings of Cloud Thermodynamics," *Amer. Metl. Soc., Bulletin* 44 (1963): 634–41; and Elizabeth Garber, "Thermodynamics and Meteorology (1850–1900)," *Annals of Science* 33 (1976): 53–

55, accurate except for Espy's "clash" with Loomis; on Espy's legacy see Kutzbach, *Thermal Theory;* on his rainmaking see Clark C. Spence, *The Rainmakers: American "Pluviculture" to World War II* (Lincoln, Neb., 1980), 9–21.

13. Excess vapor will produce rain, frozen rain will produce hail, and excess steam power will produce tornadoes. See James P. Espy, *The Philosophy of Storms* (Boston, 1841), viii–ix. Earlier accounts are Espy, "Theory of Rain, Hail, Snow, and the Waterspout, Deduced from the Latent Caloric of Vapour and the Specific Caloric of Atmospheric Air," *Geological Society of Pennsylvania, Transactions* 1, pt. 2 (Philadelphia, 1835): 342–43; Espy, "Essays on Meteorology," Nos. I and II. "Theory of Hail," *Journ. Frankl. Inst.* n.s. 17 (1836): 240–46, 309–16; No. III. "Examination of Hutton's, Redfield's and Olmsted's Theories," No. IV. "North East Storms, Volcanoes, and Columnar Clouds," *Journ. Frankl. Inst.* n.s. 18 (1836): 100–108, 239–46; and "Brief Synopsis of the Principles of James P. Espy's Philosophy of Storms," *Amer. Journ. Sci.* 39 (1840): 120–23.

14. On Dalton and Gay-Lussac, see W. E. Knowles Middleton, *A History of the Theories of Rain and Other Forms of Precipitation* (New York, 1965), 132–46, although he misses Espy's earlier influence on Ferrel. On Brandes (1777–1834), see Schneider-Carius, *Weather Science, Weather Research,* 179–87; and Heinrich W. Brandes, *Beiträge zur Witterungskunde* (Leipzig, 1820).

15. Michael T. Meier, "Caleb Goldsmith Forshey, 1812–1881: Engineer of the Old Southwest" (Ph.D. dissertation, Memphis State University, 1982).

16. Redfield to Forshey, Sept. 12, 1840, Redfield Letterbooks.

17. Forshey, "Natchez Tornado of May 7, 1840," in Espy, *Philosophy of Storms,* 337–45. Redfield hints that Espy may have lectured in Natchez in 1840 (Redfield to Matthew Henry Webster, June 30, 1840, Redfield Letterbooks).

18. Redfield to Forshey, Sept. 12, 1840, Redfield Letterbooks.

19. Forshey's Journal (#6A), Cover Inscription: "Philadelphia August 1840 to October 6-1840," Mississippi Valley Collection, Memphis State University, Memphis, Tenn.

20. Robert Hare, "Additional Objections to Redfield's Theory of Storms," *Amer. Journ. Sci.* 43 (1842): 138–39.

21. Robert Hare, "On the Causes of the Tornado, or Water Spout," *Amer. Phil. Soc., Transactions* n.s. 5 (1837): 376.

22. Robert Hare, "Objections to Mr. Redfield's Theory of Storms with Some Strictures upon His Reasoning," *Amer. Journ. Sci.* 42 (1841): 140–47; *Phil. Mag.* 23 (1843): 92–106; *Amer. Phil. Soc., Proceedings* (1841–43). An earlier version is Hare, "On the Theory of Storms, with Reference to the Views of Mr. Redfield," *Phil. Mag.* 19 (1841): 423–32.

23. Hutton (1726–97); James Hutton, "The Theory of Rain," *Royal Society of Edinburgh, Transactions* 1 (1788): 41–86.

24. Robert Hare, "On the Gales Experienced in the Atlantic States of North-America," *Amer. Journ. Sci.* 5 (1822): 352–56. See also Hare, "On the Causes of the Tornado," 380.

25. Hare, "Additional Objections to Redfield," 138.

26. *Amer. Phil. Soc., Transactions* n.s. 5 (1837): 375–84, 407–20, 421–26; *Journal of the Philadelphia Academy of Natural Sciences* 7 (1834–37): 269–81; *Amer. Journ. Sci.* 35 (1839): 207.

27. *Amer. Phil. Soc., Early Proceedings,* 674–75, 687.

28. James P. Espy, "Deductions from Observations Made, and Facts Collected on the Path of the Brunswick Spout of June 19th, 1835," *Amer. Phil. Soc., Transactions* n.s. 5 (1837): 423.

29. Bache to Hare, Mar. 17, 1836, Edgar Fahs Smith Memorial Collection in the History of Chemistry, Special Collections, Van Pelt Library, University of Pennsylvania.

30. A. D. Bache, "Notes and Diagrams, Illustrative of the Directions of the Forces Acting

at or Near the Surface of the Earth, in Different Parts of the Brunswick Tornado of June 19, 1835," *Amer. Phil. Soc., Transactions* n.s. 5 (1837): 407–20.

31. Hare, "On the Causes of the Tornado," 375–81. The facts are presented in Hare, "Facts and Observations Respecting the Tornado Which Occurred at New Brunswick, New Jersey, in June Last, Abstracted from a Written Statement made by James P. Espy, M.A.P.S.," *Amer. Phil. Soc., Transactions* n.s. 5 (1837): 381–84.

32. Section titled "New Jersey Tornado of 1835" in William Redfield, "On the Courses of Hurricanes; with Notices of the Tyfoons of the China Sea, and Other Storms," *Amer. Journ. Sci.* 35 (1839): 207.

33. Henry to Bache, July 9, 1840, *JHP* 4:417–18.

34. Elias Loomis, "Observations on a Hurricane which Passed over Stow, in Ohio, October 20th, 1837," *Amer. Journ. Sci.* 33 (1838): 371. Loomis (1811–89); *DSB*. See also H. A. Newton, "A Memorial to Professor Elias Loomis," *Amer. Metl. Journ.* 7 (1890): 97–117; Newton, *Elias Loomis, 1811–1889: Memorial Address* (New Haven, 1890); and Newton, "Biographical Memoir of Elias Loomis, 1811–1889," *Natl. Acad. Sci., Biog. Mem.* 3 (1895): 213–52.

35. Espy to Loomis, Apr. 30, 1842, Loomis Papers.

36. Elias Loomis, "On a Tornado which Passed over Mayfield, Ohio, February 4th, 1842, with Some Notices of Other Tornadoes," *Amer. Journ. Sci.* 43 (1842): 278–300.

37. James P. Espy, "Notes of an Observer—Meteorology," *Journ. Frankl. Inst.* n.s. 13 (1834): 9–11.

38. James P. Espy, "Observations on the Importance of Meteorological Observations Particularly as Regards the Dew Point; and Also on the Several Fluctuations of the Barometer," *Journ. Frankl. Inst.* n.s. 8 (1831): 389–406.

39. William Redfield, "On the Evidence of Certain Phenomena," *Journ. Frankl. Inst.* n.s. 15 (1835): 372–79; *Amer. Journ. Sci.* 28 (1835): 310–18.

40. Espy, "Essays on Meteorology," No. I, 240–46.

41. Espy, "Essays on Meteorology," No. III, 100–108; Denison Olmsted, "Of the Phenomena and Causes of Hail Storms," *Amer. Journ. Sci.* 18 (1830): 1–11.

42. Espy, "Essays on Meteorology." No. III, 107–8.

43. William Redfield, "In Reply to Mr. Espy, on the Whirlwind Character of Certain Storms," *Journ. Frankl. Inst.* n.s. 19 (1837): 112–27.

44. Reid to Redfield, Feb. 1, 21, 1838, Reid's Correspondence.

45. Redfield to Reid, Oct. 30, 1838, ibid.

46. Redfield to Reid, Mar. 26, Oct. 30, 1838, Aug. 25, 1841, April 9, 1838; Reid to Redfield, May 24, 1838, ibid.

47. Reid to Redfield, Aug. 20, Oct. 11, 1838, ibid; William Reid, "A Report Explaining the Progress Made towards Developing the Law of Storms; and of What Seems Further Desirable to Be Done, to Advance Our Knowledge of the Subject," *Athenaeum* (1838): 594–95.; *BAAS Annual Report, 1838* pt. 2, *Notices and Abstracts of Communications* (London, 1839), 21–25.

48. "An Account of the Proceedings of the Eighth Meeting of the British Association for the Advancement of Science," *Amer. Journ. Sci.* 35 (1839): 275–321.

49. Reid, "Report Explaining the Progress," 594–95.

50. Redfield to Reid, Oct. 2, 1838, Reid's Correspondence.

51. James P. Espy, "Examination of Col. Reid's Work on the Law of Storms," *Journ. Frankl. Inst.* n.s. 23 (1839): 38–50, 149–58, 217–31, 289–98, and Espy, "Objections to Mr. Espy's Theory of Rain, Hail, &c., in a Letter to Mr. Espy, by Mr. Graham Hutchinson of Glasgow, with Replies by James P. Espy," ibid., 80–90.

52. Espy, "Col. Reid's Work," 38.

53. William Redfield, "Remarks on the Prevailing Storms of the Atlantic Coast, of the North American States," *Amer. Journ. Sci.* 20 (1831): 39; Reid to Redfield, Feb. 1, 1838, Reid's Correspondence.

54. Reid to Redfield, Dec. 27, 1838, Reid's Correspondence.

55. Davis, "Redfield, Reid, Espy and Loomis on the Theory of Storms," 344–54.

56. Espy to Redfield, July 18, 1836, Redfield Papers. Joseph Henry accepted Espy's invitation (Espy to Henry, July 19, 1836, *JHP* 3:77–79).

57. Espy to Redfield, Nov. 5, 1839, Redfield Papers; Redfield to Espy, Nov. 7, 1839, Redfield Letterbooks.

58. Redfield to Henry, May 31, 1837, *JHP* 3:367–71.

59. Redfield to Henry, June 3, 1839, *JHP* 4:229–30.

60. Redfield asked Reid to insert "*New Jersey Tornado* of 1835 & *Test of Mr Espy's Theory* in the Athenaeum and the Literary Gazette," noting, "Some of my friends appear to think this notice too lenient" (Redfield to Reid, Nov. 14, 1838, Reid's Correspondence). The American article was Redfield, "On the Courses of Hurricanes."

61. Espy, "Col. Reid's Work," 229–30.

62. William Redfield, "Remarks on Mr. Espy's Theory of Centripetal Storms, Including a Refutation of His Positions Relative to the Storm of September 3rd, 1821: with Some Notice of the Fallacies which Appear in His Examination of Other Storms," *Journ. Frankl. Inst.* n.s. 23 (1839): 323–36, 363–78.

63. William Redfield, "Further Notice of the Fallacies of Mr. Espy's Examination of Storms," *Journ. Frankl. Inst.* n.s. 24 (1839): 1–4.

64. Redfield, "Remarks on Mr. Espy's Theory," 326; Espy, *Philosophy of Storms*, 249–53.

65. Other texts in Henry's library also bear the clear mark of Espy's hand. See the Bell-Henry Library copy of William Reid, *The Progress of the Development of the Law of Storms and of the Variable Winds* (London, 1849), especially the wind arrows in the maps, redrawn in pencil to illustrate Espy's centripetal storm theory; and Robert Russell, *North America, Its Agriculture and Climate* (Edinburgh, 1857), marginalia on pp. 304, 306, 315, and 333.

66. Donald M. Scott, "The Popular Lecture and the Creation of a Public in Mid-Nineteenth-Century America," *Journal of American History* 66 (1980): 791–809; Carl Bode, *The American Lyceum: Town Meeting of the Mind* (New York, 1956); and Cecil B. Hayes, *The American Lyceum: Its History and Contribution to Education* (Washington, D.C., 1932); Spence, *Rainmakers*.

67. *Pennsylvania Senate Journal* (1837–38), 1:197.

68. "Memorials of James P. Espy and of the Philadelphia Lyceum, Praying the Aid of the General Government in Making a Series of Meteorological Observations, May 2, 1838," U.S. Senate, Doc. 414, 25th Cong., 2d sess. See also "Memorial from the Pennsylvania Lyceum, etc. Asking an Appropriation to Aid in the Advancement of Meteorology, Apr. 20, 1838," U.S. House, Doc. 344, 25th Cong., 2d sess. Signatories include John Locke, Walter R. Johnson, Henry Vethake, Richard Harlan, Paul B. Goddard, Isaiah Lukins, Joseph Henry, Roswell Park, Charles D. Meigs; and a letter of support signed by Benjamin Peirce, Daniel Tredwell, John Ware, J. W. Webster, Nathaniel Bowditch, Joseph Hopkinson, Peter S. DuPonceau, and N. Chapman.

69. See *Pennsylvania Senate Journal* (1838–39), 1:20, 56, 243; *Journal of the House of Representatives, Pennsylvania* (1838–39), 1:106; and (1840), 1:802.

70. *Journal of the House of Representatives, Pennsylvania* (1840), 2, pt. 2, 273–75.

71. Ibid. (1838–39), 1:50; and (1840), 1:1112.

72. *Genesee Farmer*, Mar. 3, 1839, p. 99.

73. Bache to Humphrey Lloyd, May 4, 1839, Henry Papers.

74. Walker to W. H. White, July 10, 1839, Royal Meteorological Society Papers, Harry

Ransom Humanities Research Center, University of Texas at Austin.

75. Henry to Redfield, Dec. 17, 1838, *JHP* 4:165.

76. Redfield to Reid, April 17, 1839, Reid's Correspondence.

77. Redfield to Reid, June 25, Nov. 25, Dec. 9, 1839, ibid.

78. "Espy's Theory of Storms," from the *New York Express* and the *Whig*, quoted in *Niles' National Register* 56 (June 22, 1839): 268; *Nantucket Inquirer*, July 13, 1839; Spence, *Rainmakers.*

79. *Bangor Advertizer*, cited in Spence, *Rainmakers*, 12; *Colonial Sentinel*, Jan. 4, Feb. 5, 1840.

80. Denison Olmsted, ["Ten Lectures on Meteorology"], *Merchants' Mag.* 1 (1839): 535.

81. James P. Espy, "Notes of an Observer. Remarks on Professor Olmsted's Theory of the Meteoric Phenomenon of November 12th, 1833, Denominated Shooting Stars, with Some Queries toward Forming a Just Theory," *Journ. Frankl. Inst.* n.s. 15 (1835): 90; Espy, "Essays on Meteorology," No. III, 100–108; and Espy, "Professor Olmsted's Objections, with Replies," *Philosophy of Storms*, 429–44.

82. *Boston Evening Mercantile Journal*, Jan. 21, 23, 1840; R. T. Paine to Redfield, Jan. 1, 1840, Redfield Papers.

83. Redfield to Reid, Nov. 25, 1839, Reid's Correspondence.

84. Jack Morrell and Arnold Thackray, *Gentlemen of Science: The Early Years of the BAAS* (Oxford, 1981), 416.

85. Espy to Bache, Sept. 18, 1836, Rhees Collection.

86. Dalton (1766–1844), *DSB;* quote from Bache's European Diary, Oct. 27, 1837, Bache Papers; John Dalton, *Meteorological Observations and Essays*, 2d ed. (London, 1834).

87. Henry's tour of Britain, France, and Belgium is documented in *JHP* 3 and Bruce Sinclair, "Americans Abroad: Science and Cultural Nationalism in the Early Nineteenth Century," in Nathan Reingold, ed., *The Sciences in the American Context: New Perspectives* (Washington, D.C., 1979), 35–53.

88. Henry to Forbes, June 30, 1834, *JHP* 2:204–5; James David Forbes, "Report on the Recent Progress and Present State of Meteorology," *BAAS, Report of the First and Second Meetings* (London, 1833), 196–258, esp. 217; see Henry to Forbes, June 7, 1836, *JHP* 3:74.

89. Harriet Henry to Espy, *JHP* 3:161.

90. Henry's European Diary, [Mar.] 23, [1837], ibid., 190, 192–93.

91. Ibid., Apr. 18, 1837, p. 286; Aug. 2, 1837, p. 435.

92. Babbage (1792–1871), *DSB; Amer. Journ. Sci.* 11 (1826): 59–66. Henry's European Diary, Apr. 3, 1837, *JHP* 3:226. See also *JHP* 1:342–43.

93. Magendie (1783–1855) and Gay-Lussac (1788–1850), both *DSB*. Babinet was *rapporteur* of the French Academy's report on Espy's theory of storms. Arago and Pouillet were the other members of the committee.

94. Melloni (1798–1854) was an exiled Italian physicist studying calorific radiation; *JHP* 3:394–95n. Henry, "Oral Communication on the Application of Melloni's Thermo-electric Apparatus to Meteorological Purposes, Nov. 3, 1843," *Amer. Phil. Soc., Proceedings* 4 (1847): 22, and Henry, "On Heat, and on a Thermal Telescope," paper read to the Association of American Geologists and Naturalists, reported in *Amer. Journ. Sci.* 2d ser. 5 (1848): 113–14.

95. Adolphe Quetelet (1796–1874), *DSB; Amer. Journ. Sci.* 11 (1826): 59–66. *JHP* 3:417.

96. Adie (1774–1858), *JHP* 3:440.

97. Ibid., 441.

98. Henry, "Visit to Melrose Abbey," ibid., 475–77.

99. Ibid., 473–77., esp. n. 15. Brewster, *Life of Newton*, 2:400–406.

100. Henry to Henry James, Aug. 22, 1843, *JHP* 5:387. See also Henry, "Syllabus of a Course of Lectures on Physics."

101. John Cawood, "Terrestrial Magnetism and the Development of International Collaboration in the Early Nineteenth Century," *Annals of Science* 34 (1977): 581, 586; Walter F. Cannon, "Humboldt or Baconianism? A Defense of American Science in the First Half of the 19th Century," paper presented at the History of Science Society, Dec. 1969, filed as item 231 in the reprint file of the Joseph Henry Papers, Smithsonian Institution. Significant extracts appear in Cannon, "Humboldtian Science," in *Science in Culture: The Early Victorian Period* (New York, 1978), 73–110. See also *JHP* 2:106.

102. *JHP* 3:430, 438, 237, 422, 507; see Morrell and Thackray, *Gentlemen of Science*, 417.

103. Dove (1803–79), *DSB*. Henry to Quetelet, May 1, 1844, *JHP* 5:23–24, 235. Other scientists receiving reports were James Clark Ross, *JHP* 4:236; Hans Christian Oersted, *JHP* 5:28; and J. J. Berzelius, *JHP* 5:30.

104. Taylor, "Scientific Work of Henry," 260; [Humboldt], "Letter from Baron von Humboldt," *Phil. Mag.* 3d ser. 9 (1836): 42–53; see *JHP* 3:303–4n. Herschel to the Albany Academy, July 1839, Records of the Albany Academy, Albany, N.Y.

105. Adams, "Magnetic and Meteorological Observations," U.S. House of Representatives, Rept. 630, 26th Cong., 1st sess.

106. Charles A. Schott, "Magnetic Survey of North America," *U.S. Weather Bureau, Bulletin* 11 (1893): 460–64; A. D. Bache, *Observations at the Magnetic and Meteorological Observatory, at the Girard College, Phila.* (Washington, D.C., 1847).

107. *Niles' National Register* 58 (June 13, 1840): 240.

108. The two papers were "On Storms," discussed below, and "On the Four Daily Fluctuations of the Barometer," report from the *Athenaeum* reprinted in Espy, *Philosophy of Storms*, 543–44. Joseph Henry, "Report of the Tenth Meeting of the British Association for the Advancement of Science, Held at Glasgow, Sept. 1840," *Biblical Repertory and Princeton Review* 13 (1841): 132–49; and J. Babinet, "Rapport sur les travaux de M. Espy relatifs aux tornados, lu a l'Academie des Sc. de Paris dans sa seance du 15 Mars 1841, par M. Babinet, rapporteur. Comm. MM. Arago, Pouillet, et Babinet," *Comptes rendues hebdomadaires de l'Academie des Sciences* 12 (1841): 34, 139, 454–62; translated and reprinted in Espy, *Philosophy of Storms*, Introduction, xxxi–xxxix. The rest of the documentation from this tour is sketchy; e.g., Faraday to Espy, Dec. 3, 1840, Osborn Files, Beinecke Rare Book and Manuscript Library, Yale University, and Espy to Quetelet, Dec. 26, 1840, Quetelet Correspondence.

109. Henry, "Report of the Tenth Meeting," 143.

110. James P. Espy, "On Storms," *Athenaeum* (1840): 794–98; also *BAAS, Annual Report, 1840*, pt. 2, *Notices and Abstracts of Communications* (1841): 30–40; reprinted in Espy, *Philosophy of Storms*, Introduction, v–xxxi. Stevelly (ca. 1795–ca. 1867), Forbes (1809–68), Smith (unidentified), Osler (1798–1863), Phillips (1800–1874).

111. Henry, "Report of the Tenth Meeting," 148.

112. Reid to Redfield, Nov. 7, 1840, Reid's Correspondence.

113. *Niles' National Register* 60 (May 1, 1841): 144.

114. Ibid. (Apr. 17, 1841): 112.

115. Morehead, *Life of Espy*, 17.

116. François Arago, *Oeuvres complètes* (Paris, 1859), 12:280–86; author's translation.

117. Espy, *Philosophy of Storms*, xxxi–xxxix, quote from xxxviii–xxxix.

118. Robert Hare, *Of the Conclusion Arrived at by a Committee of the Academy of Sciences of France . . . with Objections to the Opinions of Peltier and Espy* (Philadelphia, 1842; rev. 1852). The dispute, is mentioned in Hare's manuscript "For the New York Herald," n.d., Hare Papers.

119. *Hare's Letter to Arago, April 24, 1840*, Pamphlet Collection, Bell-Henry Library, Smithsonian Institution.

120. Bache to Redfield, Oct. 22, 1841, Redfield Papers.

121. Hare to Arago, Sept. 16, 1841, Hare Papers; author's translation.

122. Hare to Redfield, Nov. 28, 1841, Redfield Papers.

123. Redfield to the Secretary of the Academy of Sciences at Paris, for Messrs Babinet, Arago and Pouillet, Committee of the Academy, n.d. [follows the entry for Nov. 25, 1841], Redfield Letterbooks.

124. Hare to Redfield, Nov. 24, 1841, Redfield Papers; Redfield to Hare, Nov. 25, 1841, Redfield Letterbooks.

125. Hare, "Objections to Mr. Redfield," 140–47.

126. Redfield, "Reply to Dr. Hare's Objections to the Whirlwind Theory of Storms," *Amer. Journ. Sci.* 42 (1842): 299–316, expanded and issued as a pamphlet: Redfield, *On Whirlwind Storms; with Replies to the Objections and Strictures of Dr. Hare* (New York, 1842).

127. Redfield, "Reply to Dr. Hare," 299.

128. Hare, "Additional Objections to Redfield," 122–40.

129. Espy to Loomis, Jan. 11, 1843, Loomis Papers; Espy to Quetelet, Nov. 27, 1842, Quetelet Correspondence.

130. Espy to Loomis, Feb. 16, 1843, Loomis Papers.

131. Benjamin Peirce, "On Espy's Theory of Storms," *Cambridge Miscellany of Mathematics, Physics and Astronomy,* no. 3 (Oct. 1842): 141–44. Cf. Davis, "Redfield, Reid, Espy and Loomis on the Theory of Storms," 335–71.

132. Peirce, "On Espy," 144.

133. Espy to Loomis, Dec. 29, 1842, Jan. 11, Mar. 4, 1843, Loomis Papers.

134. H. W. Dove, "On the Law of Storms," *Cambridge Miscellany of Mathematics, Physics and Astronomy,* no. 2 (July 1842): 94–96, and no. 3 (Oct. 1842): 138.

Chapter 3. Observational Horizons in the 1830s and 1840s

1. Armand N. Spitz, "Meteorology in the Franklin Institute," *Journ. Frankl. Inst.* 237 (1944): 271–87, 331–57; [Frankl. Inst.], "Minutes of the Board of Managers," Sept. 21, 1842, Frankl. Inst. Archives, Philadelphia.

2. Henry to Bache, Nov. 25, 1835; *JHP* 2:478–82. Espy, "Remarks on Olmsted's Theory," 9–19, 85–92, 158–65, 234–38.

3. Espy, "Remarks on Olmsted's Theory," 90; Espy, "On the Importance of Hygrometric Observations in Meteorology, and the Means of Making Them with Accuracy." *Journ. Frankl. Inst.* n.s. 7 (1831): 221–29, 361–64; Henry to James Henry, Dec. 25, 1835, *JHP* 2:489; see also *JHP* 2:239–40.

4. Herrick to Redfield, Apr. 2, 1839, Edward Claudius Herrick Papers, Manuscripts and Archives, Yale University Library. Herrick is referring to Bache's paper on the progress of meteorology in the United States, commissioned by the BAAS but never completed.

5. Elias Loomis, "Memorial of a Committee of the Western Literary Institute and College of Professional Teachers—To the Honorable the General Assembly of the State of Ohio," clipping from the *Hudson Observer,* Jan. 24, 1842, Local Meteorology, vol. 2, United States, "Scrap Book," RG 27, NA. See also G. Emerson to Loomis, Aug. 24, 1841, Loomis Papers.

6. Loomis, "Memorial to the State of Ohio."

7. Loomis to Peirce, Nov. 3, 1845, Benjamin Peirce Papers, quoted by permission of the Houghton Library, Harvard University.

8. Espy, "On the Importance of Hygrometric Observations," 221–29, 361–64.

9. Espy, "Observations on the Importance of Meteorological Observations," 389.

10. *JHP* 3:369n.

11. Gross limits his analysis to the published record in his article "The American Philosophical Society and the Growth of Meteorology in the United States, 1835–1850," *Annals of*

Science 29 (1972): 321–38; the Espy manuscript, one of many, is entitled "Meteorology: James P. Espy to Committee of the American Philosophical Society, Oct. 17, 1836," MS, Amer. Phil. Soc. Library, Philadelphia.

12. "Circular of the Joint Committee on Meteorology of the American Philosophical Society and Franklin Institute," Mar. 20, 1838, copy in Amer. Phil. Soc. Library, Philadelphia. Another copy, addressed to Senator Nathaniel P. Tallmadge of New York, is in the collections of the Albany Institute of History and Art. Pages 72 to 172 in Espy's *Philosophy of Storms,* entitled "Labors of the Joint Committee," reproduces the circular and subsequent reports.

13. Joint Committee, "Circular in Relation to Meteorological Observations, by the Joint Committee of the American Philosophical Society, and of the Franklin Institute," *Journ. Frankl. Inst.* n.s. 14 (1834): 382–85.

14. Franklin Institute, Committee on Meteorology, "Report of the Committee on Meteorology to the Board of Managers of the Franklin Institute, Embodying the Facts Collected by the Meteorologist Relative to the Storm of the 16th, 17th and 18th March, 1838," *Journ. Frankl. Inst.* n.s. 22 (1838): 161.

15. Bruce Sinclair, "Gustavus A. Hyde, Professor Espy's Volunteers, and the Development of Systematic Weather Observation," *Amer. Metl. Soc., Bulletin* 46 (1965): 779–84.

16. Joint Committee, "First Report of the Joint Committee of the American Philosophical Society, and Franklin Institute, on Meteorology," *Journ. Frankl. Inst.* n.s. 16 (1835): 4.

17. Joint Committee, "Second Report of the Joint Committee on Meteorology, of the American Philosophical Society and Franklin Institute," *Journ. Frankl. Inst.* n.s. 17 (1836): 386, 391. See Bruce Sinclair, *Philadelphia's Philosopher Mechanics: A History of the Franklin Institute, 1824–1865* (Baltimore, 1974), 152–59.

18. Spitz, "Meteorology in the Franklin Institute," 281.

19. Joint Committee, "Third Report of the Joint Committee on Meteorology, of the American Philosophical Society and Franklin Institute of the State of Pennsylvania, for the Promotion of the Mechanic Arts," *Journ. Frankl. Inst.* n.s. 19 (1837): 17–21.

20. Espy to Bache, Sept. 18, 1836, Rhees Collection. See also "Fifty-fourth Quarterly Report of the Board of Managers of the Franklin Institute," *Journ. Frankl. Inst.* n.s. 20 (1837): 130.

21. Espy to Henry, Dec. 24, 1837, *JHP* 3:535, 535–36n.

22. Emerson to Loomis, Nov. 24, 1841, Loomis Papers; Gouverneur Emerson, "The Part Taken by the American Philosophical Society and Franklin Institute in Establishing Stations for Meteorological Observations," *Amer. Phil. Soc., Proceedings* 11 (1871): 516–20.

23. Emerson to Loomis, Aug. 24, 1841, Loomis Papers. A reproduction of one of the early forms printed for the observers is found in Spitz, "Meteorology in the Franklin Institute," 284. Michael Jacobs (1801–71), *DAB,* Lutheran clergyman and educator, made observations for the Franklin Institute at Gettysburg, Pennsylvania, from 1839 to 1865. One series was made for the Smithsonian Institution. His observations are in the Amer. Phil. Soc. Library. See Catlett, ed., *New Guide,* entry 567.

24. Espy, "Directions.—By the Joint Committee," rpt. in *Philosophy of Storms,* 146–48.

25. James P. Espy, *Hints to Observers on Meteorology* (Philadelphia, 1837), rpt. in *Philosophy of Storms,* 148–60.

26. Franklin Institute, Committee on Meteorology, "Report," 161–75; Franklin Institute, Board of Managers, "Fifteenth Annual Report of the Board of Managers of the Franklin Institute, of the State of Pennsylvania, for the Promotion of the Mechanic Arts," *Journ. Frankl. Inst.* n.s. 23 (1839): 93.

27. Emerson to Loomis, Nov. 24, 1841, Loomis Papers.

28. Marine T. McChandler to Henry, Mar. 29, 1849, SIMC-NA.

29. Espy, *Philosophy of Storms,* 169–72.

30. [Navy Department], "*Circular* to Commanders of Squadrons and Stations, March 27, 1834," Navy Department, General Orders and Circulars, 1798–1862, M997, reel 1, NA.

31. Ruschenberger to Mahlon Dickerson, Secretary of the Navy, Nov. 20, 1834; Ruschenberger to Dr. B. Washington, Surgeon of the Medical Bureau of the Navy, Nov. 20, 1834, Naval RC.

32. Washington to [Naval Surgeons], Dec. 10, 1834, and Surgeon G. S. Sproston to Washington, Dec. 27, 1834, Naval RC.

33. Surviving navy reports are on file in the Naval RC. Stations, observers, dates, and number of reports are as follows: Navy Yards: Baltimore Naval Station, Dr. Sproston, 1834, 3 months; Boston Navy Yard, 1835, 2 months; Gosport Navy Yard, 1834–35, 10 months; New York Navy Yard, 1834–37, 25 months; Norfolk Navy Yard, 1835, 1 month; Pensacola Navy Yard, Dr. Edwards, 1835, 6 months, 1838, 3 months; Portsmouth Navy Yard, Dr. Plumstead, 1834–35, 3 months; Depot of Naval Instruments, Washington, D.C., Charles Wilkes, 1834–37, 34 months; Washington Navy Yard, Isaac Hull, 1834, 3 months. Ships at Sea: USS *Erie* along the coast of Brazil (Bahia, Montevideo, Rio de Janiero), Fleet Surgeon Chase, 1835–36, 16 months; USS *Delaware* in the Mediterranean Sea (Gibraltar, Malta), 1835, 6 months; USS *Ohio* in the Mediterranean Sea, 1838–40, 16 months; USS *North Carolina* in the Pacific Ocean, Professor of mathematics Mr. Houton, 1839, 1 month.

34. The committee members were Matthew Henry Webster (1803/4–46), Philip Ten Eyck (1802–92), John V. L. Pruyn (Albany Academy, 1824), and Horace B. Webster (1811/12–43).

35. Herschel, "Report of the Meteorological Committee of the South African Literary and Scientific Institution," *Edinburgh New Philosophical Journal* 21 (1836): 239–46, quoted in *Albany Institute, Transactions* 2 (1833–52): 97. Locations and names of observers during term days between December 21, 1835, and December 21, 1836, are published in the *Transactions* of the Albany Institute, 1835–36. North American locations included Albany, Flushing, and New York City; Baltimore; Burlington, Vermont; Cincinnati; Cape Diamond and Montreal, Quebec; Gardiner, Maine; Middletown and New Haven, Connecticut; and Williamstown, Massachusetts. Observers included the meteorological committee of the Albany Institute; John S. McCord of the Natural History Society of Montreal; Professor C. Gill of the Institute at Flushing; Professor A. W. Smith at Wesleyan University; Professor John Locke at the Medical College of Ohio; William Redfield of New York City; J. Watt, superintendent of telegraphs in Canada; R. H. Gardiner of Gardiner, Maine; E. C. Herrick and J. D. Whelpley of New Haven; Professor A. Hopkins at Williams College; Mr. Benedict in Burlington, Vermont; and a committee of the Maryland Academy of Science and Literature.

36. Daniels, *Age of Jackson*, 86–101.

37. Elias Loomis, "On the Storm Which Was Experienced throughout the United States about the 20th of December, 1836," *Amer. Phil. Soc., Transactions* n.s. 7 (1840): 125–63; rev. version, *AAAS Proceedings* 9 (1856): 179–83; see Kutzbach, *Thermal Theory*, 27–31.

38. Hopkins (1807–72). James H. Coffin, "Meteorological Observations and Researches at Williams College, No. 1," *Berkshire County Whig* 1 (Nov. 25, 1841): 1. See "Old Wooden Observatory on Greylock," *Berkshire Hills*, July 1903, p. 157. A photo of the wooden base of a self-registering wind vane, constructed by James H. Coffin in Ogdensburg, New York, in 1837 is located in the Abbe Papers, JHU.

39. Coffin (1806–73), *DSB*; John C. Clyde, *The Life of James H. Coffin, LL.D.* (Easton, Pa., 1881); Arnold Guyot, "Biographical Memoir of James Henry Coffin, 1806–1843," *Natl. Acad. Sci., Biog. Mem.* 1 (1877): 257–64. Quote from Coffin to Loomis, Sept. 2, 1840, Loomis Papers.

40. Coffin, "Meteorological Observations and Researches at Williams College, No. 2," *Berkshire County Whig* 1 (Jan. 27, 1842): 1.

41. *Courier*, Dec. 20, 1841?, clipping in Elias Loomis's Scrapbook, II, Loomis Family

Papers; Harvard Meteorological Society to Redfield, May 12, 1842, Redfield Papers. Espy was lecturing in the Boston area in 1840 and his *Philosophy of Storms* was published in Boston in 1841.

42. Olmsted's printed lecture notes on meteorology, with marginalia, are in Sterling Library, Yale University.

43. Charles I. Weiner, "Joseph Henry's Lectures on Natural Philosophy: Teaching and Research in Physics, 1832–1847" (Ph.D. dissertation, Case Institute of Technology, 1965), 251.

44. The Redfield Papers and Letterbooks contain references to individual observers, e.g., V. W. Smith to Redfield, Mar. 17, 1832; theorists, e.g., Coffin to Redfield, Nov. 13, Dec. 16, 1845, Redfield to Loomis, Dec. 2, 1837; and institutions, e.g., Redfield to Matthew Henry Webster, Apr. 3, 1837 (re: the Albany Academy), Redfield to Surgeon General Lawson, Jan. 6, 1845 (re: the Army Medical Reports), and Redfield to Matthew Fontaine Maury, Jan. 6, 1845 (re: the Naval Observatory marine observations).

45. Redfield to James D. Dana, U.S. Exploring Expedition, Aug. 11, 1838, Redfield Letterbooks.

46. Redfield to Reid, Apr. 17, 1839, Reid's Correspondence; Redfield to Dove, May 8, 1840; Redfield to Piddington, Aug. 12, 1843; Darwin to Redfield, Feb. 24, 1840, Redfield Letterbooks.

47. Coffin to Loomis, Aug. 1, 1842, Loomis Papers.

48. James H. Coffin, "The Prevailing Winds in North America," paper read at the Association of American Geologists and Naturalists meeting in New Haven, 1845.

49. Clyde, *Life of Coffin,* 167–77; Coffin to Redfield, Sept. 1, 1846, Redfield Papers.

50. James H. Coffin, ed., *Meteorological Register and Scientific Journal* (Oswego, N.Y., 1838), monthly, 16 pages, 4to, $2 per annum; notice in *Amer. Journ. Sci.* 36 (1839): 165. No listing for this journal, the first specialized meteorological serial in America, was found in the National Union Catalog, Serial Record, or British Library Catalog.

51. Redfield to Coffin, June 20, 1839, Redfield Letterbooks; Coffin to Loomis, June 25, 1839, Loomis Papers.

52. Coffin to Loomis, Sept. 2, 1840, Loomis Papers.

53. S. G. Arnold, "The Philosophy of Storms," *Merchants' Mag.* 5 (Oct. 1841): 351–55.

54. Arago, *Oeuvres complètes,* 12:285–86; author's translation.

55. "Espy's Theory of Storms," *Merchants' Mag.* 6 (Mar. 1842): 272.

56. John Quincy Adams, *Memoirs of John Quincy Adams* (Philadelphia, 1876), 11:52–53.

57. Ibid., 65–66. Richard Wylly Habersham (1786–1842), representative from Georgia.

58. Appointed May 7, 1842; see *Niles' National Register* 62 (June 11, 1842): 240. Postage was a major expense for Coffin as well; see Clyde, *Life of Coffin,* 244.

59. Espy to Loomis, Apr. 30, 1842, Loomis Papers.

60. Espy, "Friends of Science," clipping in Elias Loomis's Scrapbook, II, Mar. 19, 1842, Loomis Family Papers.

61. Espy to Henry, July 2, 1842, *JHP* 5:232–34; also *JHP* 2:196n, and *JHP* 3:369–370n, 535–36.

62. "Espy's Conical Ventilator," from the *Richmond Enquirer,* Apr. 12, 1842, republished in *Niles' National Register* 62 (Apr. 16, 1842): 99–100. See also "Espy's Ventilator," from the *Pennsylvanian,* reprinted in *Niles' National Register* 63 (Oct. 15, 1842): 100. A brief discussion of Espy's ventilator appears in *JHP* 5:233n. See also "Report on Espy's Ventilator," U.S. House of Representatives, Rept. 531, 29th Cong., 1st sess.

63. "Mr. Espy. Prediction and Invention," from the *Richmond Enquirer,* May 3, 1842, republished in *Niles' National Register* 62 (May 7, 1842): 148.

64. U.S. Army Medical Department, *Meteorological Register for the Years 1826, 1827, 1828, 1829, and 1830, from Observations made by the Surgeons of the Army and Others at the Military Posts of the United States. To Which is Appended the Meteorological Register for the Years 1822, 1823, 1824, and 1825,* compiled by Samuel Forry (Philadelphia, 1840); U.S. Army Medical Department, *Statistical Report on the Sickness and Mortality of the Army of the United States, January, 1819, to January, 1839,* compiled by Samuel Forry (Washington, D.C., 1840); Forry, *The Climate of the United States and Its Endemic Influences* (New York, 1842).

65. Forry, *Climate of the United States,* vi, 142, 127.

66. J. N. Nicollet, *Essay on Meteorological Observations* (Washington, D.C., 1839); Thomas Gardner Mower, "Notice of the Meteorological Observations Now Making at the Military Posts of the United States," *Amer. Phil. Soc., Proceedings* 3 (1843): 158–62.

67. J. C. Spencer to Espy, Aug. 26, 1842, Naval RC; Reingold, *Science in Nineteenth-Century America,* 128.

68. Espy, "To the Friends of Science, Dec. 2, 1842," *Niles' National Register* 63 (Dec. 10, 1842): 228–29.

69. Espy, "To the Friends of Science, Dec. 6, 1842," *Niles' National Register* 64 (June 24, 1843): 263–64.

70. *Amer. Journ. Sci.* 44 (1843): 212.

71. Mower, "Notice of Observations," 158–62; Clarence J. Root, "The Meteorological Service One Hundred Years Ago," *Amer. Metl. Soc., Bulletin* 11 (1930): 74–77.

72. [Lawson], U.S. Army Surgeon General, *Annual Report, 1844,* U.S. Senate, Public Doc. 1, 28th Cong., 2d sess., 151–58 and table.

73. James P. Espy, *First Report on Meteorology to the Surgeon General of the United States Army* (Washington, D.C., 1843). Public distribution of the report was delayed until 1845, according to a notice in the *Daily National Intelligencer* 33 (July 18, 1845): 3. Further evidence of the delay is found in a footnote in the report dated February 24, 1845. The *American Almanac* published Espy's twenty generalizations from his first report in 1846: "Summary of Espy's First Report to Congress," *American Almanac* (1846), 86–87.

74. Redfield to Reid, Jan. 8, 1844, Reid's Correspondence.

75. Henry to Forbes, June 3, 1845, James David Forbes Papers, St. Andrews University Library, St. Andrews, Scotland.

76. Espy, "Report on Meteorology," in U.S. Army Surgeon General, *Annual Report, 1845,* U.S. Senate, Public Doc. 5, 29th Cong., 1st sess., 1:234–35; reproduced in Thomas C. Cochran, ed., *The New American State Papers: Science and Technology* (Wilmington, Del., 1973), 2:361–62.

77. [Espy], "Lecture on Meteorology," advertisement, *Daily National Intelligencer* 33 (Jan. 9, 1845): 3; Reingold, *Science in Nineteenth-Century America,* 129; Adams, *Memoirs,* 11: 506.

78. A. P. Upshur, Secretary of the Navy, to Espy, Mar. 6, 1843, Naval RC; Espy to Maury, Mar. 21, 1843, LR, NOR-NA.

79. Espy to George Bancroft, Secretary of the Navy, June 30, 1845, Naval RC. See E. K. Rawson, Library and Naval War Records, Navy Department to L. C. Ferrell, Superintendent of Documents (response to inquiry), Mar. 19, 1898, Naval RC: "Prof. Jas. P. Espy, whose reports are referred to in the foregoing letter, was employed to perform meteorological work in the War Department and was appointed by the Secretary of War under act of Congress of August 23, 1842. He was a Professor of Mathematics in the Navy, and served from 1842 to 1845 in that capacity. He was in the employ of the Navy Department as a meteorologist from 1848 to 1857."

Chapter 4. The Structure of the Smithsonian Meteorological Project

1. Henry to Coffin, Dec. 31, 1846, Henry Papers.

2. *Smithsonian Report, 1859;* "Extract from the Report of the Board of Officers Convened under the Direction of the Hon. Thomas Corwin, Secretary of the Treasury, on the Light House System of the United States Coast," *Amer. Journ. Sci.* 2d ser. 13 (1852): 318–36.

3. "Circular" issued Nov. 20, 1852, copy in Henry Papers. Responses to the circular are listed in "Diary Smithsonian," Smithsonian Lists and References, RG 27, NA. The responses, extending as far back as 1785, included observations taken by Thomas Jefferson and Josiah Meigs.

4. Henry, "Report of the Secretary for 1849," reprinted in *Smithsonian Report, 1853,* 160.

5. Joseph Henry, "Programme of Organization of the Smithsonian Institution, Dec. 8, 1847," *Smithsonian Report, 1848,* 6, 25.

6. Ibid., 25.

7. Loomis, "Report on the Meteorology of the United States," *Smithsonian Report, 1848,* App. 2, pp. 28, 30, 36, 41, 46; see also Loomis to Sabine, "On the Determination of Differences of Longitude Made in the United States by Means of the Electric Telegraph, and on the Projected Observations for Investigating the Laws of the Great North American Storms," *Phil. Mag.* 81 (Aug. 1847): 338–40.

8. Espy, "Extract from a Communication from Professor Espy on the Subject of Meteorology," *Smithsonian Report, 1848,* App. 3, pp. 47–48; Sally G. Kohlstedt, *The Formation of the American Scientific Community: The American Association for the Advancement of Science, 1848–1860* (Urbana, Ill., 1965), 39, calls the Franklin Institute a "model" for the Smithsonian Institution, without mentioning the two meteorological projects.

9. This "second report" to the surgeon general was not his better-known *Second Report on Meteorology to the Secretary of the Navy* (Washington, D.C., 1851). Other short reports appeared in U.S. Navy Department, *Annual Report, 1852,* pt. 2, 336; and ibid., 1853, pt. 3, 393–94. Espy, "To the Friends of Science," *Daily National Intelligencer* 35 (Mar. 10, 1847): 3; Eric R. Miller, "New Light on the Beginnings of the Weather Bureau from the Papers of Increase A. Lapham," *Monthly Weather Review* 59 (1931): 65–70; Coffin to Henry, July 7, 1845, Henry Papers.

10. Espy to Henry, Dec. 7, 1846, Henry Papers.

11. Henry to Loomis, Apr. 22, 1847, Loomis Papers. The "club" was a scientific social group in Philadelphia, see *JHP* 2: 290–91.

12. Henry to Hon. John Y. Mason, Feb. 12, 1848, Henry Papers.

13. Henry to Bache, July 29, 1848, Henry Papers; emphasis added.

14. Ibid.; Espy to Henry, [Sept. 1848], Henry Papers.

15. Smithsonian Institution, Chief Clerk, 1846–1933, "Forms, Circulars, Announcements," Nov. 1, 1848, RU 65, SIA. Copies are also in the Library of the American Antiquarian Society, Worcester, Mass. [marked "Ansd Dec 28 1848 (affirmative)"], and in the Henry Papers. Lists of Espy's correspondents appear in his second and fourth *Meteorological Report.* In 1846 Coffin estimated that he had gathered some 190 observations from the following locations: 5 from North America; 79 from the Northeast and New York; 8 from Michigan, Iowa, and Wisconsin; 50 from New Jersey and Pennsylvania; 7 from Ohio, Indiana, Illinois, Missouri, and the Northern Territory; 6 from Delaware, Maryland, and Virginia; 5 from North Carolina, Tennessee, and Arkansas Territory; 19 from South Carolina, Georgia, Alabama, Mississippi, Louisiana, and Bermuda; 11 from Florida, Texas, and the West Indies (Coffin to Redfield, Sept. 1, 1846, Redfield Papers).

16. James P. Espy, *Second Report on Meteorology to the Secretary of the Navy,* submitted Nov.

12, 1849, communicated to the U.S. Senate, Mar. 13, 1850, bound, printed, and paginated with Espy, *Third Report on Meteorology, with directions for Mariners, etc.* (Washington, D.C., 1851).

17. Espy to Bache, Feb. 20, 1849, Bache Papers.

18. Henry Desk Diary, Jan. 15–Apr. 4, 1849, Henry Papers. The appropriation was on Espy's mind again within six months, see Espy to Henry, Oct. 23, 1849, SIMC-NA.

19. Henry Desk Diary, Apr. 27, 1849, Henry Papers: "Called on Sec. of the Navy; will allow the sum which has been kept back from Mr. Espy for expenses of Dr. Foreman etc."; Henry to J. C. Dobbin, Sept. 10, 1853, LR-Misc., Naval RC.

20. Quoted in H. T. De La Beche, *How to Observe. Geology* (London, 1835), iii–iv; see John Ruskin, "Remarks on the Present State of Meteorological Science," *Meteorological Society of London, Transactions* 1 (1839): 56–59.

21. Carlene E. Stephens, *Inventing Standard Time* (Washington, D.C., 1983); Ian R. Bartky, "Inventing, Introducing and Objecting to Standard Time," *Vistas in Astronomy* 28 (1985): 105–11.

22. Coffin to Henry, Dec. 7, 1855, Henry Papers.

23. Lapham to [Smithsonian], May 17, 1851, SIMC-NA.

24. Henry Little to Joseph Henry, July 22, 1856, "Meteorological Scraps," Smithsonian Meteorological Project, RG 27, NA, contains "Meteorological Notes, taken during the March of Colonel Morrison's command from Fort Gibson C.N. to Big Timbers on the Upper Arkansas and back from 22nd of June to 25th of October, 1855 by Captain Henry Little, 7th Infantry."

25. Ovid Plumb to Hon. Truman Smith, M[ember of] C[ongress], June 1, 1849, (forwarded to the secretary of the navy and on to the Smithsonian), SIMC-NA.

26. S. R. Frierson, Columbus, Mississippi, to Henry, Dec. 21, 1848, SIMC-NA.

27. John Newton to Smithsonian, Jan. 31, 1850, SIMC-NA; Milo G. Williams, University of Urbana, Illinois, to Henry, Jan. 2, 1853, SIMC-SIA; Rev. Jason F. Walker, Troy Conference Academy, Vermont, to Arnold Guyot, Jan. 7, 1853, SIMC-SIA.

28. Long series of observations published in the *Smithsonian Contributions to Knowledge* include those of Alexis Caswell (1860), Nathan D. Smith (1860), Parker Cleaveland (1867), and Samuel P. Hildreth (1867).

29. Chappelsmith (1807–ca. 1883), *Dictionary of Artists in America* (New Haven, 1957).

30. Chapplesmith to Foreman, Nov. 25, 1852, Rain, Wind, General Meteorology, Smithsonian Meteorological Records, RG 27, NA; John Chapplesmith, "Account of a Tornado Near New Harmony, Indiana, April 30, 1852, with a Map of the Track, Diagrams, and Illustrative Sketches," *Smithson. Contrib.* 7, Article II (1855), also issued as a pamphlet.

31. Lawrence Stone, "Prosopography," *Daedalus* 100 (Winter 1971): 46–79.

32. Lewis Pyenson, "'Who the Guys Were': Prosopography in the History of Science," *History of Science* 15 (1977): 178–79. Pyenson's article contains numerous references to the prosopography of elites.

33. The relative precision of the sampling was calculated using the formulas given by Morris H. Hansen, William N. Hurwitz, and William G. Madow, *Sample Survey Methods and Theory*, Vol. 1, *Methods and Applications* (New York, 1953), 124–29.

34. Clement L. Martzolff, "Ohio University—The Historic College of the Old Northwest," *Ohio Archaeological and Historical Quarterly* 19 (1910): 430–38. The college closed its doors in 1845, reopened them in 1848, and graduated a class in 1850. By the 1870s, however, matters were again at "low ebb." Finances were still a problem, salaries were low, and a stable faculty was hard to maintain.

35. Nason (1831–95), *BDAS;* Harris (1832–1920), *WWW* 1. Edward Dwight Eaton, *Historical Sketches of Beloit College* (New York, 1928), 211–13. Henry B. Nason, ed., *Elderhorst's*

Manual of Quantitative Blowpipe Analysis and Determinative Mineralogy (Philadelphia, 1865), five editions, and *Hand-book of Mineral Analysis* (Philadelphia, 1871), a translation of Friedrich Wöhler, *Practische Übungen in der chemischen Analyse;* Elijah P. Harris, *A Manual of Qualitative Chemical Analysis* (Amherst, Mass., 1875), ten editions. *Annual Catalogue of Beloit College,* 1869–1870 (Beloit, Wisc., 1869), 4.

36. Bruce, *Launching of Modern American Science,* 187–200.

37. Loomis, "Report on the Meteorology of the United States," 28–46.

Chapter 5. Stormy Relations among Theorists and Administrators

1. Henry, *Smithsonian Report, 1865,* 52.

2. Henry to Redfield, Jan. 11, 1848, Redfield Papers; Redfield to Henry, Sept. 15, 1848, Redfield Letterbooks.

3. Piddington (1797–1858), *DNB;* Redfield to Reid, Oct. 10, 1848, Reid's Correspondence.

4. Redfield to Reid, July 5, 1849, Reid's Correspondence. Hare's paper was "On the American Climate and Theories of Its Storms."

5. Smith, *Life of Robert Hare,* 472–73. On Hare's meteorology see pp. 445–76.

6. Robert Hare, "On the Whirlwind Theory of Storms"; William Redfield, "On the Apparent Necessity of Revising the Received Systems of Dynamical Meteorology"; James P. Espy, "On the Diminution of Temperature in Air by Expansion, and Permeation of Air by Moisture," all in *AAAS, Proceedings* 4 (1851): 231–42, 366–69, 371–74.

7. Arnold Guyot, "On the Progress of the System of Meteorological Observations Conducted by the Smithsonian Institution, and the Propriety of Its Immediate Extension Throughout the American Continent," *AAAS, Proceedings* 6 (1852): 167–68.

8. Robert Hare, "Concerning the Making of Meteorological Observations All Over the United States Proposed by Guyot," n.d., MS, Hare Papers.

9. Redfield to Secretary of the Navy Wm. B. Preston, Nov. 1, 1849, Redfield Letterbooks. The references are to Henry Piddington, *The Sailor's Horn-Book* (London, 1848), and Reid, *Progress of the Development of the Law of Storms.*

10. Preston to Redfield, Dec. 18, 1849, Redfield Papers.

11. Espy, *Second Report on Meteorology to the Secretary of the Navy,* 7, 9–10.

12. Redfield to Hon. Frederick P. Stanton, Chair, Naval Committee, Feb. 12, 1852, Redfield Letterbooks.

13. Espy to Henry, May 29, 1849, SIMC-NA.

14. Henry to Espy, Apr. 23, 1849, SIMC-NA; Espy, *Third Report on Meteorology,* 17–19.

15. Espy to Henry, Mar. 20, 1850, Henry Papers.

16. James P. Espy, *Fourth Meteorological Report,* U.S. Senate, Ex. Doc. 65, 34th Cong., 3d sess., 23.

17. Espy to Henry, Apr. 30, 1849, SIMC-NA.

18. Espy to Henry, May 29, 1849, ibid.

19. Henry to Foreman, Aug. 10, 1849, Henry Papers.

20. Henry Desk Diary, July 21, 1849, ibid.

21. Henry to Foreman, Sept. 9, 1849, ibid.

22. [Henry] to Bache, July 10, 1851, ibid.

23. *Journ. Smithsonian Regents,* 467, 472–73, 478.

24. Hare, ["Gift of apparatus to the Smithsonian Institution"], n.d., MS, Hare Papers; See also Hare, "Propositions Relative to Dr. Hare's Apparatus, To Be Submitted to the Executive Committee of the Smithsonian Institution," n.d., MS, ibid.

25. Henry Desk Diary for 1852, Henry Papers. See Espy, *Third Report on Meteorology,* 27.

26. Robert Hare, "Strictures on Professor Espy's Report on Storms to the Secretary of the Navy, as Respects the Theoretical Inferences," *AAAS, Proceedings* 6 (1852): 152–60; Hare, "Dr. Hare on the Law of Storms," *Merchants' Mag.* 26 (1852): 190–95; Hare, *Queries and Strictures by Dr. Hare, Respecting Professor Espy's Meteorological Report to the Naval Department. Also, the Conclusion Arrived at by a Committee of the Academy of Sciences of France, Agreeably to Which, Tornadoes Are Caused by Heat; While, Agreeably to Peltier's Report to the Same Body, Certain Insurers Had Been Obliged to Pay for a Tornado as an Electrical Storm. With Abstract from Peltier's Report* (Philadelphia, 1852); also *Journ. Frankl. Inst.* 3d ser. 23 (1852): 325–31 and 3d ser. 24 (1852): 28–39.

27. *Journ. Frankl. Inst.* 3d ser. 24 (1852): 29–30.

28. J. C. Dobbin to Stanton, Dec. 19, 1853, Hare Papers, indicates that the government had spent $10,882.66 on Espy's salary and expenses through December 15, 1853.

29. Hare, "To the Senate and House of Representatives of the United States Congress Assembled," n.d. [ca. 1853], MS, Hare Papers.

30. Espy to Quetelet, and Espy to Flügel (forwarded to Quetelet), Apr. 30, 1855, Quetelet Correspondence; A. Quetelet, "Sur la theorie des ourages de M. Espy," *Annuaire météorologique de la France* 4 (1852), and *Sur le climat de la Belgique*, pt. 4 (Brussels, 1851), 99.

31. Espy to Quetelet, Apr. 30, 1855, Quetelet Correspondence.

32. Henry to Alexander, July 16, 1852, Henry Papers.

33. Notice of Espy's appropriations during his period of cooperation with the Smithsonian Institution appear in William Jones Rhees, ed., *The Smithsonian Institution: Documents Relative to Its Origin and History, 1835–1899*, vol. 1, 1835–1887, *Smithson. Misc. Coll.* 42 (1901). The annual Act for Naval Service for 1853–56 and 1859, and the Legislative, Executive and Judicial Act for 1858 provided for Espy's salary and back pay. The final bill stipulated "That the employment of a meteorologist under the contract of the Secretary of the Navy, shall cease on and after June 30, 1859." According to the Report of the Secretary of the Navy, Dec. 2, 1859, U.S. Senate, Exec. Doc. 2, 36th Cong., 1st sess., 1377, the balance remaining in the meteorological fund on June 30, 1859, was $1,225. It is not known how this final balance was spent.

34. Redfield, "Summary Statements of Some of the Leading Facts in Meteorology," 12.

35. William Redfield, "On the First Hurricane of September 1853, in the Atlantic; with a Chart; and Notices of Other Storms," *Amer. Journ. Sci.*, 2d ser., 18 (1854): 1–18, 176–90; section entitled "What Are Cyclones?" 188–89. See James H. Coffin, "An Investigation of the Storm Curve, Deduced from the Relation Existing between the Direction of the Wind, and the Rise and Fall of the Barometer," *AAAS, Proceedings* 7 (1853): 83–101.

36. Redfield to Stoddard, Feb. 19, 1855, Redfield Letterbooks.

37. Redfield to George Engleman[n], Sept. 26, Oct. 30, 1854; and Redfield to Prof. A. Winchell, Oct. 14, 1854, Redfield Letterbooks.

38. Hare, "On . . . cyclones," n.d., MS, Hare Papers.

39. Ibid.

40. Robert Hare, "Illustration of Cycloidal Curvature as involved in the Suppositious Traveling Whirlwinds, and some New Demonstrations of the Impossibility of Storms being Whirlwinds, unless on a limited Scale, as the Consequence of In-blowing Winds," paper presented at the eighth meeting of the AAAS, Washington, D.C., 1854; and Hare, "A Brief Exposition of the Absurdity of the Doctrine That Any Storm Can Be a Travelling Whirlwind, Unless as a Consequence of Centripetal Currents, or That if due to Such Currents, It Can Have a Diameter Much More Than Eighty-one Times the Height of the Stratum of the Atmosphere within Which It May Be Generated," paper presented at the tenth meeting of the AAAS, Albany, N.Y., 1856.

41. Hare's manuscript is in the Bell-Henry Library, Smithsonian Institution.

42. Henry to Hare, Feb. 18 and Apr. 3, 1856, Hare Papers.

43. William Redfield, "On the Spirality of Motion in Whirlwinds and Tornadoes," *Amer. Journ. Sci.* 2d ser. 23 (1857): 23–24; Redfield, "On Various Cyclones or Typhoons of the North Pacific Ocean; with a Chart, Showing Their Courses or Progression," *AAAS, Proceedings* 10 (1856): 214–22; Redfield, "Observations in Relation to the Cyclones of the Western Pacific, Embraced in a Communication to Commodore Perry," in Matthew C. Perry, *United States Japan Expedition*, vol. 2, comp. Francis L. Hawkes (Washington, D.C., 1856), 331–59.

44. Olmsted, "Biographical Memoir of Redfield."

45. Espy, *Fourth Meteorological Report*, 13, 151.

46. Ibid., 97–111, quote, 105.

47. Espy to Quetelet, Mar. 30, 1858, Quetelet Correspondence.

48. Henry to Hare, Feb. 3, 1857, Hare Papers.

49. Henry to Herschel, Feb. 22, 1858, Herschel Papers. The Hare Papers contain numerous spiritualist manuscripts. See also Robert Hare, *Experimental Investigation of the Spirit Manifestations, Demonstrating the Existence of Spirits and Their Communion with Mortals* . . . (New York, 1855). For background information, see Mary Farrell Bednarowski, "Nineteenth Century Spiritualism: An Attempt at a Scientific Religion" (Ph.D. dissertation, University of Minnesota, 1973; Ann Arbor: University Microfilms, 73-25581).

50. Espy's Smithsonian Lectures, Feb. 22, 24, 26, 1858, reported in the *Washington Evening Star* (Feb. 24, 26, March 1, 1858). See Henry Desk Diary, 1858, Henry Papers: "*Sat. Feb. 20*, Prepared letter to Sir John Herschel relative to meteorology; *Mon. Feb. 22*, First lecture on Meteorology by Mr. Espy. Tolerable good house—Lecture not well delivered; *Wed. Feb. 24*, Mr. Espy gave his 2nd lecture to tolerable good audience."

51. Espy, "To the Committee on Science and the Arts of the Franklin Institute of the State of Pennsylvania, For the Promotion of the Mechanic Arts," June 22, 1858, MS, Frankl. Inst. Archives, Philadelphia, Pa.

52. Espy to Frazer, Feb. 11, 1859, John Fries Frazer Papers, Amer. Phil. Soc. Library. Espy's petition to the Franklin Institute is marked "Discharged June 12/62."

53. Sinclair, *Philosopher Mechanics*, 158; De Young, "Storm Controversy," 657.

54. Bache, in *Journ. Smithsonian Regents*, 108.

55. Joseph Henry, "Meteorology in Its Connection with Agriculture," U.S. Patent Office, *Annual Report, Agricultural, 1858*, 483.

56. Henry, *Smithsonian Report, 1859*, 31, commenting on Elias Loomis, "On Certain Storms in Europe and America, December, 1836," *Smithson. Contrib.* 11, Article VIII (1859); James H. Coffin, "On the Currents of the Atmosphere," *AAAS, Proceedings* 12 (1859): 205.

57. Secretary of the Navy Upshur to Maury, July 11, 1842, LR, NOR-NA; a similar program was followed at the Greenwich Observatory under the director G. B. Airy (A. J. Meadows, "Airy and After," *Nature* 225 [1975]: 592–95).

58. The preferred biography of Maury (1806–73) is Francis Leigh Williams, *Matthew Fontaine Maury: Scientist of the Sea* (New Brunswick, N.J., 1963). Although on matters of interpretation Williams is Maury's advocate, his biography is based on extensive research in original sources. See also Lewis J. Darter, Jr., "The Federal Archives Relating to Matthew Fontaine Maury," *American Neptune* 1 (1941): 149–58. On the history of the Depot of Charts and Instruments see Stephen J. Dick, "How the U.S. Naval Observatory Began, 1830–65," in Dick and LeRoy E. Doggett, eds., *Sky with Ocean Joined* (Washington, D.C., 1983), 167–81.

Marc Rothenberg, "Observers and Theoreticians: Astronomy at the Naval Observatory, 1845–1861," in Dick and Doggett, eds., *Sky with Ocean Joined*, 29–43; observes that Maury diffused his resources "in a broad spectrum of the physical sciences," which included meteorology and hydrography. Rothenberg's critique of Maury as a leader of the "Naval Observatory" hinges on the observation that "Maury was a man who seemed out of synchronization with the rest of the astronomical community" at a time when astronomy was beginning a period of rapid development (36).

The official designation "U.S. Naval Observatory and Hydrographic Office" was ordered on December 12, 1854; the Hydrographic Office was separated from the Naval Observatory five years after Maury's tenure ended in 1861.

59. Because of his theoretical calculations, credit for the discovery of Neptune was given to French astronomer and meteorologist Urbain Jean Joseph Le Verrier (1811–77), *DSB*. Morton Grosser, *The Discovery of Neptune* (Cambridge, Mass., 1962), 58–69; "Le Verrier's Planet," *Amer. Journ. Sci.* 2d ser. 3 (1847): 128–32. Using Le Verrier's predictions, Johann G. Galle and Heinrich d'Arrest at the Berlin Observatory were the first to observe and identify the planet.

60. *Astronomische Nachrichten*, no. 605 (Aug. 1847): 66–79.

61. Maury to Lucien Minor, May 21, 1847, Maury Papers, LC; Maury to Henry, May 25, 1847, LS, NOR-NA.

62. Maury to Henry, Sept. 20, 1847, LS, NOR-NA.

63. Henry to Maury, Oct. 11, Nov. 9, 1847, NOR-LC.

64. Maury to Henry, Nov. 15, 1847, LS, NOR-NA; for more on the Walker incident, see Williams, *Maury*, 167–72.

65. Maury to Henry, Nov. 27, 1850, LS, NOR-NA; Henry to Bache, Nov. 30, 1850, Bache Papers; Bache to Henry, Dec. 26, 1850, Henry Papers; Maury to Bache, Dec. 26, 1850, Rhees Collection.

66. Maury to Henry and Charles C. Jewett, Dec. 2, 1848, LS, NOR-NA.

67. Henry Desk Diary, Feb. 9, 12, 1849, Henry Papers.

68. Maury to George Manning, July 6, 1849, WCC Correspondence.

69. Maury to Humboldt, Apr. 1, 1851, LS, NOR-NA; Maury to Manning, May 8, 1848, WCC Correspondence.

70. WCC Correspondence, typescript finding aid, n.d. E.g., "Maury's Wind and Current Chart," *Merchants' Mag.* 18 (1848): 516–17. To help promote the sailing directions, a petition in praise of Maury's work—written in Maury's hand—was sent to Manning for circulation to be signed by ship owners, masters, and merchants (Maury to Manning, Jan. 3, 1849, WCC Correspondence).

71. Edward Sabine, president of the Royal Society, received thirty-one sets (Maury to Manning, Dec. 20, 1849, WCC Correspondence).

72. Maury to Henry, Nov. 29, 1850, LS, NOR-NA.

73. [Matthew F. Maury], *On the Establishment of an Universal System of Meteorological Observations by Sea and Land* (Washington, D.C., 1851).

74. John F. Crampton, Britain's Chargé d'Affairs to Daniel Webster, Secretary of State, Nov. 13, 1851; transmitted to Secretary of the Navy William A. Graham, Nov. 14, 1851; Charles Morris, Chief of Bureau of Ordnance and Hydrography to Maury, Nov. 19, 1851; Maury to Morris, Nov. 21, 1851, all letters quoted, ibid.

75. Maury to the Foreign Ministers of Belgium, Austria, Netherlands, Sweden and Norway, Two Sicilies and Parma, Sardinia, Guatemala, Argentina, Chile, Mexico, Nicaragua, Venezuela, and Peru, Dec. 23, 1851, LS, NOR-NA.

76. Matthew F. Maury, "On the General Circulation of the Atmosphere," *AAAS, Proceedings* 3 (1850): 126–47, chart from 137. Maury, *The Physical Geography of the Sea* (1855), rpt. ed. John Leighly (New York, 1963). Leighly's "Introduction" to the volume is uniformly critical of Maury.

77. Maury to Henry, Jan. 14, 1852, LS, NOR-NA. Similar letters were sent to Professors Alexis Caswell, Benjamin Silliman, Sr., Olmsted, Redfield, Hare, Guyot, and Espy, and to bureau chiefs Bache of the Coast Survey; Surgeon General Lawson; General Joseph G. Totten, chief of the Bureau of Engineers, U.S. Army; Colonel J. J. Abert, chief of the Bureau of Topographical Engineers; and to the American Philosophical Society, the Franklin Institute, and numerous college presidents.

78. Abert to Maury, Jan. 24, 1852, LR, NOR-NA.

79. Webster to Maury, July 8, 1852, ibid.

80. Maury to Henry, Jan. 22, 1852, LS, NOR-NA.

81. Henry to Maury, Jan. 23, 1852, LR, NOR-NA; Henry Desk Diary, Jan. 24, Mar. 2, 1852, Henry Papers.

82. Henry to Maury, May 31, 1852, LR, NOR-NA; Williams, *Maury*, 205.

83. Maury to Henry, June 1, 1852, LS, NOR-NA.

84. U.S. Navy to Maury, July 8, 1853, LR, NOR-NA. Maury's biographer, Williams, *Maury*, 547, n. 129, refers to the proceedings of the Brussels Conference (printed in both French and English) but does not provide a reference to the document.

85. Maury to Lieut. Marin Jansen, Royal Dutch Navy, Nov. 8, 1853, Sept. 2, 1858, Box KN, Naval RC; only the Dutch navy followed the conference suggestions.

86. Lamont [Bavaria] to Maury, n.d., [ca. 1853], LR, NOR-NA. According to Lamont, the same noncooperation happened after the "Magnetic Conference" in England in 1845.

87. Maury to Secretary of the Navy James C. Dobbin, Nov. 7, 1854, LS, NOR-NA, cited in Williams, *Maury*, 312. Williams devotes an entire chapter to Maury's crusade for a United States Weather Bureau, the details of which are not repeated here.

88. *U.S. Agricultural Society Journal* 3 (1856): pt. 1, "Journal of the 4th Annual Meeting, January, 1856," 48–51, cited in Williams, *Maury*, 317.

89. U.S. Senate, Rept. 292, 34th Cong., 3d sess. See also *Congressional Globe*, Dec. 18, 1856, p. 149.

90. Quoted in Charles Lee Lewis, "Maury—First Meteorologist," *Southern Literary Messenger* 3 (1941): 483.

91. Henry to G. T. Kingston, Aug. 3, 1858, AES File 1855b: Incoming Letters to the Toronto Observatory, 1855–69, AES, Downsview, Ontario, Canada; Henry to Lapham, Dec. 31, 1858, Lapham Papers.

92. Henry to Sabine, dated March and July 8, 1861, Sabine Papers.

93. Wotherspoon to Henry, Oct. 30, 1849, SIMC-NA.

94. "No. of Military Registers (returned) for 1851," Mar. 14, 1854, Smithsonian Lists and References, RG 27, NA.

95. Lawson, U.S. Army Surgeon General, *Annual Report, 1852*, U.S. Senate, Ex. Doc. 1, 32d Cong., 2d sess., pt. 2, p. 138.

96. Lawson, U.S. Army Surgeon General, *Annual Report, 1853*, U.S. Senate, Ex. Doc. 1, 33d Cong., 1st sess., pt. 2, p. 144.

97. Blodget (1823–1901), *BDAS*; J. K. McGuire, "The Father of American Climatology," *Weatherwise* 10 (1957): 92–94, 97.

98. Jewett (1816–68) and Choate (1799–1859), both *DAB*. [Asa Gray?], "The Smithsonian Institution," *Amer. Journ. Sci.* 2d ser. 20 (1855): 1–21, reports that one of Choate's suggestions was for the Smithsonian Institution to be administered as a bureau of the Department of the Interior. See also *Six Articles upon the Smithsonian Institution, from the Boston Post, together with the Letters of Professors Peirce and Agassiz* (Boston, 1855). A recent study of the two factions is Joel J. Orosz, "Disloyalty, Dismissal, and a Deal: The Development of the National Museum at the Smithsonian Institution, 1846–1855," *Museum Studies Journal* 2 (1986): 22–33.

99. Blodget to Henry and Espy, Jan. 4, 1849, SIMC-NA.

100. Henry Desk Diary, Jan. 6, 26, 31, 1852, Henry Papers; Henry's statement to Congress is found in U.S. House of Representatives, ["Report of the Select Committee on the Management of the Smithsonian Institution]," U.S. House of Representatives, Rept. 141, 33d Cong., 2d sess., 96–113. See also U.S. Senate, ["Resolution of the Senate to Inquire whether any Action is Necessary and Proper in Regard to the Smithsonian Institution"], U.S. Senate, Rpt. 484, 33d Cong., 2d sess.

101. Henry Desk Diary, Feb. 27, Mar. 2, 1852, Henry Papers.

102. Blodget's instructions were to credit the institution for the facts but to assume personal responsibility for his theories or speculations. For more details and the testimony of Blodget before Congress, see U.S. House of Representatives, ["Report of the Select Committee"], 57–76.

103. Henry Desk Diary, July 28, Aug. 12, Nov. 6, 15, 18, Dec. 23, 1852, Henry Papers. Blodget joined Henry and Coffin in planning the proper way to calculate the force of the wind as reported by the observers. He also requested, unsuccessfully, that army post surgeons change their times of observations to fit the Smithsonian system.

104. Blodget to Choate, May 8, 1854, Brown Collection; quoted by permission of the Houghton Library, Harvard University.

105. Ibid.

106. U.S. Census Office, "Meteorology of the United States," *Seventh Census of the United States, 1850*, U.S. House of Representatives, Misc. Doc., 32d Cong., 2d sess., xcii–xciii.

107. The first letter signed this way is Blodget to Dr. A. S. Baldwin, July 15, 1853, Lorin Blodget's Outgoing Correspondence, 1853–54, Meteorological Project Records, RU 60, SIA.

108. Blodget to Choate, May 8, 1854, Brown Collection; quoted by permission of the Houghton Library, Harvard University.

109. Orosz, "Disloyalty, Dismissal, and a Deal," 22–33.

110. Ibid.

111. Blodget to Choate, May 8, 1854, Brown Collection; quoted by permission of the Houghton Library, Harvard University; U.S. House of Representatives, Rept. 141, 33d Cong., 2d sess., 87.

112. Henry to Guyot, July 29, 1854, Guyot Papers.

113. Henry to Lapham, May 29, 1854, Lapham Papers.

114. Lorin Blodget, "Agricultural Climatology of the United States Compared with that of Other Parts of the Globe," in U.S. Patent Office, *Annual Report, 1853*, pt. 2, *Agriculture*, U.S. Senate, Ex. Doc. 27, 33d Cong., 1st sess., 328–432.

115. Mason (1804–82), *DAB*, commissioner of patents, 1853–57. Mason to Henry, June 19, 1854, and Henry to Mason, June 21, 1854, in U.S. Patent Office, *Annual Report, 1853*, 327.

116. Lawson, U.S. Army Surgeon General, *Annual Report, 1855*, U.S. Senate, Ex. Doc. 1, 33d Cong., 2d sess., pt. 2, p. 82.

117. Henry, *Smithsonian Report, 1855*, 27.

118. Wotherspoon to Blodget, Apr. 10, 1854, quoted in Lawson to Henry, Apr. 5, 1856, Surgeon General's Letterbook 25, RG 112, NA.

119. Henry to Lawson, Apr. 1, 1856, quoted, ibid. Blodget's essay was "Report on the Prominent Features of General Climate in the United States, as Exhibited in the Distribution of Temperature and of Rain, and in Explanation of the Illustrative Charts," in U.S. Army Medical Department, *Army Meteorological Register for Twelve Years, from 1843 to 1854 Inclusive* (Washington, D.C., 1855), 681–763. Forry and Dove had also used army data in constructing their charts.

120. Lawson to Henry, Apr. 5, 1856, Surgeon General's Letterbook 25, RG 112, NA.

121. Lawson to Jefferson Davis, May 10, 1856, Letters to the Secretary of War, vol. 2, Office of the Surgeon General, RG 112, NA.

122. Coffin's testimony is found in U.S. House of Representatives, ["Report of the Select Committee"], 91ff., quote from 92.

123. Ibid., 61. The issue of official versus personal mail was clarified by Henry in his letter to Blodget dated March 31, 1854 reproduced in this report.

124. Henry to Gray, Nov. 13, 1856, Historic Letter File, Gray Herbarium Library, Harvard University.

125. Henry to Guyot, Dec. 22, 1856, Guyot Papers.

126. Lorin Blodget, *Climatology of the United States and the Temperate Latitudes of the North American Continent: Embracing a Full Comparison of These with the Climatology of the Temperate Latitudes of Europe and Especially in Regard to Agriculture, Sanitary Investigations, and Engineering with Isothermal and Rain-Charts for Each Season, the Extreme Months, and the Year, Including a Summary of the Statistics of Meteorological Observations in the United States, Condensed from Recent Scientific and Official Publications* (Philadelphia, 1857). See Robert De C. Ward, "Lorin Blodget's 'Climatology of the United States': An Appreciation," *Monthly Weather Review* 42 (1914): 23–27.

127. Henry to Coffin, Nov. 29, 1855, Henry Papers.

Chapter 6. Cooperative Observations and Contributions to Knowledge

1. Beck to Henry, May 10, 1849, SIMC-NA.

2. Regents of the State of New York, *Annual Report, 1850*, 278–86. Guyot (1807–84), physical geographer and meteorologist, *DSB;* James D. Dana, "Memoir of Arnold Guyot, 1807–1884," *Natl. Acad. Sci., Biog. Mem.* 2 (1886): 309–47; Donald Drew Egbert, *Princeton Portraits* (Princeton, 1947), 113–15.

3. Guyot to Henry, Jan. 12, 1850, Guyot Letters.

4. Guyot to Henry, Nov. 7, 1849, ibid.

5. Guyot to Henry, Jan. 15, 1850, ibid.

6. Guyot to Henry, Dec. 5, 1850, ibid.

7. Arnold Guyot, *Directions for Meteorological Observations, Intended for the First Class of Observers* (Washington, D.C., 1850); Guyot and Joseph Henry, "Directions for Meteorological Observations," *Smithsonian Report, 1855*, 215–44; Guyot, "Directions for Meteorological Observations and the Registry of Periodical Phenomena," *Smithson. Misc. Coll.* 1, Article I (1860).

8. Henry, *Smithsonian Report, 1851*, rpt. (1854), 222.

9. Arnold Guyot, *A Collection of Meteorological Tables, with Other Tables Useful in Practical Meteorology*, 1st ed. (Washington, D.C., 1852); 2d ed. (Washington, D.C., 1857); 3d ed. titled *Tables, Meteorological and Physical, Prepared for the Smithsonian Institution* (Washington, D.C., 1859), *Smithson. Misc. Coll.* 1, Article III (1862); 4th ed. (Washington, D.C., 1884); Baird, "Advertisement," to 4th ed., *Smithson. Misc. Coll.* 23 (1884): iii. Quote from Guyot, "Preface to the Second Edition," vii–ix.

10. Franklin B. Hough, comp., *Results of a Series of Meteorological Observations Made under Instructions from the Regents of the University at Sundry Stations in the State of New York. Second Series from 1850 to 1863, Inclusive, with Record of Rainfall and Other Phenomena to 1871* (Albany, 1872), vi. The observations are published *in extenso*.

11. *Amer. Acad., Proceedings* 2 (1850): 198, 220.

12. Guyot to Henry, Dec. 5, 1850, Guyot Letters.

13. Ibid.

14. Henry to Guyot, Mar. 26, 1851, Guyot Papers.

15. Ibid., list of stations, "as designated by Governor N. Briggs." A list of the twelve

observers of "The Meteorological System of the State of Massachusetts," is given by Espy, *Fourth Meteorological Report,* 8.

16. "List of Meteorological Registers Received from the Stations of the Massachusetts State System," Smithsonian Lists and References, RG 27, NA.

17. Eric R. Miller, "The Evolution of Meteorological Institutions in the United States," *Monthly Weather Review* 59 (1931): 1–6.

18. "Resolution of the Illinois State Board of Education, 1855," in *Smithsonian Report, 1855,* 81–82.

19. *Smithsonian Report, 1858,* 34.

20. *Acts and Resolves Passed by the 36th Legislature of the State of Maine,* Chapter 42 (Augusta, 1857), 22; Willis to Henry, July 16, 1857, SIMC-SIA.

21. Willis to Henry, Apr. 3, July 16, 1857, SIMC-SIA.

22. LeFroy (1817–90), *DNB.* Henry, *Smithsonian Report, 1849,* 176. LeFroy said Henry had come "principally to examine our Photographic apparatus, but took the opportunity of making himself acquainted with our whole System of observation, particularly in meteorology" (LeFroy to Edward Sabine, Sept. 26, 1849, PRO document reference BJ 3/39, Edward Sabine Papers, Records of the Kew Observatory, Public Record Office, London; Crown copyright). The most recent publication on meteorology (and magnetism) in Canada is Suzanne Zeller, *Inventing Canada: Early Victorian Science and the Idea of a Transcontinental Nation* (Toronto, 1987), 115–80. See also Morley K. Thomas, "A Brief History of Meteorological Services in Canada, Part 1: 1839–1930," *Atmosphere* 9 (1971): 3–15; E. S. Hallman, "110 Years Ago," *Weather* 5 (1950): 155–58; and A. D. Thiessen, *The Founding of the Toronto Magnetic Observatory and the Canadian Meteorological Service,* 400–403, reprinted from the *Journal of the Royal Astronomical Society of Canada* 34–40 (1940–46).

23. The locations were Queenston, Montreal, Kingston, Toronto, London, Fredricton, Halifax, and Newfoundland (Hallman, "110 Years Ago," 156). "Lieutenant Colonel Sabine had already promised cooperation in Canada" (Sabine to Henry, Oct. 17, 1847, quoted in Henry to Loomis, Oct. 21, 1847, Loomis Papers).

24. *Smithsonian Report, 1852,* rpt. (1854), 246. For Hudson Bay see D. W. Moodie and A. J. W. Catchpole, *Environmental Data from Historical Documents by Content Analysis: Freeze-up and Break-up of Estuaries on Hudson Bay, 1714–1871, Manitoba Geographical Studies* vol. 5 (Winnipeg, 1975); for Canada in general see Morley K. Thomas, *Pre-Confederation Climate Data, Canadian Meteorological History,* No. 1 (Downsview, Ontario, 1984).

25. "Meteorological Stations Connected with the Senior County Grammar Schools," *Upper Canada Annual School Report for 1858,* rpt. in *Smithsonian Report, 1858,* 417–20.

26. "Departmental Regulations for the Meteorological Stations of U.C.," Local Meteorology 2, British America, RG 27, NA. A discussion of these regulations appears in Zeller, *Inventing Canada,* 143–66.

27. *Smithsonian Report, 1858,* 420.

28. *Smithsonian Report, 1873,* 84–85.

29. *Smithson. Misc. Coll.* 18 (1880): 259–60; Morley K. Thomas, "Professor Kingston's Scheme: Founding the Meteorological Service," *Chinook,* Summer 1986, pp. 51–55.

30. Lapham to Henry, Dec. 25, 1858 [not found], cited in Henry to Lapham, Dec. 31, 1858, Lapham Papers.

31. Meade to Henry, Jan. 27, Feb. 7, 1859, LS, Lake Survey Records.

32. Henry to Meade, Feb. 11, 1859, Feb. 13, 1861, LR, Lake Survey Records.

33. U.S. Army Topographical Engineers, *Annual Report, 1860,* U.S. Senate, Ex. Doc. 2, 36th Cong., 1st sess., Appendix B, p. 715.

34. Henry, *Smithsonian Report, 1860,* 55.

35. U.S. Army Topographical Engineers, *Annual Report, 1859,* U.S. Senate, Ex. Doc. 1, 35th Cong., 2d sess., Appendix 1, p. 1276.

36. U.S. Army Topographical Engineers, "Report of the Topographical Bureau, Nov. 14, 1860," U.S. Senate Ex. Doc. 1, vol. 2, 36th Cong., 2d sess., in *Meteorology of the Lakes, 1860,* 310, separately bound volume in the Bell-Henry Library, Smithsonian Institution. These results are marked in the margin, possibly by Henry.

37. "History of the Weather Service in Detroit," typescript finding aid (Detroit, 1970), RG 27, NA.

38. Volume 79 of the *Smithson. Misc. Coll.* contains the Lake System data from August 1862 to October 1870. Information for Table 6.1 is taken from U.S. Army Topographical Engineers, "Report of the Topographical Bureau, Nov. 14, 1860," U.S. Senate Ex. Doc. 1, vol. 2, 36th Cong. 2d sess., bound separately as *Meteorology of the Lakes, 1860,* in the Bell-Henry Library, Smithsonian Institution.

39. The original act of Congress was dated March 3, 1839.

40. U.S. Patent Office, *Annual Report, 1843,* 1–17, and *1848,* Appendix 3, p. 339.

41. Mason, U.S. Patent Office, *Annual Report, 1856,* quoted in *Smithsonian Report, 1860,* 54.

42. Henry to Mason, July 2, 1855, Records of the Agricultural Division of the Patent Office, RG 16, NA; Mason to Henry, July 3, 1855, SIMC-SIA.

43. Henry to Coffin, Nov. 29, 1855, Henry Papers; see Green, "Price List of Standard Meteorological Instruments," n.d. but ca. 1860, Forms, Circulars, Announcements, RU 65, SIA.

44. Henry, *Smithsonian Report, 1856,* 26–27.

45. James H. Coffin, *Psychrometrical Tables for Determining the Force of Aqueous Vapor, and the Relative Humidity of the Atmosphere from Indications of the Wet and the Dry Bulb Thermometer, Fahrenheit* (Washington, D.C., 1856), *Smithson. Misc. Coll.* 1, Article III (1862).

46. Henry, *Smithsonian Report, 1857,* 27.

47. Henry to Coffin, Jan. 15, June 15, 1855; and Coffin to Henry, Mar. 1, July 12, 1855, Henry Papers.

48. *Smithsonian Meteorological Observations for the Year 1855* (Washington, D.C., 1857); MS entitled "Summaries of Meteorological Observations 1855," in Smithsonian General Meteorology, RG 27, NA. The pamphlet recorded observations from 91 locations with barometer, 155 with thermometer, 51 with psychrometer, 141 of "face of the sky," and 102 of rain and snow. Mean values, maxima, and minima, at 7:00 A.M., 2:00 P.M., and 9:00 P.M. were given (U.S. Patent Office, *Results of Meteorological Observations Made under the Direction of the United States Patent Office and the Smithsonian Institution from the Year 1854 to 1859, Inclusive,* 2 vols. U.S. House of Representatives, Ex. Doc. 55, 36th Cong., 1st sess.

49. A. Hall to Henry, Nov. 15, 1862, SIMC-NA.

50. Coffin to Loomis, Sept. 2, 1840, Loomis Papers; Loomis, "Report on the Meteorology of the United States," 28.

51. Joseph Henry, "Meteorology in Its Connection with Agriculture," U.S. Patent Office, *Annual Report, Agricultural, 1855,* 357–94; *1856,* 455–92; *1857,* 419–506; *1858,* 429–93; *1859,* 461–524.

52. Henry to Herschel, Feb. 22, 1858, Herschel Papers.

53. Henry to Asa Gray, May 22 and 31, 1858, Historic Letter File, Gray Herbarium Library, Harvard University.

54. The turnover rate was high: Commissioner of Patents William D. Bishop, 1859–60, was succeeded by Philip F. Thomas and Superintendent of Agriculture Thomas G. Clemson, both in 1860–61. Clemson was soon replaced by Isaac Newton, 1861–62, who served as commissioner of agriculture from 1862 to 1867.

55. James S. Grinnell, Chief Clerk, Department of Agriculture, to Baird, June 29, 1863, OSI-SIA; John Johnson to Henry, July 20, 1863, SIMC-NA.

56. Newton (1800–1867), *DAB;* Grinnell to Henry, Aug. 4, 1863, SIMC-NA.

57. Henry, *Smithsonian Report, 1863*, 32.

58. Henry to G. T. Kingston, Dec. 6, 1856, AES File 1855b: Incoming Letters to the Toronto Observatory, 1855–69, Atmospheric Environment Service, Downsview, Ontario, Canada.

59. Helen T. Finneran, "List of Locations, with Dates Covered by Coast Survey Notebooks, 1843–90," in "Preliminary Inventory of Operational and Miscellaneous Meteorological Records of the Weather Bureau," Appendix 2, typescript (May 1965), RG 27, NA; Henry to Guyot, Oct. 13, 1853, Guyot Papers.

60. Schott (1826–1901); Cleveland Abbe, "Biographical Memoir of Charles Anthony Schott, 1826–1901," *Natl. Acad. Sci., Biog. Mem.* 8 (1915): 87–133. Elisha K. Kane, *Meteorological Observations in the Arctic Seas . . . 1853, 1854, and 1855*, comp. C. A. Schott, *Smithson. Contrib.* 11, Article V (1859); Francis L. M'Clintock, *Observations in the Arctic Seas . . . 1857, 1858, 1859*, comp. C. A. Schott, *Smithson. Contrib.* 13, Article III (1862); Isaac I. Hayes, *Physical Observations in the Arctic Seas . . . during 1860 and 1861*, comp. C. A. Schott, *Smithson. Contrib.* 15, Article V (1867).

61. Henry to Sabine, Dec. 2, 1853, Sabine Papers; William J. Rhees, *An Account of the Smithsonian Institution, Its Founder, Building, Operations, etc.* (Washington, D.C., 1859; rpt. Philadelphia, 1864); A. D. Bache, "On the Winds of the Western Coast of the United States, from Observations in Connection with the U.S. Coast Survey," *AAAS, Proceedings* 11 (1858): 183–90.

62. Ludlum, *Early American Tornadoes*, 124–31.

63. Henry, *Smithsonian Report, 1860*, 57; Henry to Bache, June 18, 1860, Records of the Coast and Geodetic Survey, RG 23, NA. Maps of Nicholson's proposed routes are in the archives.

64. Charles A. Schott, "Tables and Results of the Precipitation, in Rain and Snow, in the United States and at Some Stations in Adjacent Parts of North America and in Central and South America," *Smithson. Contrib.* 18, Article II (1872); Schott, "Tables, Distribution, and Variations of the Atmospheric Temperature in the United States and Some Adjacent Parts of North America," *Smithson. Contrib.* 21, Article V (1876); Schott's MS is in RG 27, NA.

65. Benjamin Peirce, "Criterion for the Rejection of Doubtful Observations," *Astronomical Journal* 2 (1852): 161–63; Benjamin Gould, "On Peirce's Criterion for the Rejection of Doubtful Observations, with Tables for Facilitating Its Application," *Astronomical Journal* 4 (1858): 81–87; Charles A. Schott, "Letter on Peirce's Criterion, Nov. 22, 1877," *Amer. Acad., Proceedings* 13 (1878): 350–51.

66. Henry, *Smithsonian Report, 1873*, 31.

67. Schott, "Tables, Distribution, and Variations of the Atmospheric Temperature," 311.

68. Baird to Henry, Nov. 3, 1849, Baird Papers; quoted in William H. Dall, *Spencer Fullerton Baird, A Biography* (Philadelphia, 1915), 190–93.

69. William A. Deiss, "Spencer F. Baird and His Collectors," *Journal of the Society for the Bibliography of Natural History* 9 (1980): 635–45.

70. Henry Desk Diary, Mar. 31, 1849, Henry Papers.

71. Henry's European Diary, *JHP* 3:463.

72. Henry to Loomis, Apr. 22, 1847, Loomis Papers.

73. Henry to Baird, July 5, 1850, Baird Papers.

74. [Diary of Edward Foreman?], Sent off to Prof. Coffin &c., June 26, 1856, Smithsonian Lists and References, RG 27, NA.

75. Henry, *Smithsonian Report, 1864*, 25.

76. See Bigelow, "Facts."

77. J. G. Kohl, "Substance of a Lecture Delivered at the Smithsonian Institution on a Collection of the Charts and Maps of America," *Smithsonian Report, 1856*, 132.

78. Henry Desk Diary, Mar. 31, 1849, Henry Papers; See also Deiss, "Baird and His Collectors."

79. Eugene Coan, *James Graham Cooper, Pioneer Western Naturalist* (Moscow, Idaho, 1981). His father, William Cooper (of Cooper's hawk fame), was the founder of the New York Lyceum of Natural History and the first American member of the Zoological Society of London.

80. Edward Foreman, "Report of the General Assistant with Reference to the Meteorological Correspondence," *Smithsonian Report, 1851*, 68–78.

81. Kirtland (1793–1877) and Kennicott (1835–66), both *BDAS;* William Healy Dall (1845–1927), *DSB.*

82. Englemann (1809–84), *BDAS*; George Englemann, "Meteorological Table for St. Louis," *St. Louis Medical and Surgical Journal* 10 (1852): 297–302; copy in Meteorological Abstracts and Valuable Tables, Smithsonian Lists and References, RG 27, NA.

83. Englemann to Baird, Apr. 25, 1851, RU 305, SIA: "I am sorry that I can not give you much in making collections as my professional engagements keep me almost entirely chained to within the city—but what comes in my way I shall cheerfully and gladly preserve for you."

84. Englemann to Baird, May 28, 1853. By "Serpents" and "Guyot's tables," Englemann meant the Smithsonian natural history and meteorology publications S. F. Baird and C. Girard, *Catalog of North American Reptiles in the Museum of the Smithsonian Institution,* Part 1, *Serpents* (Washington, D.C., 1853), and Guyot, *Collection of Meteorological Tables.*

85. On the emergence of ecology see Frank N. Egerton, "Ecological Studies and Observations before 1900," in B. J. Taylor and T. J. White, eds., *Issues and Ideas in America* (Norman, Okla., 1976), 311–51.

86. Henry to Coffin, Feb. 18, 1869, Henry Papers; James H. Coffin, "On the Winds of the Northern Hemisphere," *Smithson. Contrib.* 6, Article VI (1853); Coffin, "The Winds of the Globe: or the Laws of Atmospheric Circulation over the Surface of the Earth," *Smithson. Contrib.* 20 (1875). A list of published international works sent to Coffin by the Smithsonian is found in "Sent off to Prof. Coffin, &c," Smithsonian Lists and References, RG 27, NA.

87. Ferrel (1817–91), *DSB;* Cleveland Abbe, "Memoir of William Ferrel, 1817–1891," with Ferrel's "Autobiographical Sketch," *Natl. Acad. Sci., Biog. Mem.* 3 (1895): 265–309. See also the tributes to Ferrel in *Amer. Metl. Journ.* 4 (1888): 441–49, and 8 (1891–92): 337–69, 441; also "William Ferrel: Complete List of His Publications," *Naturalist* 4, no. 5 (1889): 2–3.

88. William M. Davis, "Ferrel's Contributions to Meteorology," *Amer. Metl. Journ.* 8 (1891–92): 354.

89. Ferrel read the translation of Laplace by Nathaniel Bowditch, *Mécanique céleste by the Marquis de La Place,* 4 vols. (Boston, 1829–39). The equations of motion of a fluid at the surface of the earth are found in Book 4, chapter 1. Coriolis (1792–1843), *DSB;* Gaspard Coriolis, *Traité de la méchanique des corps solides et du calcul de l'effet des machines,* 2d ed. (Paris, 1844); Poisson (1781–1840), *DSB;* Siméon Poisson, *Recherches sur le mouvement des projectiles dans l'air en ayant égard à leur figure et leur rotation, et à l'influence du mouvement diurne de la terre* (Paris, 1839).

90. William Ferrel, "An Essay on the Winds and Currents of the Ocean," *Nashville Journal of Medicine and Surgery* 11 (1856): 287–301, 375–89, rpt. in *Professional Papers of the Signal Service* 12 (Washington, D.C., 1882), 1–19; Ferrel, "The Influence of the Earth's Rotation upon the Relative Motion of Bodies Near Its Surface," *Astronomical Journal* 5 (1858): 97–100, quote from 99; Ferrel, "The Motions of Fluids and Solids Relative to the Earth's Surface,"

Mathematical Monthly 1 (1859): 140–48, 210–16, 300–307, 366–73, 397–406, and 2 (1859–60): 89–97, 339–46, 374–90, rpt. in *Professional Papers of the Signal Service* 8 (Washington, D.C., 1882).

91. On Ferrel and Coriolis see J. L. Jordan, "On Coriolis and the Deflective Force," *Amer. Metl. Soc., Bulletin* 47 (1966): 401–3; and Harold L. Burstyn, "The Deflecting Force and Coriolis," *Amer. Metl. Soc., Bulletin* 47 (1966): 887–91.

92. *DSB;* Kutzbach, *Thermal Theory,* 35.

93. Ferrel, "Essay on the Winds." On gyratory storms see Charles Tracy, "On the Rotary Action of Storms," *Amer. Journ. Sci.* 45 (1843): 65–73.

94. Henry to Ferrel, Jan. 2, 1857, William Ferrel Papers, William R. Perkins Library, Duke University.

95. Espy to Henry, Jan. 4, 1857 [received Jan. 28, 1857], Henry Papers.

96. [Cleveland Abbe], "William Ferrel, the Signal Service and the Weather Bureau," dated Oct. 10–11, 1891, typescript in Abbe Papers, JHU, subsequently published as Cleveland Abbe, "Ferrel's Influence in the Signal Office," *Amer. Metl. Journ.* 8 (1891–92): 342–48. Abbe (1838–1916), *DSB;* William J. Humphreys, "Biographical Memoir of Cleveland Abbe, 1838–1916," *Natl. Acad. Sci., Biog. Mem.* 8 (1919): 469–508.

97. Henry, *Smithsonian Report, 1856,* 26.

Chapter 7. Weather Telegraphy

1. Two standard histories of the Weather Bureau are Gustavus A. Weber, *The Weather Bureau: Its History, Activities, and Organization* (New York, 1922); and Donald R. Whitnah, *A History of the United States Weather Bureau* (Urbana, Ill., 1961). An anniversary volume, Patrick Hughes, *A Century of Weather Service: A History of the Birth and Growth of the National Weather Service, 1870–1970* (New York, 1970), is a valuable source of early photographs. See also Joseph M. Hawes, "The Signal Corps and Its Weather Service, 1870–1890," *Military Affairs* 30 (1966): 68–76.

2. Redfield, in *Amer. Journ. Sci.* (1846), quoted in Cleveland Abbe, "Historical Notes on the Systems of Weather Telegraphy and Especially Their Development in the United States," *Amer. Journ. Sci.* 3d ser. 2 (1871): 83.

3. Loomis, "Report on the Meteorology of the United States," 41.

4. George J. Symons, "The First Daily Weather Map," *Metl. Mag.* 32 (1897): 133–35; Bernard Ashley, *Weather Men* (London, 1974), 39–41, says it was August 31, 1848, but provides no references.

5. A copy of this map is published in *Metl. Mag.* 31 (1896): 113.

6. Henry Pocket Notebook, 1848 and Henry Desk Diary, Feb. 9, 1848, both in *Journ. Smithsonian Regents,* 471.

7. O'Reilly (1806–86), Kendall (1789–186·, both *DAB;* Henry Desk Diary, Jan. 30, 31, 1849, and passim, Henry Papers; *Journ. Smithsonian Regents,* 471.

8. Henry Desk Diary, May 1, 1849, June 5, 1850, Henry Papers; Glaisher to Henry, July 8, 1850, SIMC-NA.

9. Jones (1802–63), *DAB;* Jones & Company advertisement in *Amer. Journ. Sci.* 2d ser. 5 (1848): 297.

10. Jones to Henry, July 21, 1849, SIMC-NA.

11. David Brooks to Abbe, Jan. 6, 1899, quoted in J. Cecil Alter, "National Weather Service Origins," *Bulletin of the Historical and Philosophical Society of Ohio* 7 (1949): 144–45.

12. Wade (1811–90), *DAB;* Wade to Abbe, Nov. 27, 1888, Abbe Papers, LC.

13. Joseph Brooks, *Boston Evening Star,* Nov. 20, 1882; W. H. Prescott, *History of the American Telegraph,* both quoted in Alter, "Weather Service Origins," 147.

14. Henry to Abbe, Oct. 16, 1869, Abbe Papers, LC.

15. Henry, *Smithsonian Report, 1858,* 32; Alter, "Weather Service Origins," 151.

16. Henry, *Smithsonian Report, 1857,* 26; *Washington Evening Star,* May 1, 7, 1857; Espy to J. H. Berryhill, Dec. 28, 1857, Rhees Collection.

17. Henry, *Smithsonian Report, 1857,* 26.

18. Henry, *Smithsonian Report, 1860,* 35; *1861,* 35.

19. Henry to Sabine, dated March and July 8, 1861, Sabine Papers.

20. H. W. Ravenel, *Charleston Courier,* July 1861, cited in David M. Ludlum, *The Weather Factor* (Boston, 1984), 69.

21. Henry to Coffin, Dec. 10, 1861, Henry Papers.

22. W. G. Fuller to Henry, Aug. 9, 1862; Lucius E. Ricksecker to Hon. Commissioner of Agriculture, Sept. 18, 1862, both in SIMC-NA.

23. Finley (1798–1879), Hammond (1828–1900), Barnes (1817–83), all in Hume, "Foundation of American Meteorology," 230–32.

24. Abbe, "Weather Telegraphy," 85.

25. "Report of the Special Committee of the Board of Regents of the Smithsonian Institution Relative to the Fire," *Journ. Smithsonian Regents,* 236–39. See also "Origin of the Fire at the Smithsonian Institution, Loss Occasioned, and Nature and Cost of Repairs Necessary," U.S. Senate, Report 129, 38th Cong., 2d sess.

26. "Meteorological Material Saved and Lost at the Fire," Jan. 24, 1865, MS; and "List of Meteorological Material Saved from the Fire of 1865," MS, both in Smithsonian Lists and References, RG 27, NA.

27. Henry to R. Lachlan, July 17, 1865, OSO-SIA.

28. "Origin of the Fire at Smithsonian Institution," 24–26.

29. Henry, *Smithsonian Report, 1865,* 50, 57.

30. Smithsonian Institution, "Circular and Blank Form to Telegraph Operators, Dec. 1866," Smithsonian Lists and References, RG 27, NA.

31. Omaha Office, American Telegraph Co. to Henry, Jan. 2, 1864, Smithsonian Clippings, RG 27, NA.

32. Henry to O. H. Palmer, Jan. 21, 1867, OSO-SIA.

33. "List of Meteorological Stations, 1866," Smithsonian Lists and References, RG 27, NA.

34. Jos. P. Martin to Horace Capron, July 25, 1868, forwarded to the Smithsonian, OSI-SIA.

35. Henry to Chester Dewey, Jan. 16, 1866, Chester Dewey Papers, Department of Rare Books and Special Collections, Rush Rhees Library, University of Rochester, Rochester, N.Y.

36. Henry to J. L. Lippencott, Apr. 4, 1866, OSO-SIA.

37. Henry, *Smithsonian Report, 1863,* 52; U.S. Department of Agriculture, *Monthly Report,* copies from 1865 to 1872 in Smithsonian General Meteorology, RG 27, NA.

38. William R. Hopkins to Henry, June 4, 1867, OSI-SIA.

39. Henry Desk Diary, Nov. 19, 1868, Henry Papers.

40. *Journ. Smithsonian Regents,* 386.

41. For details see Truman Abbe, *Professor Abbe and the Isobars: The Story of Cleveland Abbe, America's First Weatherman* (New York, 1955).

42. Abbe to Lapham, Jan. 7, 1870, Lapham Papers. See also Cleveland Abbe, "How the United States Weather Bureau Was Started," *Scientific American* n.s. 114 (1916): 529; William J. Humphreys, "Origin and Growth of the Weather Service of the United States, and Cincinnati's Part Therein," *Scientific Monthly* 18 (1924): 372–82; and Miller, "New Light," 65–70.

43. Abbe to Mr. Orton, President of Western Union, July 29, 1868, quoted in Alter,

"Weather Service Origins," 157–58. Alter's long article reproduces many original documents.

44. Abbe to Gano, May 7, 1869, quoted, ibid., 159.

45. The Western Meteorological Society was formed on July 20, 1869, to support Abbe's forecasts at Cincinnati. See Walter Alden to Abbe, July 21, 1869, with enclosed news clipping, "Meteorological Society Formed—Constitution and By-Laws Adopted—A Bulletin to be Published at Once," Abbe Papers, LC; *Invitation to a Meeting of Those Interested in the Establishment of a Meteorological Association,"* ([Cincinnati], 1869).

46. L. Young to Abbe, July 26, 1869, Abbe Papers, LC.

47. Edward Goldsmith to Abbe, Sept. 3, 1869, Abbe Papers, LC.

48. Abbe, "How the Weather Bureau Started," 529. Alter, "Weather Service Origins," 165, reproduces the "Bulletin" for Sept. 2, 1869. The forecasts seem to have been the exception, not the rule.

49. Abbe, "How the Weather Bureau Started," 529.

50. Quoted in Alter, "Weather Service Origins," 168.

51. Quoted in Abbe, "How the Weather Bureau Started," 529.

52. *Proceedings of the Second Annual Meeting of the National Board of Trade* (Richmond, Va., Dec. 1869), quoted in R. H. Weightman, "Establishment of a National Weather Service—Who Was Responsible for It," typescript (Washington, D.C., 1952), 27, in General Correspondence, 1951–55, RG 27, NA.

53. Abbe to Secretary of War, Mar. 21, 1870 (copy), Abbe Papers, LC. See Abbe, "Meteorology—1859 to 1909," in Reminiscences of Employees, 1907–46, RG 27, NA; and Abbe, "How the Weather Bureau Started," 529.

54. The full text of Lapham's memorial to Congress is given in Miller, "New Light," 67. See also Increase A. Lapham, "The Great Storms," Letter to the Editor of the *Bureau: A Chronicle of the Commerce and Manufactures of Chicago and the Northwest,* Dec. 1869, p. 59, copy in Abbe Papers, JHU. The article also appeared in the *New York Times,* Dec. 8, 1869.

55. Increase A. Lapham, "Memorial of Professor J. [sic] A. Lapham to H. E. Paine, Dec. 8, 1869," in "Disasters on the Lakes," U.S. House of Representatives, Misc. Doc. 10, 41st Cong., 2d sess. See also Lapham, "Will It Pay?" in *Bureau,* Jan. 1870, pp. 83–84, copy in Abbe Papers, JHU. Also *Milwaukee Sentinel,* Dec. 8, 1869.

56. Increase A. Lapham, "Meteorological Observations," *Bureau,* Jan. 1870, p. 91, map on p. 93, copy in Abbe Papers, JHU. His data came from Coffin, "Observations Relative to Three Storms in the Year 1859," in U.S. Patent Office, *Results of Meteorological Observations 1854 to 1859,* vol. 2. Average wind directions were taken from Coffin, "Winds of the Northern Hemisphere."

57. Loomis to Paine, Jan. 10, 1870, "Disasters on the Lakes," U.S. House of Representatives, Ex. Doc. 10, pt. 2, 41st Cong., 2d sess.

58. Henry to Paine, Jan. 10, 1870, ibid.

59. J. K. Barnes to Paine, Jan. 5, 1870, ibid.

60. *Congressional Globe,* 41st Cong., 2d sess., 90, pt. 1, p. 177. Weightman, "Establishment of Weather Service," 31–33; Cleveland Abbe, "The Meteorological Work of the U.S. Signal Service, 1870–1893," *U.S. Weather Bureau, Bulletin* 11 (1893): 236.

61. Myer (1828–80), graduate of Hobart College (1847) and Buffalo Medical College (1851), assistant surgeon in the Army Medical Department (1854), and army signal officer (1860). Paul J. Scheips, "Albert James Myer, Founder of the Army Signal Corps: A Biographical Study" (Ph.D. dissertation, American University, 1966); Paul J. Scheips, "'Old Probabilities': A. J. Myer and the Signal Corps Weather Service," *Arlington Historical Magazine* 5 (1974): 29–43; and George M. Kober, "General Albert J. Myer and the United States Weather Bureau," *Military Surgeon* 65 (1929): 65.

62. Paine to Duane Mowry, Milwaukee, Wisc., Oct. 8, 1903, quoted in Miller, "New Light," 68; cited in Lewis J. Darter, Jr., "The Origin and Development of the Weather Bureau," in Darter, comp., *List of Climatological Records in the National Archives* (Washington, D.C., 1942), 9.

63. A copy of the order, dated June 27, 1860, and signed by President Buchanan the following day, is in the Myer Papers, microfilm reel 1.

64. Copy in Myer Papers, microfilm reel 2.

65. Miller, "New Light," 69. More general information is found in G. L. Van Deusen, *Historical Sketch of the Signal Corps (1860–1928)*, Signal School Pamphlet 32 (Fort Monmouth, N.J., 1929).

66. Myer to Paine, Jan. 18, 1870, "Disasters on the Lakes [pt. 2]," David J. Marshall, "Myer's Triumph and the Postwar Signal Corps," in Max L. Marshall, ed., *The Story of the U.S. Army Signal Corps* (New York, 1965), 83.

67. *U.S. Statutes at Large* 16 (1871): 369, 90.

68. According to Alter, who reproduces both reports and maps, the new series began on February 9, 1870, the day the weather service bill was signed into law ("Weather Service Origins," 179). According to Humphreys, however, the "Weather Bulletin of the Western Union Telegraph Co., compiled at the Cincinnati Observatory" began in December 1869, when Abbe's initial experiment ended ("Biographical Memoir of Cleveland Abbe," 490).

69. Abbe to Secretary of War, Mar. 24, 1870 (copy); Abbe to Buchanan, President of the Board of Control of the Cincinnati Astronomical Society, Oct. 8, 1870 (copy); Thomas O. Selfridge to Abbe, Oct. 10, 1870, all in Abbe Papers, LC; Humphreys, "Origins and Growth," 380.

70. A. Wessel to L. B. Harrison, Oct. 21, 1870; H. W. Howgate to Abbe, Oct. 21, 1870; Myer to Abbe, Dec. 27, 30, 1870, all in Abbe Papers, LC.

71. Myer to William W. Belknap, Secretary of War, Feb. 17, 1870, LS–Misc., Signal Office Records; Belknap to Myer, Feb. 28, 1870, LR–War Dept., ibid. Formal assignment was made by General Order No. 29, Mar. 15, 1870.

72. Abbe to Lapham, Jan. 7, 1870, Lapham Papers.

73. Lapham to Abbe, Jan. 12, 1870, Abbe Papers, LC.

74. Elias Loomis, *A Treatise on Meteorology: With a Collection of Meteorological Tables* (New York, 1868); Alexander Buchan, *Handy Book of Meteorology*, 2d ed. (Edinburgh, 1869); Albert J. Myer, *Signal Office Report, 1871*, 321.

75. U.S. Signal Office, "Practical Use of Meteorological Reports and Weather-Maps" (1871) and "Instructions to Observer-Sergeants" (1871); Guyot and Henry, "Directions for Meteorological Observations"; Espy, *Philosophy of Storms*, and *First* and *Fourth Meteorological Reports;* Piddington, *Sailor's Horn-Book;* Robert Fitzroy, *The Weather-Book: A Manual of Practical Meteorology* (London, 1863); F. P. B. Martin, *A Memoir on the Equinoctial Storms of March–April, 1850; an Inquiry into the Extent to Which the Rotatory Theory May Be Applied* (Kingston, Eng., 1852); William C. Ley, *Laws of the Winds Prevailing in Western Europe*, pt. 1 (London, 1872); Myer, *Signal Office Report, 1873*, 303.

76. Myer, *Signal Office Report, 1871*, 321, 335, 354–59.

77. H. W. Howgate to Abbe, Oct. 21, 1870; Myer to Abbe, Oct. 27, 1870, Abbe Papers, LC.

78. Myer's correspondence after 1870, preserved in the National Archives, is dominated by telegrams received rather than by letters. *Signal Office Report, 1873*, 306; Scheips, "Old Probabilities," 35; Darter, "Origin and Development," 14. Stations were established on Mount Washington, N.H. (1870); Mount Mitchell, N.C.; and Pikes Peak, Colo. (1873).

79. Miller, "New Light," 69.

80. L. B. Norton to Abbe, Jan. 12, 1871, Abbe Papers, LC.

81. Alter, "Weather Service Origins," 182–83.

82. The first was Elias Loomis, "Results Derived from Examination of the United States Weather Maps for 1872 and 1873," *Amer. Journ. Sci.* 3d ser. 8 (1874): 1–15.

83. U.S. Army Engineers, *Annual Report, 1872,* 1032, 1068.

84. Barnes to Myer, June 19, 1874, in LR–Misc., Signal Office Records; Myer to Barnes, June 26, 1874, LS–Misc., ibid., both in *Signal Office Report, 1875,* 362; see also Darter, "Origin and Development," 19–20.

85. Cleveland Abbe, "The Origin of the Weather Bureau," *Independent,* May 20, 1897, p. 5; John Grable to William J. Rhees, Jan. 27, 1872, SIMC-NA.

86. Henry, *Smithsonian Report, 1870,* 44.

87. Henry to Myer, Dec. 5, 1873, LR–Misc., Signal Office Records; Myer to Henry, Jan. 9, 1874, LS–Misc., ibid., see *Signal Office Report, 1874,* 704–5.

88. Joseph Henry, "To the Voluntary Meteorological Observers Who Have Reported to the Smithsonian Institution," Feb. 2, 1874, Henry Papers; copies in *Smithsonian Report, 1873,* 31; *Signal Office Report, 1874,* 704–5.

89. Myer, "To the Voluntary Meteorological Observers Who Have Reported to the Smithsonian Institution," Feb. 16, 1874, *Signal Office Report, 1874,* 705.

90. Darter, "Origin and Development," 19–20.

91. As Weightman pointed out, the foundation of a national storm warning system was not the responsibility of a single person or institution. Credit must be distributed among those who organized observational systems to collect data on storms—Espy, Henry, Guyot, Abbe, and Myer; those who studied such observations and developed the laws of storms and storm movement—Loomis, Espy, and Coffin; those who demonstrated that weather predictions were possible and called for support—Henry and Lapham; those who were members of Congress and who had the vision and ability to draft and support legislative measures—Paine and Wilson; and those who developed telegraphic facilities without which forecasts and warnings would have been impossible—Morse, Western Union, and the other telegraph companies (Weightman, "Establishment of a National Weather Service," 30–35).

Chapter 8. The Worldwide Horizon

1. Henry Desk Diary, May 17, 1870; Henry Pocket Notebook, Sept. 2, 1870, both in Henry Papers.

2. Henry to Baird, June 25, 1870, Baird Papers.

3. Henry to Royal Commission on Science, June 27, 1870, Henry Papers.

4. Henry to Rhees, Nov. 12, 1870, ibid.

5. Henry to Baird, July 7, 1870, Baird Papers (visit with Sir Henry James, director-general of the ordnance survey); Henry to Harriet Henry, Aug. 9, 1870 (visit to International Commission of Measures, Paris); Henry Pocket Notebook, Aug. 30, 1870 (discussion on testing meteorological instruments with Col. A. Stange of Lambeth Observatory, India); Henry Pocket Notebook, Aug. 29, 1870 (visit to Kew Observatory); Henry to John Tyndall, Nov. 9, 1870 (on the charting of an Atlantic storm); all items except the first are in Henry Papers.

6. Baird to Henry, July 11, 1870, Baird Papers.

7. Buchan, *Handy Book of Meteorology,* quoted in *Metl. Mag.* 3 (1869): 197.

8. Kutzbach, *Thermal Theory,* 12–13.

9. Quoted in Khrgian, *Meteorology,* 1:115.

10. Louis Dufour, "Sketch History of Meteorology in Belgium," *Weather* 6 (1951): 359–64. See also Dufour, *Ésquisse d'une histoire de la météorologie en Belgique* (Brussels, 1953), issued

as *Institut Royal Météorologique de Belgique, Miscellanées* 40 (Brussels, 1953); and Gustav Hellmann, "The Organization of the Meteorological Service in Some of the Principal Countries of Europe," trans. and ed. J. S. Harding, *Metl. Mag.* 16 (1881): 51–57.

11. Hellmann, "Umriss einer Geschichte"; data from the Verein appear in *Annalen für Meteorologie, Erdmagnetismus und verwandte Gegenstände* beginning in 1842. See also Hellmann, "Die Entwicklung der meteorologischen Beobachtungen in Deutchland von der ersten Anfängen bis zur Einrichtung staatlicher Beobachtungsnetze," *Abhandlungen der Preussische Akademie der Wissenschaften, Physikalisch-mathematische Klasse,* no. 1 (Berlin, 1926).

12. Mahlman (1812–48); Gustav Hellmann, *Geschichte des Königlich Preussischen Meteoroloigischen Instituts von seiner Gründing im Jahre 1847 bis zu seiner Reorganisation im Jahre 1885* (Berlin, 1887). See also Hellmann, "Katalogen der Schriften und Erfindungen," *Repertorium,* 1–744. A bibliography of 208 of Dove's works appears on pages 93–103.

13. E. Doublet, "La météorologie en France et en Allemagne," *Revue philomathique de Bordeaux et sud Ouest* 15 (1912): 173; See also Heinrich Seilkopf, "Zur Geschichte der meteorologischen Arbeit an der Deutchen Seewarte, Hamburg," *Annalen der Meteorologie* 3 (1950): 53–56.

14. George J. Symons, "History of English Meteorological Societies, 1823 to 1880," *Quarterly Journal of the Royal Meteorological Society* 7 (1881): 75. See also Robert Watson-Watt, "The Evolution of Meteorological Institutions in the United Kingdom," *Quarterly Journal of the Royal Meteorological Society* 76 (1950): 115–24; and Richard Corless, "A Brief History of the Royal Meteorological Society," *Weather* 5 (1950): 78–83.

15. George Birkbeck, President of the Meteorological Society of London, to J. G. Tatem, quoted in Symons, "English Meteorological Societies," 70.

16. Ibid., 76. Symons provides a useful summary of the minutes of the proceedings of the society for 1823–24 and 1836–43.

17. Ruskin, "Remarks on the Present State of Meteorological Science."

18. The *Quarterly Journal of Meteorology* was published from 1841 to 1843.

19. Symons, "English Meteorological Societies," 93, 88.

20. See chapter 7, nn. 4 and 5.

21. Among the numerous short articles on the British Meteorological Office are the following: "Meteorological Office Centenary, 1855–1955," *Metl. Mag.* 84 (1955): 161–98; David Brunt, "A Hundred Years of Meteorology, 1851–1951," *Advancement of Science* 8 (1951): 114–24; Brunt, "The Centenary of the Meteorological Office: Retrospect and Prospect," *Science Progress* 44 (1956): 193–207; G. A. Bull, "Short History of the Meteorological Office," *Metl. Mag.* 83 (1955): 163–67; R. P. W. Lewis, "The Founding of the Meteorological Office, 1854–55," *Metl. Mag.* 110 (1981): 221–27; and Oliver Graham Sutton, "The Meteorological Office, 1855–1955," *Nature* 175 (1955): 963–65. More substantial and more recent is Jim Burton, "Robert Fitzroy and the Early History of the Meteorological Office," *British Journal for the History of Science* 19 (1986): 147–76.

22. Alfred Angot, "Premier catalogue des observations météorologiques faites en France depuis l'origine jusqu'en 1850," in *Annales du Bureau Central Météorologique de France, 1895,* vol. 1, *Mémoires* (Paris, 1897), 89–146. The rain stations established in the 1830s were located in Baye, Corbigny, Decize, and Laroche.

23. Doublet, "La météorologie en France et en Allemagne." See also Météorologie Nationale, [France], *Ce qu'est la météorologie française* (Paris, 1952); and John L. Davis, "Weather Forcasting and the Development of Meteorological Theory at the Paris Observatory, 1853–1878," *Annals of Science* 41 (1984): 359–82.

24. Alexander Woeikof, "Meteorology in Russia," *Smithsonian Report, 1872,* 267–98; F. Clawer, *Catalog der Meteorologischen Beobachtungen im Russichen Reich Zusammengestelt,* in H.

Wild, ed., *Repertorium Für Meteorologie* vol. 2 (St. Petersburg, 1872), contains information on meteorological observations at 330 locations in Russia from 1726. See also Pavel Nikolaevich Tverkoi, *Razvitie Meteorologii v U.S.S.R.* (Development of meteorology in the USSR) (Leningrad, 1949).

25. See Henry to Coffin, Sept. 28, Nov. 1, 1849, SIMC-SIA.

26. Woeikof, "Meteorology in Russia."

27. Cleveland Abbe, "Meteorology in Russia," *Monthly Weather Review* 27 (1899): 106.

28. A. Kh. Khrgian, "The History of Meteorology in Russia," *Actes du VIIIe Congrès International d'Histoire des Sciences* (Paris, 1958): 446. A book-length treatment by the same author with international comparisons is Khrgian, *Meteorology.*

29. Myer, *Signal Office Report, 1874,* 505; on the international congresses see H. G. Cannegieter, "The History of the International Meteorological Organization, 1872–1951," *Annalen der Meteorologie* n.s., 1 (1963): 7–280, esp. 7–12; and World Meteorological Organization, *One Hundred Years of International Co-operation in Meteorology (1873–1973), A Historical Review* (Geneva, 1973).

30. Latour, "Visualization and Cognition."

31. David Elliston Allen, *The Naturalist in Britain: A Social History* (London, 1976), 109.

32. Bruce, *Launching of Modern American Science,* 66.

33. Peter Dobkin Hall, *The Organization of American Culture, 1700–1900: Private Institutions, Elites, and the Origins of American Nationality* (New York, 1982), 242, 227–39.

34. Allan Nevins, *The War for the Union,* vol. 1, *The Improvised War, 1861–1862* (New York, 1959), v.

35. Allen, *Naturalist in Britain,* 176–94; Elizabeth B. Keeny, "The Botanizers: Amateur Scientists in Nineteenth-Century America" (Ph.D. dissertation, University of Wisconsin, 1985), 19.

36. John Lankford, "Amateurs versus Professionals: The Controversy over Telescope Size in Late Victorian Science," *Isis* 72 (1981): 28.

37. John C. Deane to Cyrus Field, Jan. 31, 1868; Field to Henry, Feb. 3, 1868, OSI-SIA; Henry to Field, Henry Desk Diary, Feb. 6, 1868, Henry Papers.

38. Henry, *Smithsonian Report, 1863,* 33–34.

39. Quoted in Reingold, "American Indifference," 57.

BIBLIOGRAPHY

Archival and Manuscript Collections

Albany Academy Archives, Albany, N.Y.
  Records and Trustees' Minutes
American Philosophical Society Library, Philadelphia
  James P. Espy MSS
  John Fries Frazer Papers
  Robert Hare Papers
  Joseph LeConte Papers
  Manuscript Communications, Natural Philosophy
  Adolphe Quetelet, selected correspondence (microfilm), H.S. Films 11
Army Medical Library, Bethesda, Md.
  Benjamin Waterhouse, weather diary, 1816
Atmospheric Environment Service, Downsview, Ontario, Canada
  AES File 1855b: Incoming Letters to the Toronto Observatory, 1855–69
Bell-Henry Library, Smithsonian Institution, Washington, D.C.
  Robert Hare MS
  Marginalia in Joseph Henry's books and pamphlets
Duke University, Durham, N.C., Manuscript Department, William H. Perkins Library
  William Ferrel Papers
Franklin Institute Archives, Philadelphia
  James P. Espy MSS
  Minutes of the Board of Managers
Harvard University, Cambridge, Mass.
  Gray Herbarium Library
    Historic Letter File: Asa Gray
  Houghton Library
    Samuel Gilman Brown Collection
    Benjamin Peirce Papers
Historical Society of Princeton, N.J.
  Arnold Guyot Papers
Huntington Library, San Marino, Calif.
  William J. Rhees Collection
Johns Hopkins University, Baltimore, Md., Special Collections, Milton S. Eisenhower Library
  Cleveland Abbe Papers, MS 60
Library of Congress Manuscript Division, Washington, D.C.
  Cleveland Abbe Papers
  Matthew F. Maury Papers
  Albert J. Myer Papers
  Redfield Family Papers
  U.S. Naval Observatory Records
    Superintendent's Office
    Correspondence of Wind and Current Chart Agents

Memphis State University, Memphis, Tenn., Mississippi Valley Collection
Caleb G. Forshey Papers
National Archives and Records Administration, Washington, D.C.
Microfilm M997, U.S. Navy Department, General Orders and Circulars, 1798–1862
Microfilm Records of the U.S. Census, 1850, 1860, 1870
Microfilm T907, U.S. Weather Bureau Climatological Records, 1819–92. 564 reels
RG 16, Records of the Office of the Secretary of Agriculture
Records of the Agricultural Division of the Patent Office
RG 23, Records of the Coast and Geodetic Survey
RG 27, Records of the Weather Bureau
General Correspondence of the Weather Bureau, 1951–55
Meteorological Correspondence of the Signal Office, 1870–93
Meteorological Correspondence of the Smithsonian Institution, 1847–67
Arnold Guyot Letters, Clippings, General Meteorology, Lists and References, Local Meteorology
Other Records, 1907–46
RG 45, Office of Naval Records and Library: Naval Records Collection
RG 77, Records of the Office of the Chief of Engineers
U.S. Lake Survey, 1845–1913
RG 78, Records of the Naval Observatory
RG 111, Records of the Office of the Chief Signal Officer
RG 112, Records of the Office of the Surgeon General (Army)
NOAA Library, Rockville, Md.
[C. Abbe?], Detailed Weather Synopses, 1871
Public Record Office, London
Records of the Kew Observatory: Edward Sabine Papers
Royal Society Library, London
John F. W. Herschel Papers
St. Andrews University Library, St. Andrews, Scotland
James David Forbes Papers
Smithsonian Institution Archives, Washington, D.C.
RU 26, Office of the Secretary (Joseph Henry, Spencer F. Baird), Incoming Correspondence, 1863–79
RU 33, Office of the Secretary (Joseph Henry, Spencer F. Baird, Samuel P. Langley), Outgoing Correspondence, 1865–91
RU 60, Meteorological Project Records
Lorin Blodget's Outgoing Correspondence, 1853–54
Incoming Correspondence
Miscellaneous Correspondence
Notes and Reports
Published Meteorological Maps
RU 65, Chief Clerk, 1846–1933
Forms, Circulars, Announcements
RU 305, Accession Records of the U.S. National Museum
RU 7001, Joseph Henry Collection
Including a group of letters to and from James H. Coffin
RU 7002, Spencer F. Baird Papers
RU 7053, Alexander D. Bache Papers
RU 7060, James Henry Coffin Papers

State Historical Society of Wisconsin, Madison
Increase A. Lapham Papers
University of Pennsylvania, Philadelphia, Special Collections, Van Pelt Library
Edgar Fahs Smith Memorial Collection in the History of Chemistry
University of Rochester, Rochester, N.Y., Department of Rare Books and Special Collections, Rush Rhees Library
Chester Dewey Papers
University of Texas at Austin, Harry Ransom Humanities Research Center
Royal Meteorological Society Papers
Yale University, New Haven, Conn.
Beinecke Rare Book and Manuscript Library
Elias Loomis Papers
Osborn Files
William C. Redfield Letterbooks, 3 vols.
William C. Redfield Papers
William Reid's Correspondence with W. C. Redfield, 3 vols.
Manuscripts and Archives, Yale University Library
Elias Loomis Family Papers
Edward Claudius Herrick Papers

Miscellaneous Journals and Annual Reports

Acts and Resolves Passed by the 36th Legislature of the State of Maine, 1857.
Albany Institute, Transactions, 1830–55.
American Almanac, 1846.
American Academy of Arts and Sciences, Memoirs, 1818–21, and *Proceedings*, 1850–78.
Annuaire météorologique de la France, 1852.
Boston Evening Mercantile Journal, 1840.
Colonial Sentinel, 1840.
Congressional Globe, 1856, 1869.
Daily National Intelligencer, 1820, 1845–47.
Dial (Cincinnati), 1860.
Genesee Farmer, 1839.
Journal of the House of Representatives, Pennsylvania, 1838–40.
Literary and Philosophical Repertory, 1812–17.
Medical and Agricultural Register, 1806–7.
Medical Repository, 1797–1800, 1821–22.
Milwaukee Sentinel, 1869.
Nantucket Inquirer, 1839.
New York Times, 1869.
Niles' National Register, 1839–43.
Niles' Weekly Register, 1817–19.
North American Review, 1815–18.
Pennsylvania Senate Journal, 1837–39.
Regents of the State of New York, *Annual Report*, 1832–50.
Societatis Meteorologicae Palatinae, *Ephemerides*, 1783–95.
Society for the Promotion of Agriculture, Arts and Manufactures, Transactions, 1801.
U.S. Army Engineers, *Annual Report*, 1872.
U.S. Army Surgeon General, *Annual Report*, 1844–45, 1852–55.
U.S. Army Topographical Engineers, *Annual Report*, 1859–60, 1872.

U.S. Coast Survey, *Annual Report,* 1859–63, 1878.
U.S. Department of Agriculture, *Monthly Report,* 1865–72.
U.S. House of Representatives: Executive Documents, Reports, 1840–70.
U.S. Navy, *Annual Report,* 1852–59.
U.S. Patent Office, *Annual Report,* 1843, 1848–59.
U.S. Senate: Executive Documents, Miscellaneous Documents, Public Documents, Reports, 1838–74.
U.S. Statutes at Large, 1871–72.
Washington Evening Star, 1857–58.

Reference Works and Bibliographies

American Meteorological Society. Cumulated Bibliography and Annual Indexes to *Meteorological and Geo-astrophysical Abstracts.* Boston, 1950–85.

———. "A Selective Bibliography in Meteorology." *Weatherwise* 10 (1957): 47–49, 95–97; 12 (1961): 73–75; 14 (1963): 187–91, 199–205; 16 (1965): 175–87.

American Philosophical Society. *Catalog of Books and Manuscripts in the American Philosophical Society Library.* 38 vols. Westport, Conn., 1970.

———. *Classified Index to the Publications of the American Philosophical Society . . . , 1769–1940.* Philadelphia, 1940.

———. "Guide to the Archives and Manuscript Collections of the American Philosophical Society." *Amer. Phil. Soc., Memoirs* 66 (1966).

Angot, Alfred. "Premier catalogue des observations météorologiques faites en France depuis l'origine jusqu'en 1850." *Annales du Bureau Central Météorologique de France, 1895,* vol. 1 *Mémoires* (Paris, 1897): 89–146.

Arksey, Laura, Nancy Pries, and Marcia Reed. *American Diaries: An Annotated Bibliography of Published American Diaries and Journals.* Vol. 1, *1492–1844.* Detroit, 1983.

Brush, Stephen, and Helmut Landsberg. *The History of Geophysics and Meteorology: An Annotated Bibliography.* New York, 1985.

Catlett, Stephen J., ed. *A New Guide to the Collections in the Library of the American Philosophical Society.* Philadelphia, 1987.

Clawer, F. "Catalog der Meteorologischen Beobachtungen im Russichen Reich, Zusammengestelt." In Heinrich Wild, ed., *Repertorium für Meteorologie,* vol. 2. St. Petersburg, 1872.

Cochran, Thomas C., ed. *The New American State Papers.* Vol. 2, *Science and Technology.* Wilmington, Del., 1973.

Darter, Lewis J. Jr., comp. *List of Climatological Records in the National Archives.* Washington, D.C., 1942. Reprint. Washington, D.C., 1981.

Debus, Allen G., ed. *World Who's Who in Science: From Antiquity to the Present.* Chicago, 1968.

Elliott, Clark A. "Models of the American Scientist: A Look at Collective Biography." *Isis* 73 (1982): 77–93.

Finneran, Helen T. "List of Locations, with Dates Covered by Coast Survey Notebooks, 1843–90." In "Preliminary Inventory of Operational and Miscellaneous Meteorological Records of the Weather Bureau," Appendix 2. Typescript (May 1965), RG 27, NA.

Fleming, James R. *Guide to Historical Resources in the Atmospheric Sciences: Archives, Manuscripts, and Special Collections in the Washington, D.C. Area.* National Center for Atmospheric Research, NCAR Technical Note 327+IA. Boulder, Colo., 1989.

Forbes, Harriette M., comp. *New England Diaries, 1602–1800: A Descriptive Catalogue of Diaries, Orderly Books, and Sea Journals.* Topsfield, Mass., 1923.

Greely, A. W. *Index of Meteorological Observations in the United States from the Earliest Records to January 1890.* Washington, D.C., 1891.

Havens, James M., ed. *An Annotated Bibliography of Meteorological Observations in the United States, 1731–1818.* Florida State University Department of Meteorology, Technical Report No. 5. Tallahassee, 1956.

———. "A Note on Early Meteorological Observations in the United States with Reference to the Germantown Temperature Records of 1731–32." *Amer. Metl. Soc., Bulletin* 39 (1958): 211–16.

Humphreys, William J. "A Review of Papers on Meteorology and Climatology Published by the American Philosophical Society prior to the Twentieth Century." *Amer. Phil. Soc., Proceedings* 86 (1942): 29–33.

Isis Cumulative Bibliography. A Bibliography of the History of Science Formed from Isis Critical Bibliographies . . . Vols. 1–6, *1913–65,* edited by Magda Whitrow; Vols. 1–2, *1966–75,* and Vol. 1, *1976–85,* edited by John Neu. London, 1971–89.

Kämtz, L. F. *Repertorium für Meteorologie.* 3 vols. Dorpat, 1862.

Matthews, William. *American Diaries: An Annotated Bibliography of American Diaries Prior to . . . 1861.* Berkeley, 1945.

Middleton, W. E. Knowles. *Catalog of Meteorological Instruments in the Museum of History and Technology at the Smithsonian Institution.* Smithsonian Studies in History and Technology, no. 2. Washington, D.C., 1969.

National Oceanographic and Atmospheric Administration. *Catalog of the Atmospheric Sciences Collection.* 24 vols. 1978–present.

Oxford English Dictionary. Oxford, 1888–1928.

Pinkett, Harold T., Helen T. Finneran, and Katherine H. Davidson, comps. *Preliminary Inventory of the Climatological and Hydrological Records of the Weather Bureau.* National Archives PI38. Washington, D.C., 1952.

Rothenberg, Marc. *The History of Science and Technology in the United States: A Critical and Selective Bibliography.* New York, 1982.

Royal Society Catalogue of Scientific Papers. Vols. 1–6, *1800–1863.* Vols. 7–8, *1864–1873.* London, 1867–77.

Smithsonian Institution. "Classified Record of Monthly Meteorological Reports Preserved in the Smithsonian Institution." *Smithsonian Report, 1873,* 84–131.

U.S. Army Signal Corps. *Bibliography of Meteorology: A Classed Catalogue of the Printed Literature of Meteorology from the Origin of Printing to the Close of 1881; with a Supplement to the Close of 1887, and an Author Index.* Edited by Oliver L. Fassig. Vol. 1, *Temperature;* Vol. 2, *Moisture;* Vol. 3, *Winds;* Vol. 4, *Storms.* Washington, D.C., 1889–91.

———. "Publications of the U.S. Signal Service from 1861 to July 1, 1891." *Signal Office Report, 1891,* 389–409.

Primary Sources

Accademia del Cimento. "Dichiarazione d'alcuni Strumenti per Conoscer l'Alterazioni dell'Aria." *Saggi di naturali esperienze fatte nell'Accademia del Cimento.* Florence, 1666. In Gustav Hellmann, ed. *Neudrucke von Schriften und Karten über Meteorologie und Erdmagnetismus,* No. 7, pp. 8–17. Berlin, 1897.

Adams, Eliphalet. *God sometimes answers his people by terrible things in righteousness. A discourse occasioned by that awful thunder-clap which struck the meeting-house in N. London, Aug. 31st, 1735. At what time one was killed outright and diverse others much hurt and wounded, yet graciously and remarkably preserved, together with the rest of the congregation, from immediate death. As it was delivered the Lord's Day following.* New London, Conn., 1735.

Adams, John Quincy. "Magnetic and Meteorological Observations." U.S. House of Representatives, Rept. 630, 26th Cong., 1st sess.

———. *Memoirs of John Quincy Adams*. Vol. 11. Philadelphia, 1876.

American Association for the Advancement of Science. "Members of the American Association for the Advancement of Science." *AAAS, Proceedings* 19 (1871): xxiii–xl.

Arago, François. *Oeuvres complètes*. 17 vols. Paris, 1854–62.

Arbuthnot, John. *An Essay Concerning the Effects of Air on Human Bodies*. London, 1733.

Arnold, S. G. "The Philosophy of Storms." *Merchants' Mag.* (1841): 351–55.

Babinet, J. "Rapport sur les travaux de M. Espy relatifs aux tornados, lu a l'Academie des Sc. de Paris dans sa seance du 15 Mars 1841, par M. Babinet, rapporteur. Comm. MM. Arago, Pouillet, et Babinet." *Comptes rendues hebdomadaires de l'Academie des Sciences* 12 (1841): 34, 139, 454–62.

Bache, A. D. "Notes and Diagrams, Illustrative of the Directions of the Forces Acting at or Near the Surface of the Earth, in Different Parts of the Brunswick Tornado of June 19, 1835." *Amer. Phil. Soc., Transactions* n.s. 5 (1837): 407–20.

———. *Observations at the Magnetic and Meteorological Observatory, at the Girard College, Phila.* Washington, D.C., 1847.

———. "On the Winds of the Western Coast of the United States, from Observations in Connection with the U.S. Coast Survey." *AAAS, Proceedings* 11 (1858): 183–90.

———. "J. P. Espy." Necrology. *Smithsonian Report, 1859*, 108–11.

Baird, Spencer F., and Charles Girard. *Catalog of North American Reptiles in the Museum of the Smithsonian Institution*. Part 1, *Serpents*. Washington, D.C., 1853. *Smithson. Misc. Coll.* 2.

Beck, T. Romeyn. "Address Delivered before the Lyceum of Natural History (Now the Second Department of the Institute) at Its First Anniversary, March 1, 1824." *Albany Institute, Transactions* 1 (1830): 145.

———. *Eulogium on the Life and Services of Simeon DeWitt, Surveyor General of the State of New York, Chancellor of the University, &c.* Albany, 1835.

———, and Joseph Henry. "Abstract of the Returns of Meteorological Observations Made to the Regents of the University by Sundry Academies in the State." *Albany Institute, Transactions* 3 (1855): 234.

Beloit College. *Annual Catalogue of Beloit College, 1869–1870*. Beloit, Wisc., 1869.

Bigelow, Jacob. "Facts Serving to Show the Comparative Forwardness of the Spring Season in Different Parts of the United States." *Amer. Acad., Memoirs* 4 (1818): 77–85.

———. (report by B. Silliman). "On the Comparative Forwardness of the Spring, in Different Parts of the United States, in 1817." *Amer. Journ. Sci.* 1 (1818): 76–77.

Blake, John Lauris. *Wonders of the Earth*. Cazenovia, N.Y., 1845.

Blodget, Lorin. "Agricultural Climatology of the United States Compared with That of Other Parts of the Globe." U.S. Patent Office, *Annual Report, 1853*, pt. 2: *Agriculture*. U.S. Senate, Ex. Doc. 27, 33d Cong., 1st sess., 328–432.

———. "Report on the Prominent Features of General Climate in the United States, as Exhibited in the Distribution of Temperature and of Rain, and in Explanation of the Illustrative Charts." In U.S. Army Medical Department, *Army Meteorological Register for Twelve Years, from 1843 to 1854 Inclusive*, pp. 688–768. Washington, D.C., 1855.

———. *Climatology of the United States and the Temperate Latitudes of the North American Continent: Embracing a Full Comparison of These with the Climatology of the Temperate Latitudes of Europe and Especially in Regard to Agriculture, Sanitary Investigations, and Engineering with Isothermal and Rain-Charts for Each Season, the Extreme Months, and the Year, Including a Summary of the Statistics of Meteorological Observations in the United States, Condensed from Recent Scientific and Official Publications*. Philadelphia, 1857.

Brandes, Heinrich W. *Beiträge zur Witterungskunde*. Leipzig, 1820.

Brewster, Sir David. *Memoirs of the Life, Writings, and Discoveries of Sir Isaac Newton*. Vol. 2. 1855. Reprint. New York, 1965.

"Brief Synopsis of the Principles of James P. Espy's Philosophy of Storms." *Amer. Journ. Sci.* 39 (1840): 120–32.
Buchan, Alexander. *Handy Book of Meteorology.* 2d ed. Edinburgh, 1869.
Chalmers, Lionel. *An Account of the Weather and Diseases of South Carolina.* 2 vols. London, 1776.
Chapplesmith, John. "Account of a Tornado Near New Harmony, Indiana, April 30, 1852, with a Map of the Track, Diagrams, and Illustrative Sketches." *Smithson. Contrib.* 7, Article II (1855); also issued as a pamphlet.
Coffin, James Henry. "Meteorological Observations and Researches at Williams College (Nos. 1 and 2)." *Berkshire County Whig* 1 (Nov. 25, 1841, and Jan. 27, 1842).
______. "An Investigation of the Storm Curve, Deduced from the Relation Existing between the Direction of the Wind, and the Rise and Fall of the Barometer." *AAAS, Proceedings* 7 (1853): 83–101.
______. "On the Winds of the Northern Hemisphere." *Smithson. Contrib.* 6, Article VI (1853).
______. *Psychrometrical Tables for Determining the Force of Aqueous Vapor, and the Relative Humidity of the Atmosphere from Indications of the Wet and the Dry Bulb Thermometer, Fahrenheit.* Washington, D.C., 1856. *Smithson. Misc. Coll.* 1, Article II (1862).
______. "On the Currents of the Atmosphere." *AAAS, Proceedings* 12 (1859): 200–205.
______. "The Winds of the Globe: or the Laws of Atmospheric Circulation over the Surface of the Earth." *Smithson. Contrib.* 20 (1875).
______, ed. *Meteorological Register and Scientific Journal.* Oswego, N.Y., 1838.
Collin, Nicholas. "Observations Made at an Early Period on the Climate of the Country Along the River Delaware, Collected from the Records of the Swedish Colony." *Amer. Phil. Soc., Transactions* n.s. 1 (1818): 340–52.
Columbus, Ferdinand. *The Life of the Admiral Christopher Columbus by His Son Ferdinand.* Translated by Benjamin Keen. New Brunswick, N.J., 1959.
Condorcet, Jean-Antoine-Nicolas de Caritat, Marquis de. *Oeuvres complètes.* 21 vols. (Paris, 1804).
Cooper, James. "On the Distribution of the Forests and Trees of North America, with Notes on Its Physical Geography." *Smithsonian Report, 1858,* 246–80.
Cotte, Louis. *Traité de métérologie.* Paris, 1774.
______. *Mémoires sur la métérologie pour servir et de supplément au Traité de Météorologie publie en 1774.* 2 vols. Paris, 1788.
Dalton, John. *Meteorological Observations and Essays.* 2d ed. London, 1834.
Davis, Charles H. "Redfield, Reid, Espy and Loomis on the Theory of Storms." *North American Review* 58 (1844): 335–71.
De La Beche, H. T. *How to Observe. Geology.* London, 1835.
Dewey, Professor [Chester]. "An Attempt to Ascertain at Which Three Hours of the Day the Thermometer Will Give Nearly the Mean Temperature." *North American Review* 6 (1818): 436–3[7].
DeWitt, Simeon. "Respecting a Plan of a Meteorological Chart, for Exhibiting a Comparative View of Climates of North America, and the Progress of Vegetation." *Society for the Promotion of Agriculture, Arts and Manufactures, Transactions* 1, 2d ed. rev. (Albany, 1801): 88–92.
______. "Result of Thermometrical Observations for the Years 1795 and 1796 Made at the City of Albany." *Society for the Promotion of Agriculture, Arts and Manufactures, Transactions* 1, 2d ed. rev. (Albany, 1801): 287–88.
"Disasters on the Lakes." U.S. House of Representatives, Misc. Doc. 10, and Ex. Doc. 10, pt. 2, 41st Cong., 2d sess.
Dove, H. W. "On the Law of Storms." *Cambridge Miscellany of Mathematics, Physics and Astronomy,* no. 2 (July 1842): 94–96, and no. 3 (Oct. 1842): 137–40.
Dunbar, William. "Meteorological Observations." *Amer. Phil. Soc., Transactions* 4 (1809): 48.

Emerson, Gouverneur. "The Part Taken by the American Philosophical Society and the Franklin Institute in Establishing Stations for Meteorological Observations." *Amer. Phil. Soc., Proceedings* 11 (1871): 516–20.

Emerson, Ralph Waldo. "The American Scholar." In Alfred R. Ferguson, ed., *The Collected Works of Ralph Waldo Emerson.* Vol. 1, *Nature, Addresses, and Lectures,* pp. 52–70. Cambridge, Mass., 1971.

Englemann, George. "Meteorological Table for St. Louis." *St. Louis Medical and Surgical Journal* 10 (July 1852): 297–302.

Espy, James P. "On the Importance of Hygrometric Observations in Meteorology, and the Means of Making Them with Accuracy." *Journ. Frankl. Inst.* n.s. 7 (1831): 221–29, 361–64.

———. "Observations on the Importance of Meteorological Observations, Particularly as Regards the Dew Point; and Also on the Several Fluctuations of the Barometer." *Journ. Frankl. Inst.* n.s. 8 (1831): 389–406.

———. "Notes of an Observer—Meteorology." *Journ. Frankl. Inst.* n.s. 13 (1834): 9–11.

———. "Notes of an Observer. Remarks on Professor Olmsted's Theory of the Meteoric Phenomenon of November 12th, 1833, Denominated Shooting Stars, with Some Queries toward Forming a Just Theory." *Journ. Frankl. Inst.* n.s. 15 (1835): 9–19, 85–92, 158–65, 234–38.

———. "Theory of Rain, Hail, Snow, and the Waterspout, Deduced from the Latent Caloric of Vapour and the Specific Caloric of Atmospheric Air." *Geological Society of Pennsylvania, Transactions* 1, pt. 2 (Philadelphia, 1835): 342–43.

———. "Essays on Meteorology," Nos. I. and II. "Theory of Hail." *Journ. Frankl. Inst.* n.s. 17 (1836): 240–46, 309–16; No. III. "Examination of Hutton's, Redfield's and Olmsted's Theories," No. IV. "North East Storms, Volcanoes, and Columnar Clouds." *Journ. Frankl. Inst.* n.s. 18 (1836): 100–108, 239–46.

———. *Hints to Observers on Meteorology.* Philadelphia, 1837.

———. "Deductions from Observations Made, and Facts Collected on the Path of the Brunswick Spout of June 19th, 1835." *Amer. Phil. Soc., Transactions* n.s. 5 (1837): 421–26.

———. "Examination of Col. Reid's Work on the Law of Storms." *Journ. Frankl. Inst.* n.s. 23 (1839): 38–50, 149–58, 217–31, 289–98.

———. "Objections to Mr. Espy's Theory of Rain, Hail, &c., in a Letter to Mr. Espy, by Mr. Graham Hutchinson of Glasgow, with Replies by James P. Espy." *Journ. Frankl. Inst.* n.s. 23 (1839): 80–90.

———. "On Storms." *Athenaeum* (1840): 794–98; also *BAAS, Annual Report, 1840.* Pt. 2, *Notices and Abstracts of Communications* (1841): 30–40.

———. *The Philosophy of Storms.* Boston, 1841.

———. *First Report on Meteorology to the Surgeon General of the United States Army.* Washington, D.C., 1843.

———. "Report on Meteorology." In U.S. Army Surgeon General, *Annual Report, 1845,* U.S. Senate, Public Doc. 5, 29th Cong., 1st sess., 1:234–35.

———. "Extract from a Communication from Professor Espy on the Subject of Meteorology." *Smithsonian Report, 1848,* Appendix 3, pp. 47–48; also in *Report of the Board of Regents of the Smithsonian Institution, January 6, 1848,* U.S. Senate, Misc. Doc. 23, 30th Cong., 1st sess., 207–8.

———. "On the Diminution of Temperature in Air by Expansion, and Permeation of Air by Moisture." *AAAS, Proceedings* 4 (1851): 371–74.

———. *Second Report on Meteorology to the Secretary of the Navy.* Submitted Nov. 12, 1849; communicated to the U.S. Senate, Mar. 13, 1850; and printed, paginated, and bound with his *Third Report on Meteorology.* Washington, D.C., 1851. U.S. Senate, Ex. Doc. 39, 31st Cong., 1st sess.

———. *Third Report on Meteorology, with Directions for Mariners, etc.* Printed, paginated, and bound with Espy's *Second Report on Meteorology to the Secretary of the Navy.* Washington, D.C., 1851. U.S. Senate, Ex. Doc. 39, 31st Cong., 1st sess.

———. *Fourth Meteorological Report.* Washington, D.C., 1857. U.S. Senate, Ex. Doc. 65, 34th Cong., 3d sess.

Falconer, William. *Remarks on the Influence of Climate, Situation, Nature of Country, Population, Nature of Food, and Way of Life, on the Disposition and Temper, Manners and Behavior, Intellects, Laws and Customs, Form of Government, and Religion, of Mankind.* London, 1781.

Farrar, John. "An Account of the Violent and Destructive Storm of the 23rd of September 1815." *Amer. Acad., Memoirs* 4 (1821): 92–97.

Ferrel, William. "An Essay on the Winds and Currents of the Ocean." *Nashville Journal of Medicine and Surgery* 11 (1856): 287–301, 375–89. Reprinted in *Professional Papers of the Signal Service* 12 (Washington, D.C., 1882): 1–19.

———. "The Influence of the Earth's Rotation upon the Relative Motion of Bodies Near Its Surface." *Astronomical Journal* 5 (1858): 97–100.

———. "The Motions of Fluids and Solids Relative to the Earth's Surface." *Mathematical Monthly* 1 (1859): 140–48, 210–16, 300–307, 366–73, 397–406; 2 (1859–60): 89–97, 339–46, 374–90.

———. "William Ferrel: Complete List of His Publications." *Naturalist* 4, no. 5 (1889): 2–3.

———. "Autobiographical Sketch." *Natl. Acad. Sci., Biog. Mem.* 3 (1895): 287–99.

Fitzroy, Robert. *The Weather-Book: A Manual of Practical Meteorology.* London, 1863.

Forbes, James David. "Report on the Recent Progress and Present State of Meteorology." *BAAS, Report of the First and Second Meetings* (1833): 196–258.

Foreman, Edward. "Report of the General Assistant, with Reference to the Meteorological Correspondence." *Smithsonian Report, 1851,* 68–78.

Forry, Samuel. *The Climate of the United States and Its Endemic Influences.* New York, 1842.

Fothergill, John. "Some Remarks on the Bills of Mortality in London with an Account of a Late Attempt to Establish an Annual Bill for this Nation." In John C. Lettsom, ed., *The Works of John Fothergill, M.D.* 3 vols. 2:107–13. London, 1783.

Franklin Institute, Board of Managers. "Fifty-fourth Quarterly Report of the Board of Managers of the Franklin Institute." *Journ. Frankl. Inst.* n.s. 20 (1837): 130.

———. "Fifteenth Annual Report of the Board of Managers of the Franklin Institute, of the State of Pennsylvania, for the Promotion of the Mechanic Arts." *Journ. Frankl. Inst.* n.s. 23 (1839): 92–95.

Franklin Institute, Committee on Meteorology. "Report of the Committee on Meteorology to the Board of Managers of the Franklin Institute, Embodying the Facts Collected by the Meteorologist Relative to the Storm of the 16th, 17th and 18th March, 1838." *Journ. Frankl. Inst.* n.s. 22 (1838): 161–75.

Goad, John. *Astro-meteorologica, or Aphorisms and Discourses of the Bodies Coelestial, Their Nature and Influences . . . Collected from the Observations at Leisure Times, of Above Thirty Years.* London, 1686.

Gould, Benjamin. "On Peirce's Criterion for the Rejection of Doubtful Observations, with Tables for Facilitating Its Application." *Astronomical Journal* 4 (1858): 81–87.

Grant, John. *Natural and Political Observations . . . upon the Bills of Mortality.* London, 1662.

[Gray, Asa?]. "The Smithsonian Institution." *Amer. Journ. Sci.* 2d ser. 20 (1855): 1–21.

Griscomb, John. "Hints Relative to the Most Eligible Method of Conducting Meteorological Observations." *Literary and Philosophical Society of New York, Transactions* 1 (1815): 341–54.

Guyot, Arnold. *Directions for Meteorological Observations, Intended for the First Class of Observers.* Washington, D.C., 1850.

———. "On the Progress of the System of Meteorological Observations Conducted by the

Smithsonian Institution, and the Propriety of Its Immediate Extension Throughout the American Continent." *AAAS, Proceedings* 6 (1852): 167–68.

———. *A Collection of Meteorological Tables, with Other Tables Useful in Practical Meteorology.* 1st ed. Washington, D.C. 1852; 2d ed. Washington, D.C., 1857; 3d ed. titled *Tables, Meteorological and Physical, Prepared for the Smithsonian Institution.* Washington, D.C., 1859, *Smithson. Misc. Coll.* 1, Article III (1862); 4th ed. Washington, D.C., 1884.

———. "Directions for Meteorological Observations and the Registry of Periodical Phenomena." *Smithson. Misc. Coll.* 1, Article I (1860).

———, and Joseph Henry. "Directions for Meteorological Observations." *Smithsonian Report, 1855,* 215–44.

Hare, Robert. "On the Gales Experienced in the Atlantic States of North-America." *Amer. Journ. Sci.* 5 (1822): 352–56.

———. "On the Causes of the Tornado, or Water Spout." *Amer. Phil. Soc., Transactions* n.s. 5 (1837): 375–81.

———. "Facts and Observations Respecting the Tornado Which Occurred at New Brunswick, New Jersey, in June Last, Abstracted from a Written Statement made by James P. Espy, M.A.P.S." *Amer. Phil. Soc., Transactions* n.s. 5 (1837): 381–84.

———. *Hare's Letter to Arago, April 24, 1840.* Pamphlet Collection, Bell-Henry Library, Smithsonian Institution.

———. "On the Theory of Storms, with Reference to the Views of Mr. Redfield." *Phil. Mag.* 19 (1841): 423–32.

———. "Objections to Mr. Redfield's Theory of Storms with Some Strictures upon His Reasoning." *Amer. Journ. Sci.* 42 (1841): 140–47; *Phil. Mag.* 23 (1843): 92–106.

———. "Additional Objections to Redfield's Theory of Storms." *Amer. Journ. Sci.* 43 (1842): 122–40.

———. *Of the Conclusion Arrived at by a Committee of the Academy of Sciences of France . . . with Objections to the Opinions of Peltier and Espy.* Philadelphia, 1842; rev. 1852.

———. "Strictures on Prof. Dove's Essay 'On the Law of Storms.'" *Amer. Journ. Sci.* 44 (1843): 137–46; *Phil. Mag.* 25 (1844): 94–102.

———. "On the Whirlwind Theory of Storms." *AAAS, Proceedings* 4 (1851): 231–42.

———. "Strictures on Professor Espy's Report on Storms to the Secretary of the Navy, as Respects the Theoretical Inferences." *AAAS, Proceedings* 6 (1852): 152–60.

———. "Dr. Hare on the Law of Storms." *Merchants' Mag.* 26 (1852): 190–95.

———. *Queries and Strictures by Dr. Hare, Respecting Professor Espy's Meteorological Report to the Naval Department. Also, the Conclusion Arrived at by a Committee of the Academy of Sciences of France, Agreeably to Which, Tornadoes Are Caused by Heat; While, Agreeably to Peltier's Report to the Same Body, Certain Insurers Had Been Obliged to Pay for a Tornado as an Electrical Storm. With Abstract from Peltier's Report.* Philadelphia, 1852; also *Journ. Frankl. Inst.* 3d ser. 23 (1852): 325–31, and 3d ser. 24 (1852): 28–39.

———. *The Whirlwind Theory of Storms.* New York, 1853.

———. "On Mr. John Wise's Observations and Inferences Respecting the Phenomena of a Thunder Storm, to Which He Was Exposed during an Aerial Voyage, Made by Means of a Balloon, June 3, 1852, from Portsmouth, Ohio." *Smithsonian Report, 1854,* 224–30.

———. "On the Suppositious Travelling Whirlwinds Called Cyclones." MS, n.d., Bell-Henry Library, Smithsonian Institution.

Hayes, Isaac I. *Physical Observations in the Arctic Seas . . . during 1860 and 1861.* Compiled by C. A. Schott. *Smithson. Contrib.* 15, Article V (1867).

Hemmer, J. J. "Historia Societas Meteorologicae Palatinae." Societas Meteorologicae Palatinae, *Ephemerides, 1781* (1783): 1–54.

Henry, Joseph. "Topographical Sketch of the State of New-York, Designed Chiefly to Show the General Elevations and Depressions of Its Surface." *Albany Institute, Transactions* 1 (1830): 87–112.

———. "On a Disturbance of the Earth's Magnetism, in Connexion with the Appearance of an Aurora Borealis, as Observed at Albany, April 19th, 1831." *Annual Report of the Regents of the State of New York* (Albany, 1832), 107–19; and *Amer. Journ. Sci.* 22 (1832): 143–55.

———. "Report of the Tenth Meeting of the British Association for the Advancement of Science, Held at Glasgow, Sept. 1840." *Biblical Repertory and Princeton Review* 13 (1841): 132–49.

———. "Oral Communication on the Application of Melloni's Thermo-electric Apparatus to Meteorological Purposes, Nov. 3, 1843." *Amer. Phil. Soc., Proceedings* 4 (1847): 22.

———. "Programme of Organization of the Smithsonian Institution, Dec. 8, 1847." *Smithsonian Report 1848,* 6, 25; also in *Report of the Board of Regents of the Smithsonian Institution, January 6, 1848,* U.S. Senate, Misc. Doc. 23, 30th Cong., 1st sess., 175, 190.

———. "On Heat, and on a Thermal Telescope." Paper read to the Association of American Geologists and Naturalists. Reported in *Amer. Journ. Sci.* 2d ser. 5 (1848): 113–14.

———. *An Account of the Smithsonian Institution, Presented to the American Association for the Advancement of Education . . . August 10th, 1853.* Newark, N.J., 1854.

———. "Meteorology in Its Connection with Agriculture." U.S. Patent Office, *Annual Report, Agricultural, 1855,* 357–94; *1856,* 455–92; *1857,* 419–506; *1858,* 429–93; *1859,* 461–524.

———. "Syllabus of a Course of Lectures on Physics." *Smithsonian Report, 1856,* 187–91.

Hippocrates. "On Airs, Waters, and Places," and "Of the Epidemics." *Hippocratic Writings.* Translated by Francis Adams. Chicago, 1952.

Holm, Thomas Campanaius. *Kort Beskrifning om Provincieu Nya Sverige uti America. . .* Stockholm, 1702.

Holyoke, Edward A. "Observations on Weather and Diseases at Salem, Massachusetts, for the Year 1786, Extracted from a Communication by Edward Augustus Holyoke, M.D. to the Massachusetts Medical Society." In William Currie, comp. *An Historical Account of the Climates and Diseases of the United States of America; and of the Remedies and Methods of Treatment, Which Have Been Found Most Useful and Efficacious, Particularly in Those Diseases Which Depend upon Climate and Situation.* Philadelphia, 1792.

Hough, Franklin B., comp. *Results of a Series of Meteorological Observations Made in Obedience to Instructions from the Regents of the University at Sundry Academies in the State of New York, from 1826 to 1850, Inclusive.* Albany, 1855.

———. comp. *Results of a Series of Meteorological Observations Made under Instructions from the Regents of the University at Sundry Stations in the State of New York. Second Series from 1850 to 1863, Inclusive, with Record of Rainfall and Other Phenomena to 1871.* Albany, 1872.

———. *Historical and Statistical Record of the University of the State of New York during the Century from 1784 to 1884.* Albany, 1885.

Hutton, James. "The Theory of Rain." *Royal Society of Edinburgh, Transactions* 1 (1788): 41–86.

Invitation to a meeting of those interested in the establishment of a meteorological association. [Cincinnati], 1869.

Jefferson, Thomas. *Notes on the State of Virginia.* Paris, 1785. Reprint. Gloucester, Mass., 1976.

Johnson, Walter R. "Observations on the Effects of a Remarkable Atmospheric Current or Storm as Witnessed on the Day Following Its Occurrence." *Journ. Philadelphia Academy of Natural Sciences* 7 (1834–37): 269–81.

Joint Committee on Meteorology of the American Philosophical Society and the Franklin Institute. "Circular in Relation to Meteorological Observations, by the Joint Committee of the American Philosophical Society, and of the Franklin Institute." *Journ. Frankl. Inst.* n.s. 14 (1834): 382–85.

———. "First Report of the Joint Committee of the American Philosophical Society, and

Franklin Institute, on Meteorology." *Journ. Frankl. Inst.* n.s. 16 (1835): 4–6.

———. "Second Report of the Joint Committee on Meteorology, of the American Philosophical Society and Franklin Institute." *Journ. Frankl. Inst.* n.s. 17 (1836): 386–93.

———. "Third Report of the Joint Committee on Meteorology, of the American Philosophical Society and the Franklin Institute of the State of Pennsylvania, for the Promotion of the Mechanic Arts." *Journ. Frankl. Inst.* n.s. 19 (1837): 17–21.

Jurin, James. "Invitatio ad Observationes Meteorologicas communi consilio instituendas." *Phil. Trans.* 32 (1723): 422–27.

Kane, Elisha K. *Meteorological Observations in the Arctic Seas . . . 1853, 1854, and 1855.* Compiled by C. A. Schott. *Smithson. Contrib.* 11, Article V (1859).

Kanold, Johann, Johann Christian Kundmann, and J. G. Brunschwitz, eds. *Breslauer Sammlung,* or *Sammlung von Natur- und Medicin-, wie auch hiezu gehörigen Kunst- und Literatur-Geschichten* (1718–30).

Kohl, J. G. "Substance of a Lecture Delivered at the Smithsonian Institution on a Collection of the Charts and Maps of America." *Smithsonian Report, 1856,* 132.

Lapham, Increase A. "The Great Storms." Letter to the Editor of the *Bureau: A Chronicle of the Commerce and Manufactures of Chicago and the Northwest,* Dec. 1869, p. 59. Copy in Abbe Papers, JHU. The article also appeared in the *New York Times,* Dec. 8, 1869.

———. "Memorial of Professor J. [sic] A. Lapham to H. E. Paine, Dec. 8, 1869," in "Disasters on the Lakes." U.S. House of Representatives, Misc. Doc. 10, 41st Cong., 2d sess.

———. "Will It Pay?" *Bureau,* Jan. 1870, pp. 83–84. Copy in Abbe Papers, JHU. Also *Milwaukee Sentinel,* Dec. 8, 1869.

———. "Meteorological Observations." *Bureau,* Jan. 1870, p. 91, map on p. 93. Copy in Abbe Papers, JHU.

Lavoisier, Antoine-Laurent. "Règles pour prédire le changement de temps d'après les variations du baromètre." In Etiene Chirou, ed., *Extraits des mémoires de Lavoisier concernant la météorologie et l'aéronautique,* pp. 139–45. Paris, 1926.

———. *Oeuvres de Lavoisier: Correspondance.* Edited by Réné Fric. 3 vols. Paris, 1955–64.

Ley, William C. *The Laws of the Winds Prevailing in Western Europe.* Pt. 1. London, 1872.

Lining, John. "A Description of American Yellow Fever." In *Essays and Observations,* vol. 2, pp. 404–32. Edinburgh, 1771.

Loomis, Elias. "Observations on a Hurricane Which Passed over Stow, in Ohio, October 20th, 1837." *Amer. Journ. Sci.* 33 (1838): 368–76.

———. "On the Storm Which Was Experienced throughout the United States about the 20th of December, 1836." *Amer. Phil. Soc., Transactions* 7 (1840): 125–63; rev. version *AAAS, Proceedings* 9 (1856): 179–83.

———. "On a Tornado Which Passed over Mayfield, Ohio, February 4th, 1842, with Some Notices of Other Tornadoes." *Amer. Journ. Sci.* 43 (1842): 278–300.

———. "On Two Storms Which Occurred in February, 1842." *Amer. Phil. Soc., Proceedings* 3 (1843): 50–56.

———. "On the Determination of Differences of Longitude Made in the United States by Means of the Electric Telegraph, and on the Projected Observations for Investigating the Laws of the Great North American Storms." *Phil. Mag.* 81 (1847): 338–40.

———. "Report on the Meteorology of the United States." *Smithsonian Report, 1848,* Appendix 2, 28–46; also in *Report of the Board of Regents of the Smithsonian Institution, January 6, 1848,* U.S. Senate, Misc. Doc. 23, 30th Cong., 1st sess., 193–207.

———. "On Certain Storms in Europe and America, December, 1836." *Smithson. Contrib.* 11, Article VIII (1859).

———. *A Treatise on Meteorology: With a Collection of Meteorological Tables.* New York, 1868.

______. "Results Derived from Examination of the United States Weather Maps for 1872 and 1873." *Amer. Journ. Sci.* 3d ser. 8 (1874): 1–15.

MacSparran, James. *America Dissected, Being a Full and True Account of All the American Colonies, Shewing the Intemperance of the Climates, Excessive Heat and Cold, and Sudden Violent Changes of Weather, Terrible and Mischievous Thunder and Lightning, Bad and Unwholesome Air, Destructive to Human Bodies, etc.* Dublin, 1753.

Mann, James. *Medical Sketches of the Campaigns of 1812, 13, 14. To Which are Added, Surgical Cases; Observations on Military Hospitals; and Flying Hospitals Attached to a Moving Army. . .* Dedham, Mass., 1816.

Mariotte, Edme. *Oeuvres de Mariotte.* 2 vols. in 1. Leiden, 1717.

Martin, F. P. B. *A Memoir on the Equinoctial Storms of March–April, 1850; an Inquiry into the Extent to Which the Rotary Theory May Be Applied.* Kingston, Eng., 1852.

Mather, Cotton. *The Christian Philosopher: A Collection of the Best Discoveries in Nature with Religious Improvements.* London, 1721; Charlestown, Mass., 1815.

Mather, Increase. *The Voice of God in Stormy Winds. Considered in Two Sermons Occasioned by the Dreadful and Unparallel'd Storm in the European Nations, Nov. 27, 1703.* Boston, 1704.

Maury, Matthew F. "On the General Circulation of the Atmosphere." *AAAS, Proceedings* 3 (1850): 126–47.

______. *On the Establishment of an Universal System of Meteorological Observations by Sea and Land.* Washington, D.C., 1851.

______. *The Physical Geography of the Sea and Its Meteorology.* 1855. Reprint edited by John Leighly. New York, 1963.

M'Clintock, Francis L. *Observations in the Arctic Seas . . . 1857, 1858, 1859.* Compiled by C. A. Schott. Smithson. Contrib. 13, Article III (1862).

Meigs, Josiah. "Circular to the Registers of the United States." *Niles' Weekly Register* 12 (May 10, 1817): 167–68.

______. "Geometric Exemplification of Temperature, Wind and Weather for 1820 at Washington." *Amer. Phil. Soc., Early Proceedings,* 505.

"Memorial from the Pennsylvania Lyceum, etc. Asking an Appropriation to Aid in the Advancement of Meteorology, Apr. 20, 1838." U.S. House of Representatives, Doc. 344, 25th Cong., 2d sess.

"Memorials of James P. Espy and of the Philadelphia Lyceum, Praying the Aid of the General Government in Making a Series of Meteorological Observations, May 2, 1838." U.S. Senate, Doc. 414, 25th Cong., 2d sess.

Mower, Thomas Gardner. "Notice of the Meteorological Observations Now Making at the Military Posts of the United States." *Amer. Phil. Soc., Proceedings* 3 (1843): 158–62.

Nicollet, J. N. *Essay on Meteorological Observations.* Washington, D.C., 1839.

"Notice of a Meteorological Register for the Years 1822, 1823, 1824 and 1825; from Observations Made by the Surgeons of the Army, at the Military Posts of the United States." *Amer. Journ. Sci.* 12 (1827): 149–54.

Olmsted, Denison. "Of the Phenomena and Causes of Hail Storms." *Amer. Journ. Sci.* 18 (1830): 1–11.

______. ["Ten Lectures on Meteorology"]. *Merchants' Mag.* 1 (1839): 529–35.

______. "Biographical Memoir of William C. Redfield." *Amer. Journ. Sci.* 2d ser. 24 (1857): 355–73.

______. "An Address in Commemoration of William C. Redfield, First President of the Association." *AAAS, Proceedings* 11 (1858): 9–34.

______. *Outlines of a Course of Lectures on Meteorology and Astronomy.* New Haven, 1858. Copy with marginalia in Sterling Library, Yale University.

Peirce, Benjamin. "On Espy's Theory of Storms." *Cambridge Miscellany of Mathematics, Physics and Astronomy*, no. 3 (Oct. 1842): 141–44.

———. "Criterion for the Rejection of Doubtful Observations." *Astronomical Journal* 2 (1852): 161–63.

Pickering, Roger. "A Scheme of a Diary of the Weather; Together with Draughts and Descriptions of Machines Subservient Thereunto." *Phil. Trans.* 43, no. 473 (1744–45): 1–18.

Piddington, Henry. *The Sailor's Horn-Book for the Law of Storms; Being a Practical Exposition of the Theory of the Law of Storms, and Its Uses to Mariners of All Classes in All Parts of the World, Shewn by Transparent Storm Cards and Useful Lessons.* London, 1848.

Prince, Thomas. *The Natural and Moral Government and Agency of God in Causing Droughts and Rains: A Sermon Deliver'd at the South Church in Boston, Thursday, Aug. 24, 1749, Being the Day of the General Thanksgiving in the Province of Massachusetts.* Boston, 1751.

Prout, William. *Chemistry, Meteorology, and the Function of Digestion, Considered with Reference to Natural Theology.* Bridgewater Treatise VIII, 2d ed. London, 1834.

Quetelet, Adolphe. *Sur le climat de la Belgique.* 2 vols. Brussels, 1847–57.

———. "Sur la theorie des ourages de M. Espy." *Annuaire météorologique de la France* 4 (1852).

Redfield, William. "Remarks on the Prevailing Storms of the Atlantic Coast, of the North American States." *Amer. Journ. Sci.* 20 (1831): 17–51.

———. "Summary Statements of Some of the Leading Facts in Meteorology." *Amer. Journ. Sci.* 25 (1834): 122–35.

———. "On the Evidence of Certain Phenomena." *Journ. Frankl. Inst.* n.s. 15 (1835): 372–79; *Amer. Journ. Sci.* 28 (1835): 310–18.

———. "In Reply to Mr. Espy, on the Whirlwind Character of Certain Storms." *Journ. Frankl. Inst.* n.s. 19 (1837): 112–27.

———. "On the Courses of Hurricanes; with Notices of the Tyfoons of the China Sea, and Other Storms." *Amer. Journ. Sci.* 35 (1839): 201–23.

———. "Remarks on Mr. Espy's Theory of Centripetal Storms, Including a Refutation of His Positions Relative to the Storm of September 3rd, 1821: with Some Notice of the Fallacies Which Appear in His Examination of Other Storms." *Journ. Frankl. Inst.* n.s. 23 (1839): 323–36, 363–78.

———. "Further Notice of the Fallacies of Mr. Espy's Examination of Storms." *Journ. Frankl. Inst.* n.s. 24 (1839): 1–4.

———. "Reply to Dr. Hare's Objections to the Whirlwind Theory of Storms." *Amer. Journ. Sci.* 42 (1842): 299–316.

———. *On Whirlwind Storms; with Replies to the Objections and Strictures of Dr. Hare.* New York, 1842.

———. "Notice of Dr. Hare's 'Strictures' on Prof. Dove's 'Essay on the law of storms.'" *Amer. Journ. Sci.* 44 (1843): 384–92.

———. "On the Apparent Necessity of Revising the Received Systems of Dynamical Meteorology." *AAAS, Proceedings* 4 (1851): 366–69.

———. "On the First Hurricane of September 1853, in the Atlantic; with a Chart; and Notices of Other Storms." *Amer. Journ. Sci.* 2d ser. 18 (1854): 1–18, 176–90.

———. "On the Spirality of Motion in Whirlwinds and Tornadoes." *Amer. Journ. Sci.* 2d ser., 23 (1857): 23–24.

———. "On Various Cyclones or Typhoons of the North Pacific Ocean; with a Chart, Showing Their Courses or Progression." *AAAS, Proceedings* 10 (1856): 214–22; also *Amer. Journ. Sci.* 2d ser. 24 (1857): 21–38.

———. "Observations in Relation to the Cyclones of the Western Pacific, Embraced in a Communication to Commodore Perry." In Matthew C. Perry, *United States Japan Expedition*, comp. Francis Hawkes, 2:331–57. Washington, D.C., 1856.

Reid, William. "A Report Explaining the Progress Made Towards Developing the Law of Storms; and of What Seems Further Desirable to Be Done, to Advance Our Knowledge of the Subject." *Athenaeum* (1838): 594–95; *BAAS, Annual Report, 1838*. Pt. 2, *Notices and Abstracts of Communications* (1839): 21–25.

———. *An Attempt to Develop the Law of Storms by Means of Facts, Arranged According to Place and Time; And Hence to Point Out a Cause for the Variable Winds, With a View to Practical Use in Navigation*. London, 1838.

———. *The Progress of the Development of the Law of Storms and of the Variable Winds, with the Practical Application of the Subject to Navigation*. London, 1849.

Rhees, William J. *An Account of the Smithsonian Institution, Its Founder, Building, Operations, etc.* Washington, D.C., 1859. Reprint. Philadelphia, 1864.

———. ed. *The Smithsonian Institution: Documents Relative to Its Origin and History, 1835–99*, 2 vols. *Smithson. Misc. Coll.* 42 and 43. (1901).

Rush, Benjamin. "Account of the Climate of Pennsylvania, and Its Influence upon the Human Body." *American Museum* 4 (1789): 26.

Ruskin, John. "Remarks on the Present State of Meteorological Science." *Meteorological Society of London, Transactions* 1 (1839), 56–59.

Russell, Robert. *North America, Its Agriculture and Climate; Containing Observations on the Agriculture and Climate of Canada, the United States, and the Island of Cuba*. Edinburgh, 1857.

Schott, Charles, A. "Tables and Results of the Precipitation, in Rain and Snow, in the United States and at Some Stations in Adjacent Parts of North America and in Central and South America." *Smithson. Contrib.* 18, Article II (1872).

———. "Tables, Distribution, and Variations of the Atmospheric Temperature in the United States and Some Adjacent Parts of North America." *Smithson. Contrib.* 21, Article V (1876). MS version in RG 27, NA.

———. "Letter on Peirce's Criterion, Nov. 22, 1877." *Amer. Acad., Proceedings* 13 (1878): 350–51.

Six Articles upon the Smithsonian Institution, from the Boston Post, together with the Letters of Professor Peirce and Agassiz. Boston, 1855.

Smithsonian Institution. *Smithsonian Meteorological Observations for the year 1855*. Washington, D.C., 1857.

Sprat, Thomas. *History of the Royal Society*. London, 1667. Reprint edited by Jackson I. Cope and Harold W. Jones. St. Louis, 1958.

Strachey, William. *The Historie of Travell into Virginia Britania*. 1612. Edited by Louis B. Wright and Virginia Freund. London, 1953.

Tracy, Charles. "On the Rotary Action of Storms." *Amer. Journ. Sci.* 45 (1843): 65–72.

Tichomirov, E. I. "Instructions for Russian Meteorological Stations of the 18th Century." In Russian, English summary. *Central Geophysical Observatory, Proceedings* (1932): 3–12.

U.S. Army Medical Department. "Regulations for the Medical Department." In *Military Laws and Rules and Regulations for the Army of the United States*, pp. 227–28. Washington, D.C., 1814.

———. *Regulations of the Medical Department*. Washington, D.C., 1818.

———. *Meteorological Register for the Years 1822, 1823, 1824, and 1825 from Observations made by the Surgeons of the Army at the Military Posts of the United States*. Washington, D.C., 1826.

———. *Statistical Report on the Sickness and Mortality of the Army of the United States, January, 1819, to January, 1839*. Compiled by Samuel Forry. Washington, D.C., 1840.

———. *Meteorological Register for the Years 1826, 1827, 1828, 1829, and 1830, from Observations made by the Surgeons of the Army and Others at the Military Posts of the United States. To Which is Appended the Meteorological Register for the Years 1822, 1823, 1824, and 1825*. Compiled by Samuel Forry. Philadelphia, 1840.

———. *Meteorological Register for Twelve Years, from 1831 to 1842 Inclusive. Compiled from Observations made by the Officers of the Medical Department of the Army at the Military Posts of the United States.* Washington, D.C., 1851.

———. *Army Meteorological Register for Twelve Years, from 1843 to 1854, Inclusive. Compiled from Observations made by the Officers of the Medical Department of the Army at the Military Posts of the United States.* Washington, D.C., 1855.

U.S. Army Signal Office. *Instructions to Observer-Sergeants.* Washington, D.C., 1871.

———. *Practical Use of Meteorological Reports and Weather-Maps.* Washington, D.C., 1871.

U.S. Army Topographical Engineers. "Report of the Topographical Bureau, Nov. 14, 1860." U.S. Senate Ex. Doc. 1, vol. 2, 36th Cong., 2d sess. Bound separately as *Meteorology of the Lakes, 1860* in the Bell-Henry Library, Smithsonian Institution.

U.S. Census Office. "Meteorology of the United States." In *The Seventh Census of the United States, 1850,* pp. xcii–xciii. U.S. House of Representatives, Misc. Doc. 32nd Cong., 2d sess.

U.S. House of Representatives. ["Report of the Select Committee on the Management of the Smithsonian Institution]," U.S. House of Representatives, Rpt. 141, 33d Cong., 2d sess.

U.S. Patent Office. *Results of Meteorological Observations Made under the Direction of the United States Patent Office and the Smithsonian Institution from the Year 1854 to 1859, Inclusive.* 2 vols. U.S. House of Representatives, Ex. Doc. 55, 36th Cong., 1st sess.

Webster, Noah. *A Brief History of Epidemic and Pestilential Diseases with the Principal Phenomena of the Physical World, Which Precede and Accompany Them, and Observations Deduced from the Facts Stated.* 2 vols. Hartford, Conn., 1799.

Whewell, William. *History of the Inductive Sciences, from the Earliest to the Present Time.* 3 vols. London, 1837.

Williamson, Hugh. "An Attempt to Account for the Change of Climate, Which Has Been Observed in the Middle Colonies in North-America." *Amer. Phil. Soc., Transactions* 1 (1771): 272–78.

———. *Observations on the Climate of Different Parts of America, Compared with the Climate in Corresponding Parts of the Other Continent. To Which Are Added, Remarks on the Different Complexions of the Human Race, with some Account of the Aborigines of America. Being an Introductory Discourse to the History of North-Carolina.* New York, 1811.

Wilson, Job. "A Meteorological Synopsis, in Connection with the Prevailing Diseases for Sixteen Years, as They Occurred at Salisbury, Massachusetts." *Medical Repository* 22 (1822): 409–13.

Wood, William. *New England's Prospect. A True, Lively, and Experimental Description of that Part of America, Commonly called New England: Discovering the State of that Countrie, both as It Stands to New-come English Planters; and to the Old Native Inhabitants. Laying Downe that which May both Enrich the Knowledge of the Mind-travelling Reader, or Benefit the Future Voyager.* London, 1634.

Secondary Sources

Abbe, Cleveland. "Historical Notes on the Systems of Weather Telegraphy and Especially Their Development in the United States." *Amer. Journ. Sci.* 3d ser. 2 (1871): 81–88.

———. "Ferrel's Influence in the Signal Office." *Amer. Metl. Journ.* 8 (1891–92): 342–48.

———. "The Meteorological Work of the U.S. Signal Service, 1870–1893." *U.S. Weather Bureau, Bulletin* 11 (1893): 232–85.

———. "Memoir of William Ferrel, 1817–1891," with Ferrel's "Autobiographical Sketch." *Natl. Acad. Sci., Biog. Mem.* 3 (1895): 265–309.

———. "The Origin of the Weather Bureau." *Independent,* May 20, 1897, p. 5.

———. "Meteorology in Russia." *Monthly Weather Review* 27 (1899): 103–7.

———. "Benjamin Franklin as Meteorologist." *Amer. Phil. Soc., Proceedings* 45 (1906): 117–28.

______. "Meteorology—1859 to 1909." In Reminiscences of Employees, 1907–46. Records of the U.S. Weather Bureau, RG 27, NA.

______. "Biographical Memoir of Charles Anthony Schott, 1826–1901." *Natl. Acad. Sci., Biog. Mem.* 8 (1915): 87–133.

______. "How the United States Weather Bureau Was Started." *Scientific American* n.s. 114 (1916): 529.

Abbe, Truman. *Professor Abbe and the Isobars: The Story of Cleveland Abbe, America's First Weatherman.* New York, 1955.

Accademia del Cimento. *Celebrazione della Accademia del Cimento nel Tricentario della Fondazione.* Pisa, 1957.

Aldredge, Robert C. "Weather Observers and Observations at Charleston, South Carolina from 1670–1871." In *Year Book of the City of Charleston.* Historical Appendix, pp. 190–257. Charleston, 1940.

Allen, David Elliston. *The Naturalist in Britain: A Social History.* London, 1976.

______. "Naturalists in Britain: Some Tasks for the Historian." *Journal of the Society for the Bibliography of Natural History* 8 (1977): 91–107.

Alter, J. Cecil. "National Weather Service Origins." *Bulletin of the Historical and Philosophical Society of Ohio* 7 (1949): 139–85.

American Philosophical Society. "Commemoration of the Life and Work of Alexander Dallas Bache." *Amer. Phil. Soc., Proceedings* 84 (1941): 124–86.

Ashley, Bernard. *Weather Men.* London, 1974.

Bartky, Ian R. "Inventing, Introducing and Objecting to Standard Time." *Vistas in Astronomy* 28 (1985): 105–11.

Battan, Louis J. *Weather.* 2d ed. Englewood Cliffs, N.J., 1985.

Bednarowski, Mary Farrell. "Nineteenth Century Spiritualism: An Attempt at a Scientific Religion." Ph.D. dissertation, University of Minnesota, 1973. Ann Arbor: University Microfilms, 73-25581.

Benjamin, Marcus. "Meteorology." In G. B. Goode, ed., *The Smithsonian Institution, 1846–1896: The History of Its First Half Century,* 647–78. Washington, D.C., 1897.

Blouet, O. M. "Sir William Reid, F.R.S., 1791–1858: Governor of Bermuda, Barbados, and Malta." *Notes and Records of the Royal Society of London* 40 (1985–86): 169–91.

Bode, Carl. *The American Lyceum: Town Meeting of the Mind.* New York, 1956.

Bowden, Mary Ellen. "The Weather Observations of Johannes Kepler and David Fabricius: Astrologically Motivated Research in the Early Seventeenth Century." Paper presented at the History of Science Society, Raleigh, N.C., 1987.

Bowes, Frederick P. *The Culture of Early Charleston.* Chapel Hill, 1942.

[British Meteorological Office]. "Meteorological Office Centenary, 1855–1955." *Metl. Mag.* 84 (1955): 161–98.

Brown, Harvey E. *The Medical Department of the United States Army from 1775 to 1873.* Washington, D.C., 1873.

Brown, Ralph. "The First Century of Meteorological Data in America." *Monthly Weather Review* 68 (1940): 130–33.

Bruce, Robert V. *The Launching of Modern American Science, 1846–1876.* New York, 1987.

Brunt, David. "A Hundred Years of Meteorology, 1851–1951." *Advancement of Science* 8 (1951): 114–24.

______. "The Centenary of the Meteorological Office: Retrospect and Prospect." *Science Progress* 44 (1956): 193–207.

Bull, G. A. "Short History of the Meteorological Office." *Metl. Mag.* 83 (1955): 163–67.

Burstyn, Harold L. "The Deflecting Force and Coriolis." *Amer. Metl. Soc., Bulletin* 47 (1966): 887–91.

Burton, Jim. "Robert Fitzroy and the Early History of the Meteorological Office." *British Journal for the History of Science* 19 (1986): 147–76.

Cannegieter, H. G. "The History of the International Meteorological Organization, 1872–1951." *Annalen der Meteorologie* n.s. 1 (1963): 7–280.

Cannon, Walter F. "Scientists and Broad Churchmen: An Early Victorian Intellectual Network." *Journal of British Studies* 4 (1964): 66–72.

———. "Humboldt or Baconianism? A Defense of American Science in the First Half of the 19th Century." Paper presented at the History of Science Society, Dec. 1969, filed as item 231 in the reprint file of the Joseph Henry Papers, Smithsonian Institution.

——— [a.k.a. Susan F. Cannon]. "Humboldtian Science." In *Science in Culture: The Early Victorian Period*, pp. 73–110. New York, 1978.

Cappel, Albert. "Societas Meteorologica Palatina (1780–1795)." *Annalen der Meteorologie* n.s. 16 (1980): 10–27, 255–61.

Carter, Horace S. "Josiah Meigs, Pioneer Weatherman." *Weatherwise* 13 (1960): 166–67, 181.

Cassedy, J. H. "Meteorology and Medicine in Colonial America: Beginnings of the Experimental Approach." *Journal of the History of Medicine and Allied Sciences* 24 (1969): 193–204.

Cassidy, David C. "Meteorology in Mannheim: The Palatine Meteorological Society, 1780–1795." *Sudhoffs Archiv für Geschichte der Medizin und der Naturwissenschaften* 69 (1985): 8–25.

Cawood, John. "Terrestrial Magnetism and the Development of International Collaboration in the Early Nineteenth Century." *Annals of Science* 34 (1977): 551–87.

———. "The Magnetic Crusade: Science and Politics in Early Victorian Britain." *Isis* 70 (1979): 493–518.

Clawer, F. *Catalog der Meteorologischen Beobachtungen im Russichen Reich Zusammengestelt.* In H. Wild, ed., *Repertorium Für Meteorologie*, vol. 2. St. Petersburg, 1872.

Clyde, John C. *The Life of James H. Coffin, LL.D.* Easton, Pa., 1881.

Coan, Eugene. *James Graham Cooper, Pioneer Western Naturalist.* Moscow, Idaho, 1981.

Cohen, I. Bernard. *Revolutions in Science.* Cambridge, Mass., 1985.

Corless, Richard. "A Brief History of the Royal Meteorological Society." *Weather* 5 (1950): 78–83.

Coulson, Thomas. *Joseph Henry: His Life and Work.* Princeton, 1950.

Dall, William H. *Spencer Fullerton Baird, A Biography.* Philadelphia, 1915.

Dana, James D. "Memoir of Arnold Guyot, 1807–1884." *Natl. Acad. Sci., Biog. Mem.* 2 (1886): 309–47.

Daniels, George H. *American Science in the Age of Jackson.* New York, 1968.

———, ed. *Nineteenth-Century American Science: A Reappraisal.* Evanston, Ill., 1972.

Darter, Lewis J., Jr. "The Federal Archives Relating to Matthew Fontaine Maury." *American Neptune* 1 (1941): 149–58.

———. "The Origin and Development of the Weather Bureau." In Darter, comp., *List of Climatological Records in the National Archives*, pp. 1–32. Washington, D.C., 1942.

Davis, John L. "Weather Forecasting and the Development of Meteorological Theory at the Paris Observatory, 1853–1878." *Annals of Science* 41 (1984): 359–82.

Davis, William Morris. "Some American Contributions to Meteorology." *Journ. Frankl. Inst.* 127 (1889): 104–15, 176–91.

———. "Ferrel's Contributions to Meteorology." *Amer. Metl. Journ.* 8 (1891–92): 348–59.

———. "The Redfield and Espy Period." *U.S. Weather Bureau, Bulletin* 11 (1893): 305–16.

Deiss, William A. "Spencer F. Baird and His Collectors." *Journal of the Society for the Bibliography of Natural History* 9 (1980): 635–45.

De Young, Gregg. "The Storm Controversy (1830–1860) and Its Impact on American Science." *Eos: American Geophysical Union, Transactions* 66 (1985): 657–60.

Dick, Stephen J. "How the U.S. Naval Observatory Began, 1830–65." In Dick and LeRoy E. Doggett, eds., *Sky with Ocean Joined,* pp. 167–81. Washington, D.C., 1983.

Doublet, E. "La Météorologie en France et en Allemagne." *Revue philomathique de Bordeaux et du Sud Ouest* 14 (1911): 213–32, 250–67; 15 (1912): 103–28, 169–86.

Dufour, Louis, "Sketch History of Meteorology in Belgium." *Weather* 6 (1951): 359–64.

______. *Ésquisse d'une histoire de la météorologie en Belgique.* Brussels, 1953. Issued as *Institut Royal Météorologique de Belgique Miscellanées* 40 (1953).

Dupree, A. Hunter. *Science in the Federal Government.* Cambridge, Mass., 1957.

Eaton, Edward Dwight. *Historical Sketches of Beloit College.* New York, 1928.

Egbert, Donald Drew. *Princeton Portraits.* Princeton, 1947.

Egerton, Frank N. "Ecological Studies and Observations before 1900." In B. J. Taylor and T. J. White, eds., *Issues and Ideas in America,* pp. 311–51. Norman, Okla., 1976.

Fassig, Oliver L. "A Sketch of the Progress of Meteorology in Maryland and Delaware." *Maryland Weather Service* 1 (1899): 331–63.

Feldman, Theodore S. "The History of Meteorology, 1750–1800: A Study in the Quantification of Experimental Physics." Ph.D. dissertation, University of California, Berkeley, 1983. Ann Arbor: University Microfilms, 84-13376.

Fleming, James R. "Storms, Strikes, Indian Uprisings, and Other Threats to Domestic Tranquility: The U.S. Army Weather Service and the Telegraph, 1870–1891." Paper presented at the History of Science Society Meeting, Cincinnati, 1988.

Forbes, T. R. *Chronicle from Aldgate: Life and Death in Shakespeare's London.* New Haven, 1971.

[France], Météorologie Nationale. *Ce qu'est la météorologie française.* Paris, 1952.

Frisinger, H. H. *The History of Meteorology to 1800.* New York, 1977.

Garber, Elizabeth, "Thermodynamics and Meteorology (1850–1900)." *Annals of Science* 33 (1976): 51–65.

Gillett, Mary C. *The Army Medical Department, 1775–1818.* Washington, D.C., 1981.

Gillispie, Charles C. *Science and Polity in France at the End of the Old Regime.* Princeton, 1980.

Glassford, William A. *Historical Sketch of the Signal Corps, United States Army.* New York, ca. 1890.

Greene, John C. *American Science in the Age of Jefferson.* Ames, Iowa, 1984.

Gross, Walter E. "The American Philosophical Society and the Growth of Meteorology in the United States, 1835–1850." *Annals of Science* 29 (1972): 321–38.

Grosser, Morton. *The Discovery of Neptune.* Cambridge, Mass., 1962.

Guyot, Arnold. "Memoir of James Henry Coffin, 1806–1873." *Natl. Acad. Sci., Biog. Mem.* 1 (1877): 257–64.

Hagarty, J. H. "Dr. James Tilton, 1745–1822." *Weatherwise* 15 (1962): 124–25.

Hall, Peter Dobkin. *The Organization of American Culture, 1700–1900: Private Institutions, Elites, and the Origins of American Nationality.* New York, 1982.

Hallman, E. S. "110 Years Ago." *Weather* 5 (1950): 155–58.

Hansen, Morris H., William N. Hurwitz, and William G. Madow. *Sample Survey Methods and Theory.* Vol. 1, *Methods and Applications.* New York, 1953.

Hawes, Joseph M. "The Signal Corps and Its Weather Service, 1870–1890." *Military Affairs* 30 (1966): 68–76.

Hayes, Cecil B. *The American Lyceum: Its History and Contribution to Education.* Washington, D.C., 1932.

Hellmann, Gustav. "The Organization of the Meteorological Service in Some of the Principal Countries of Europe." Trans. in *Metl. Mag.* 16 (1881): 33–39, 51–57, 69–73, 83–89, 99–106, 122–28, 142–45, 158–61, 165–73.

______. *Repertorium der Deutchen Meteorologie.* Liepzig, 1883.

———. *Geschichte des Königlich Preussischen Meteorologischen Instituts von seiner Gründing im Jahre 1847 bis zu seiner Reorganisation im Jahre 1885*. Berlin, 1887.

———. *Neudrucke von Schriften und Karten über Meteorologie und Erdmagnetismus*, nos. 5 and 7. Berlin, 1896–97.

———. *Beiträge zur Geschichte der Meteorologie*. 3 vols. Berlin, 1914–22.

———. "Die Entwicklung der meteorologischen Beobachtungen in Deutchland von der ersten Anfängen bis zur Einrichtung staatlicher Beobachtungsnetze." *Abhandlungen der Preussische Akademie der Wissenschaften*, Physikalisch-mathematische Klasse, no. 1. Berlin, 1926.

Henry, Alfred J. "Early Individual Observers in the United States." *U.S. Weather Bureau, Bulletin* 11 (1893): 291–302.

Henry, Joseph. "Memoir of A. D. Bache, 1806–1867." *Natl. Acad. Sci., Biog. Mem.* 1 (1877): 181–212d.

Hindle, Brooke. *The Pursuit of Science in Revolutionary America, 1735–1789*. Chapel Hill, 1956.

Hughes, Patrick. *A Century of Weather Service: A History of the Birth and Growth of the National Weather Service, 1870–1970*. New York, 1970.

Hume, Edgar Erskine. "The Foundation of American Meteorology by the United States Army Medical Department." *Bulletin of the History of Medicine* 8 (1940): 202–38.

Humphreys, William J. "Biographical Memoir of Cleveland Abbe, 1838–1916." *Natl. Acad. Sci., Biog. Mem.* 8 (1919): 469–508.

———. "Origin and Growth of the Weather Service of the United States, and Cincinnati's Part Therein." *Scientific Monthly* 18 (1924): 372–82.

Jordan, J. L. "On Coriolis and the Deflective Force." *Amer. Metl. Soc., Bulletin* 47 (1966): 401–3.

Keeny, Elizabeth Barnaby. "The Botanizers: Amateur Scientists in Nineteenth-Century America." Ph.D. dissertation, University of Wisconsin, 1985.

Kemper, C. E. "Letters of Rev. James Madison, President of William and Mary College, to Thomas Jefferson." *William and Mary College Quarterly Historical Magazine* 2d ser. 5 (July 1925): 145–58.

Khrgian, A. Kh. "The History of Meteorology in Russia." *Actes du VIIIe Congrès Internatinal d'Histoire des Sciences*, pp. 445–48. Paris, 1958.

———. *Meteorology: A Historical Survey*. Vol. 1. 2d ed. Leningrad, 1959. Translated by Ron Hardin. Jerusalem, 1970.

Kington, J. A. "A Late Eighteenth-Century Source of Meteorological Data." *Weather* 25 (1970): 169–75.

Kober, George M. "General Albert J. Myer and the United States Weather Bureau." *Military Surgeon* 65 (1929): 65–83.

Kohlstedt, Sally G. *The Formation of the American Scientific Community: The American Association for the Advancement of Science, 1848–1860*. Urbana, Ill., 1965.

Kupperman, Karen Ordahl. "The Puzzle of the American Climate in the Early Colonial Period." *American Historical Review* 87 (1982): 1262–89.

Kutzbach, Gisela. *The Thermal Theory of Cyclones: A History of Meteorological Thought in the Nineteenth Century*. Boston, 1979.

Landsberg, Helmut E. "Early Stages of Climatology in the United States." *Amer. Metl. Soc., Bulletin* 45 (1964): 268–74.

———. "Early Weather Observations in America." *EDIS: Environmental Data and Information Service* 10, no. 4 (1979): 21–23.

———. "A Bicentenary of International Meteorological Observations." *World Meteorological Organization, Bulletin* 29 (1980): 235–38.

Langley, Samuel P. "The Meteorological Work of the Smithsonian Institution." *U.S. Weather Bureau, Bulletin* 11 (1893): 216–20.

Lankford, John. "Amateurs versus Professionals: The Controversy over Telescope Size in Late Victorian Science." *Isis* 72 (1981): 11–28.

Latour, Bruno. "Visualization and Cognition: Thinking with Eyes and Hands." *Knowledge and Society: Studies in the Sociology of Culture Past and Present* 6 (1986): 1–40.

Leighly, John. "Introduction." In Matthew Fontaine Maury, *The Physical Geography of the Sea and Its Meteorology*, pp. ix–xxx. Cambridge, Mass., 1963.

Lewis, Charles Lee. "Maury—First Meteorologist." *Southern Literary Messenger* 3 (1941): 482–83.

Lewis, R. P. W. "The Founding of the Meteorological Office, 1854–55." *Metl. Mag.* 110 (1981): 221–27.

Ludlam, F. H. *The Cyclone Problem: A History of Models of the Cyclonic Storm.* London, 1966.

Ludlum, David M. *Early American Hurricanes, 1492–1870.* Boston, 1963.

______. *Early American Winters, 1604–1820.* Vol. 1. Boston, 1966.

______. "Thomas Jefferson and the American Climate." *Amer. Metl. Soc., Bulletin* 47 (1966): 974–75.

______. "The Espy-Redfield Dispute." *Weatherwise* 22 (1969): 224–29, 245, 261.

______. *Early American Tornadoes, 1586–1870.* Boston, 1970.

______. *The Weather Factor.* Boston, 1984.

McAdie, Alexander. "Simultaneous Meteorological Observations in the United States during the Eighteenth Century." *U.S. Weather Bureau, Bulletin* 11 (1893): 303–4.

MacDonald, James E. "James P. Espy and the Beginnings of Cloud Thermodynamics." *Amer. Metl. Soc., Bulletin* 44 (1963): 634–41.

McGuire, J. K. "The Father of American Climatology." *Weatherwise* 10 (1957): 92–94, 97.

Marshall, David J. "Myer's Triumph and the Postwar Signal Corps." In Max L. Marshall, ed., *The Story of the U.S. Army Signal Corps*, pp. 76–89. New York, 1965.

Martzolff, Clement L. "Ohio University—The Historic College of the Old Northwest." *Ohio Archaeological and Historical Quarterly* 19 (1910): 411–45.

Meadows, A. J. "Airy and After." *Nature* 225 (1975): 592–95.

Meier, Michael T. "Caleb Goldsmith Forshey, 1812–1881: Engineer of the Old Southwest." Ph.D. dissertation, Memphis State University, 1982.

Meigs, William M. *Life of Josiah Meigs.* Philadelphia, 1887.

Middleton, W. E. Knowles. *The History of the Barometer.* Baltimore, 1964.

______. *A History of the Theories of Rain and Other Forms of Precipitation.* New York, 1965.

______. *A History of the Thermometer and Its Use in Meteorology.* Baltimore, 1966.

______. *Invention of the Meteorological Instruments.* Baltimore, 1969.

Milham, Willis I. *The History of Meteorology in Williams College.* Williamstown, Mass., 1936.

Miller, Eric R. "Tradition versus History in American Meteorology." *Monthly Weather Review* 58 (1930): 65–66.

______. "New Light on the Beginnings of the Weather Bureau from the Papers of Increase A. Lapham." *Monthly Weather Review* 59 (1931): 65–70; abstract in Amer. Metl. Soc. *Bulletin* 11 (1930): 77–78.

______. "The Evolution of Meteorological Institutions in the United States." *Monthly Weather Review* 59 (1931): 1–6.

______. "The Pioneer Meteorological Work of Elias Loomis at Western Reserve College, Hudson, Ohio, 1837–1844." *Monthly Weather Review* 59 (1931): 194–95.

______. "American Pioneers in Meteorology." *Monthly Weather Review* 61 (1933): 189–93.

Moodie, D. W., and A. J. W. Catchpole. *Environmental Data from Historical Documents by Content Analysis: Freeze-up and Break-up of Estuaries on Hudson Bay, 1714–1871.* Manitoba Geographical Studies 5. Winnipeg, 1975.

Morehead, L. M. *A Few Incidents in the Life of Professor James P. Espy.* Cincinnati, 1888.

Morrell, Jack, and Arnold Thackray. *Gentlemen of Science: The Early Years of the BAAS*. Oxford, 1981.

Mulkay, M. J., G. N. Gilbert, and S. Woolgar. "Problem Areas and Research Networks in Science." *Sociology* 9 (1975): 187–203.

Nevins, Allan. *The War for the Union*. Vol. 1, *The Improvised War, 1861–1862*. New York, 1959.

Newcomb, Simon. "Memoir of Joseph Henry." *Natl. Acad. Sci., Biog. Mem.* 5 (1905): 1–45.

Newton, H. A. "A Memorial to Professor Elias Loomis." *Amer. Metl. Journ.* 7 (1890): 97–117.

______. *Elias Loomis, 1811–1889: Memorial Address*. New Haven, 1890.

______. "Biographical Memoir of Elias Loomis, 1811–1889. *Natl. Acad. Sci., Biog. Mem.* 3 (1895): 213–52.

Odgers, Merle M. *Alexander Dallas Bache: Scientist and Educator, 1806–1867*. Philadelphia, 1947.

Oleson, Alexandra, and Sandborn C. Brown, eds. *The Pursuit of Knowledge in the Early Republic: American Scientific and Learned Societies from Colonial Times to the Civil War*. Baltimore, 1976.

Orosz, Joel J. "Disloyalty, Dismissal, and a Deal: The Development of the National Museum at the Smithsonian Institution, 1846–1855." *Museum Studies Journal* 2 (1986): 22–33.

Pomerantz, Martin A. "Benjamin Franklin—The Compleat Geophysicist." *Eos: American Geophysical Union, Transactions* 57 (1976): 492–505.

Price, Derek J. de Solla. *Little Science, Big Science*. New York, 1963.

Pueyo, M. G. "Un contintuateur des travaux concernant la métérologie agricole . . . L. Cotte." *Comptes rendus. Academie d'Agriculture de France* 68 (1982): 604–9.

______. "Les observations métérologiques des correspondants de L. Cotte," *Comptes rendus. Academie d'Agriculture de France* 68 (1982): 658–63, 1429–35.

Pyenson, Lewis. "'Who the Guys Were': Prosopography in the History of Science." *History of Science* 15 (1977): 155–88.

Randolph, Frederick J., and Frederick L. Francis. "Thomas Jefferson as a Meteorologist." *Monthly Weather Review* 23 (1895): 456–58.

Redfield, John Howard. *Recollections of John Howard Redfield*. Philadelphia, 1900.

Reingold, Nathan. "Cleveland Abbe at Pulkowa: Theory and Practice in the Nineteenth Century Physical Sciences." *Archives Internationales d'Histoire des Sciences* 17 (1964): 133–47.

______. *Science in Nineteenth-Century America: A Documentary History*. New York, 1964.

______. "American Indifference to Basic Research: A Reappraisal." In George Daniels, ed., *Nineteenth Century American Science: A Reappraisal*, pp. 38–62. Evanston, Ill., 1972.

______. "The New York State Roots of Joseph Henry's National Career." *New York History* 54 (1973): 133–44.

______, ed. *The Papers of Joseph Henry*. 5 vols. Washington, D.C., 1972–85.

Rigby, Malcolm. "Ephemerides of the Meteorological Society of the Palatinate." *Environmental Data Service*, Feb. 1973, pp. 10–16.

Root, Clarence J. "The Meteorological Service One Hundred Years Ago." *Amer. Metl. Soc., Bulletin* 11 (1930): 74–77.

Rothenberg, Marc. "The Educational and Intellectual Background of American Astronomers, 1825–1875." Ph.D. dissertation, Bryn Mawr College, 1974.

______. "Observers and Theoreticians: Astronomy at the Naval Observatory, 1845–1861." In Stephen J. Dick and LeRoy E. Doggett, eds., *Sky with Ocean Joined*, pp. 29–43. Washington, D.C., 1983.

Rudwick, Martin. *The Great Devonian Controversy*. Chicago, 1985.

Sargent, Frederick, II. *Hippocratic Heritage: A History of Ideas about Weather and Human Health*. New York, 1982.

Scheips, Paul J. "Albert James Myer, Founder of the Army Signal Corps: A Biographical Study." Ph.D. dissertation, American University, 1966.

______. "Old Probabilities: A. J. Myer and the Signal Corps Weather Service." *Arlington Historical Magazine* 5 (1974): 29–43.

Schneider-Carius, Karl. *Weather Science, Weather Research: History of Their Problems and Findings from Documents during Three Thousand Years.* Translated from the German edition, Frieburg, 1955. New Delhi, 1975.

______. "Geschichtlicher Überblick uber die Entwicklung der Meteorologie." In Franz Baur, ed., *Meteorologisches Taschenbuch begrundet von Franz Linke,* pp. 662–708. Leipzig, 1962.

Schott, Charles A. "Magnetic Survey of North America." *U.S. Weather Bureau, Bulletin* 11 (1893): 460–64.

Scott, Donald M. "The Popular Lecture and the Creation of a Public in Mid-Nineteenth-Century America." *Journal of American History* 66 (1980): 791–809.

Seilkopf, Heinrich. "Zur Geschichte der meteorologischen Arbeit an der Deutchen Seewarte, Hamburg." *Annalen der Meteorologie* 3 (1950): 53–56.

Shaw, Sir Napier. *Manual of Meteorology.* Vol. 1, *Meteorology in History.* London, 1926.

Sheynin, O. B. "On the History of the Statistical Method in Meteorology." *Archive for the History of Exact Sciences* 31 (1984): 53–95.

Shyrock, Richard H. "American Indifference to Basic Research during the Nineteenth Century." *Archives internationales d'histoire des sciences* 28 (1948): 50–65.

Sinclair, Bruce. "Gustavus A. Hyde, Professor Espy's Volunteers, and the Development of Systematic Weather Observation." *Amer. Metl. Soc., Bulletin* 46 (1965): 779–84.

______. *Philadelphia's Philosopher Mechanics: A History of the Franklin Institute, 1824–1865.* Baltimore, 1974.

______. "Americans Abroad: Science and Cultural Nationalism in the Early Nineteenth Century." In Nathan Reingold, ed., *The Sciences in the American Context: New Perspectives,* pp. 35–53. Washington, D.C., 1979.

Singer, C., and E. A. Underwood. *A Short History of Medicine.* 2d ed. London, 1962.

Smart, Charles. "The Connection of the Army Medical Department with the Development of Meteorology in the United States." *U.S. Weather Bureau, Bulletin* 11 (1893): 207–16.

Smith, Edgar F. *The Life of Robert Hare, an American Chemist, 1781–1858.* Philadelphia, 1917.

Spence, Clark C. *The Rainmakers: American "Pluviculture" to World War II.* Lincoln, Neb., 1980.

Spitz, Armand N. "Meteorology in the Franklin Institute." *Journ. Frankl. Inst.* 237 (1944): 271–87, 331–57.

Stephens, Carlene E. *Inventing Standard Time.* Washington, D.C., 1983.

Stone, Lawrence. "Prosopography." *Daedalus* 100 (Winter 1971): 46–79.

Sutton, Oliver Graham. "The Meteorological Office, 1855–1955." *Nature* 175 (1955): 963–65.

Symons, George James. "History of English Meteorological Societies, 1823 to 1880." *Quarterly Journal of the Royal Meteorological Society* 7 (1881): 65–98.

______. "The First Daily Weather Map." *Metl. Mag.* 32 (1897): 133–35.

Taylor, W. B. "The Scientific Work of Joseph Henry." In *A Memorial of Joseph Henry,* pp. 205–425. Washington, D.C., 1880.

Thiessen, A. D. *The Founding of the Toronto Magnetic Observatory and the Canadian Meteorological Service.* Reprint from the *Journal of the Royal Astronomical Society of Canada* 34–40 (1940–46).

Thomas, Morley, K. "A Brief History of Meteorological Services in Canada, Part 1: 1839–1930." *Atmosphere* 9 (1971): 3–15.

______. *Pre-Confederation Climate Data.* Canadian Meteorological History, No. 1. Downsview, Ontario, 1984.

______. "Professor Kingston's Scheme: Founding the Meteorological Service." *Chinook,* Summer 1986, pp. 51–55.

Thompson, Kenneth. "Forests and Climate Change in America: Some Early Views." *Climatic Change* 3 (1980): 47–64.

Thompson, Robert Luther. *Wiring a Continent: The History of the Telegraph Industry in the United States, 1832–1866.* Princeton, 1947.

Thoms, Herbert. *The Doctors Jared of Connecticut.* Hamden, Conn., 1958.

Traumüller, Friedrich. *Die Mannheimer meteorologische Gesellschaft (1780–1795): Ein Beitrag zur Geschichte der Meteorologie.* Leipzig, 1885.

True, Webster Prentiss. *The Smithsonian Institution.* Smithsonian Scientific Series, vol. 1. New York, 1929.

Tverkoi, Pavel Nikolaevich. *Razvitie Meteorologii v U.S.S.R.* (Development of meteorology in the U.S.S.R.). Leningrad, 1949.

U.S. Weather Bureau. "History of the Weather Service in Detroit." Detroit, 1970. National Archives Typescript Finding Aid, RG 27, NA.

Van Deusen, G. L. *Historical Sketch of the Signal Corps (1860–1928).* Signal School Pamphlet 32. Fort Monmouth, N.J., 1929.

Waldo, Frank. "Some Remarks on Theoretical Meteorology in the United States, 1855 to 1890." *U.S. Weather Bureau, Bulletin* 11 (1893): 317–25.

Walter, Emil J. "Technische Bedingungen in der historischen Entwicklung der Meteorologie." *Gesnerus* 9 (1952): 55–66.

Ward, Robert De C. "Lorin Blodget's 'Climatology of the United States': An Appreciation." *Monthly Weather Review* 42 (1914): 23–27.

Watson-Watt, Robert. "The Evolution of Meteorological Institutions in the United Kingdom." *Quarterly Journal of the Royal Meteorological Society* 76 (1950): 115–24.

Weber, Gustavus A. *The Weather Bureau: Its History, Activities, and Organization.* New York, 1922.

Weightman, R. H. "Establishment of a National Weather Service—Who Was Responsible for It?" Typescript. Washington, D.C., 1952. Copy in General Correspondence, 1951–55, RG 27, NA.

Weiner, Charles I. "Joseph Henry's Lectures on Natural Philosophy: Teaching and Research in Physics, 1832–1847." Ph.D. dissertation: Case Institute of Technology, 1965.

White, Andrew Dickson. *A History of the Warfare of Science with Theology in Christendom.* 2 vols. New York, 1896.

Whitnah, Donald R. *A History of the United States Weather Bureau.* Urbana, Ill., 1961.

Williams, Frances Leigh. *Matthew Fontaine Maury: Scientist of the Sea.* New Brunswick, N.J., 1963.

Woeikof, Alexander. "Meteorology in Russia." *Smithsonian Report, 1872,* 267–98.

Wolf, Abraham. *A History of Science, Technology, and Philosophy in the Eighteenth Century.* New York, 1939.

World Meteorological Organization. *One Hundred Years of International Co-operation in Meteorology (1873–1973), A Historical Review.* Geneva, 1973.

Wright, Sydney L. *The Story of the Franklin Institute.* Philadelphia, 1938.

Yeo, Richard. "An Idol of the Marketplace: Baconianism in Nineteenth Century Britain." *History of Science* 23 (1985): 251–98.

Zeller, Suzanne. *Inventing Canada: Early Victorian Science and the Idea of a Transcontinental Nation.* Toronto, 1987.

INDEX

AAAS (American Association for the Advancement of Science), 92
  Blodget's maps, 111
  committee on meteorology, 96, 108
  Hare's cycloidograph, 102
  Henry-Blodget confrontation, 112
  and Maury, 108
  and Redfield, 96, 103
  and Smithsonian observers, 92, 175
  and storm controversy, 95–97
Abbe, Cleveland, 173
  Cincinnati weather reports, 141, 150–52
  cooperation with Smithsonian, 151
  on Ferrel's theory, 139
  opinion of Army weather service, 157
  and Signal Office, 156, 158–59
Abbe, George, 152
Abert, J. J., 108
Accademia del Cimento, 1
Adair, William, 10
Adams, Daniel, 10
Adams, John Quincy
  opinion of Espy, 66
  report on observatories, 49
Adie, Alexander, 47
Africa, 4, 16
Agassiz, Louis, 118, 170
Alabama, 13, 43, 177–78, 180, 184
Alaska, 135
Albany Academy, 20
Albany Institute, 55
  and Henry, 20
  term day observations, 62–63
Albany Lyceum of Natural History
  and Henry, 20
Alexander, Stephen, 20, 101
Allen, David, 170
Almanacs, 8–9
Amateurs, 81, 86–89, 170–72
American Academy of Arts and Sciences, 120
American Association of Variable Star Observers, 172
American Chemical Society, 93
American Institute, 19
American Journal of Science, 19, 55, 70, 142
  and Army meteorological data, 16
  Loomis's papers, 159
  and phenological phenomena, 12
  Redfield's papers, 52
American Lyceum, 43–44
American Medical Association, 92
American Philosophical Society, xvii, xxi
  and Army observations, 57
  committee on meteorology, 57, 61, (*see also* Joint Committee)
  and Espy, 57
  Jefferson, as president of, 10
  library, 48
  Minutes, 32
  observations of Meigs, 19
  and physical observatories, 48
  Transactions, 4, 55
Anemometers, 49. *See also* Instruments
Annales de Chimie et de Physique, 168
Annales de l'observatoire physique central (Russia), 169
Annuaire magnetique et météorologique (Russia), 168
Annuaire météorologique de France, 101, 168
Antinori, Luigi, 1
Appalachian Mountains, 7, 134
Arago, François, 23, 45, 47, 52–53, 166, 168
  and Espy (*see* Espy, supporters of)
  Hare's letters to, 51
Aristotle, 1–2
Arizona, 134
Arkansas, 150, 179
  lack of observers, 64
Army Medical Department, xvii, xxii, 1, 166. *See also* Systems
  controversy with Smithsonian, 110–14
  and Espy, 25, 56
  Meteorological Registers, 15, 68, 110, 113–14
  and Meyer, 155

Army Medical Library, 14
Army Topographical Engineers, 124
Associated Press, 142, 151
Association of American Geologists and Naturalists, 65
Astrology, 8–9
Astronomical Journal, 138
Astronomische Nachrichten, 106
Astronomy, 106
Athenaeum Club, 163
Atlantic Ocean, 4, 24–25, 70
Atlas météorologique de la France, 168
Atmospheric electricity, 49, 68–69
  Espy's views, 27–28
  Forshey's questions, 27
  and Franklin, 7
  Hare's theory, 26–27, 29, 34, 51, 97, 100
  and Henry, 21
  and Peltier, 51, 100
  Redfield's views, 29
Atmospheric tides, 36
Austro-Hungary, 165

BAAS (British Association for the Advancement of Science)
  and Bache, 38
  and Blodget, 111
  and Espy, 43, 49–50, 52, 101
  and Forbes, 46
  and Henry, 48, 163
  and Herschel, 38–39, 101
  and Redfield, 23
  and Reid, 39
Babbage, Charles, 47
Babinet, Jacques, 23, 47
Bache, Alexander Dallas, 21, 32, 40, 43, 57, 79. *See also* Coast Survey
  and BAAS, 38
  Brunswick spout (*see* Tornadoes)
  debate with Olmsted, 56
  and Espy, 44, 105
  European tour, 45–46
Bacon, Francis, 47
Baconianism, xx, 24, 48, 63
Baden, 165
Baird, Spencer F., 76, 164, 170
  as assistant secretary of the Smithsonian, 133
  and Smithsonian observers, 130–33
Bancker, Charles N., 57
Bangor Advertiser, 45
Barnes, Joseph K. *See* Surgeons General
Barometers, 1, 6, 9, 12, 47, 120, 164. *See also* Instruments
  in education, 164
  gauge on nephelescope, 98
  height surveys, 56
  minimum pressure, lines of, 41, 63, 71
  minimum pressure, paths of, 159–60
  mountain, 48
  Newman, 48, 119, 147
  Smithsonian, 83, 119
Bartram, William, 10
Battan, Louis J., 31
Bauern-Praktik, 9
Beck, T. R., 19–21, 47, 118
Behring, Vitus, 1
Belgium, 163, 165. *See also* International developments
Beloit College, 93
Berkshire County Whig, 64
Bermuda, 24, 70
Bernoulli, Daniel, 1
Bigelow, Jacob, 12
Bishop, "Professor", 12
Blake, John Lauris, 8
Blodget, Lorin, 76, 82, 110–15
  Climatology of the United States, 115, 157
Bond, William Cranch, 64
Boston Evening Mercantile Journal, 45
Botanical Society of Edinburgh, 170
Bowditch, Nathaniel, 120
Bowdoin College, 12
Brandes, H. W., 26, 54
Breslauer Sammlung, 1
Brewster, David, 24, 47, 49
Bridgewater treatise, 8
Britain. *See* England; Scotland
British Board of Trade, Meteorological Department, 166–67
British Meteorological Office, 157, 166, 168
British Meteorological Society, 166–67
British Royal Engineers, 107
Brooks, David, 143
Brooks, Joseph, 143
Bruce, Robert, xix
Brunswick spout. *See* Tornadoes
Brussels Observatory, 165–66

Buchan, Alexander
Handy Book of Meteorology, 157
opinion of American meteorology, 164
Buck, Rufus, 86
Bulletin of International Simultaneous Observations, 166, 170
Bureau of Topographical Engineers, 108

California, 132, 180, 182
Cambridge Miscellany, 54
Campanius, John, 2
Canada, 13, 70, 75, 117, 123, 132, 134, 139. *See also* Systems
Cannon, Walter F., xxii, 48
Caribbean Sea, 25, 75
Carolinas, 7, 134
Carvel, Hans, 66
Cawood, John, 48
Central Physical Observatory (Russia), 169
Chalmers, Lionel, 6
Chapplesmith, John, 88–89. *See also* Tornadoes
Charts, xxi, 19, 47. *See also* Maps
Espy's, 71
Schott's, 132
temperature change, 132
Chicago Academy of Sciences, 152
China, 16
Choate, Rufus, 111
Cincinnati Chamber of Commerce, 151
Cincinnati Observatory, 141, 150–52
Civil War, 82, 120, 146–47, 155, 171
and cooperative projects, 147
and observers, 146–47
Cleaveland, Parker, 12
Climatic change, 2–5, 15
Schott's analysis, 129–32
Coast Survey, 164. *See also* Bache, Alexander Dallas; Systems
cooperation with Smithsonian, 128–32
and Nicholson, 129
Coffin, James, xix, 55, 63–64, 75–76, 114, 176
correspondence network, 64–66, 208
Meteorological Register and Scientific Journal, 65
on Smithsonian data, 83
as Smithsonian observer, 91
"Winds of the Globe", 91, 136
"Winds of the Northern Hemisphere", 91, 136, 157, 168
Coffin, S. J., 136
Collective biography. *See* Smithsonian observers
College of Physicians, 6
Collegiate associations
Harvard College, 64
New England professors, 1, 10–13
Princeton College, 64
Williams College, 63–64
Yale College, 64
Colonial Sentinel, 45
Colorado, 164
Columbia College, 119
Columbus, Christopher, 3
Columbus, Ferdinand, 3
Congress, 19, 127
appropriations, 125
appropriations, Army Medical Department, 69, 70
and Blodget, 112, 115
and Espy, 67, 72, 79, 98–99
and Hare, 100–101, 104
investigation of Smithsonian, 115
and Maury, 110
memorials, 59, 148
and weather service bill, 153–56
Connecticut, 7, 37, 39, 88, 132, 178, 180
Academy of Arts and Sciences, 17
Cooper, James, 133–34. *See also* Maps
Cooperation. *See* Smithsonian meteorological project; Systems
Coriolis, Gaspard, 138
Coriolis acceleration, 138
Correspondence Météorologique (Russia), 169
Cotte, Père Louis, 1
Crimean War, Balaclava, 168
Custis, Peter, 10
Cyclones, 102–3, 138

Dakota Territory, 134, 180
Dall, William Healy, 135
Dalton, John, 26
visit with Bache, 45
Daniell, John Frederic, 46
hygrometer, 21
Daniels, George, 24
Darwin, Charles, 64

Darwinism, 171
Data collection and reduction, 113, 134–36. *See also* Systems
  Albany Institute, 62–63
  Army Medical Department, 68–71, 114
  Babbage, 47
  and Baconianism, xx
  Blodget, 111–15
  Chapplesmith, 88
  Coffin, 55, 64, 126–27, 139
  colonial America, 6
  Cooper, 133
  England, 5
  Espy, 40
  Germany, 167
  Glaisher, 142
  Hall, 10
  international observations, 170
  Lake Survey, 125
  Lapham, 153–54
  Loomis, 159
  Maury, 107
  and meteorological theories, 173
  and mobilization of inscription, xxi
  Navy, 61
  New York, 118
  Redfield, 24, 55
  Russia, 169
  Schott, 129–32, 139
  Signal Office, 158, 161
Davis, Charles H., 23, 39
Davis, William Morris, 173
Day, Jeremiah, 10
Delaware, 2, 13
Department of Agriculture, 76, 93, 150. *See also* Smithsonian meteorological project
Depot of Charts and Instruments, 106
Deutsche Seewarte, 166–67
Devonian controversy, 23
Dewey, Chester, 10, 12
DeWitt, Simeon, 19
De Young, Gregg, 105
Dove, Heinrich Wilhelm, 48, 50–51, 54, 64, 120, 165, 167
Drake, Daniel, 17
Dunbar, William, 4, 9

Eights, Jonathan, 12
Eliot, Jared, 7
Emerson, Gouverneur, 57
England, 46, 64, 153–54, 172. *See also* International developments
Englemann, George, 135
Espy, James P., xix, xxi-xxii, 23–24, 32, 57, 76
  advice to Smithsonian, 78
  and BAAS, 49–50
  correspondence network, 208
  critics, 39, 49, 53 (*see also* Hare; Redfield)
  debate with Olmsted, 45, 56
  on Ferrel's theory, 139
  and Herschel, 39
  and Loomis, 53
  and lyceum lectures, 43–45
  marginalia, 41–42
  and Navy, 67–68, 72, 78–81, 97
  petitions to government, 43
  Philosophy of Storms, 27–28, 32, 41–42, 46, 78, 99, 152
  on plucked chickens, 35
  and Reid, 39
  reports on meteorology, 70–72, 97, 100, 104
  responses to Forshey, 27–28
  storm theory, 25–26
  tour of England and Scotland, 49
  tour of France, 50
  ventilators, 67–68
  weather maps, 71–72
Espy, James P., supporters of
  Arago, 23, 50, 66
  Army Medical Department, 69–72
  Bache, 25, 32, 36, 101
  Chapplesmith, 88
  Coffin, 105–6
  concerns of, 44
  Ferrel, 138
  Forshey, 27
  French Academy, 23, 36, 49–51, 66, 100
  Hare, 100
  Henry, 25, 36, 78–81, 101, 105–6
  Lapham, 153
  Loomis, 105–6
  Pennsylvania, 60
  Philadelphia circle, 23, 40
  Quetelet, 101
  Smithsonian, 79–81
Europe. *See* International developments

Faraday, Michael, 51
Farmer's Almanac, 9
Farrar, John, 12
Ferrel, William, xix, 76
  studies of the general circulation, 136–39
Finley, Clement A. *See* Surgeons General
Fitzroy, Robert, 165, 167
Florida, 68, 132, 134, 146, 177, 183
Forbes, James, 46–47, 49
Force, William Q., 147
Foreman, Edward, 81, 89, 142
Forry, Samuel, and Army Medical Department, 68–69
Forshey, Caleb G., questions on storms, 27–30
Fothergill, John, 6
France, 43, 153–54, 164–65, 172. *See also* International developments
Francis, L. C., 60
Franco-Prussian War, 164
Franklin, Benjamin, 58
  and climatic change, 3
  kite experiment, 8
  storm theory, 7–8
Franklin Institute, xvii, xxi, 59, 105, 204
  committee on meteorology, 57, 61, (*see also* Joint Committee)
  grant from Pennsylvania, 60
  Journal, 40, 55–56, 58
Franklin Kite Club, 39
Frazer, John Fries, 105
Freeman, Thomas, 10
French Academy
  Comptes rendus, 168
  Hare's appeal, 51–52, 100
  prize for theory of hail, 45
  Redfield's appeal, 52
  report on Peltier, 51
  and support for Espy (see Espy, James P., supporters of)
Fuller, W. G., 146

Gano, John A., 151, 153
Gardiner Lyceum, 92
Gardiner, Robert Hallowell, 92, 179
Gauss, Carl Friedrich, 48
Gay-Lussac, Joseph Louis, 26, 47
General Land Office, 1, 17–19, 21. *See also* Systems
  phenological observations, 17
Genesee Farmer, 44
Georgia, 7, 19, 132, 149
Germany, 64, 172. *See also* International developments
Girard College, 49, 119
Glaisher, James, 167
  and daily weather reports, 141
Global Atmospheric Research Programme, 173
Goddard, Paul B., 57
Goldsmith, Edward, 152
Gould, Benjamin, 129
Grant, Ulysses S., 155–56
Gray, Asa, 115, 170
Great Lakes, 124
  shipping disasters, 153
Great Northern Expedition, 1
Great September gale, 5
Green, James, 86, 119–20, 122, 125. *See also* Instruments
Greenwich Observatory, 142
Greig, John, 19
Gross, Walter E., 57
Gulf of Mexico, 4, 7, 31, 70
Gulf Stream, 4
Guyot, Arnold, 76
  as meteorological consultant, 82, 117–22
  Meteorological Tables, 119–20, 125, 128, 135
  on storm controversy, 97

Hadley, George, 136
Hadley cells, 137
Hall, Fredrick, 10
Hall, Peter, 171
Halley, Edmund, 37
Halley's comet, 78
Hammond, William A. *See* Surgeons General
Hare, Robert, xxii, 23–24, 32. *See also* Atmospheric electricity
  critics (*see* Espy, James P.; Redfield, William)
  dispute with Peltier, 51
  and French Academy, 51, 100
  and Navy, 101
  polemics, 50–53, 95–97
  responses to Forshey, 29–31
  storm theory, 26–27
Harlan, James, 109

Harris, Elijah Paddock, 93
Harvard College, 3, 12, 55, 64
  Observatory, 119
Hawley, Gideon, 19
Hayden, Ferdinand, 164
Hayes, Isaac I., 136
Heckewelder, John Gottlieb Ernestus, 10
Hemmer, Johann Jakob, 1
Henry, Joseph, xix, xxi, 2, 24, 32. *See also* Smithsonian Institution
  on Baconianism, 47, *see also* Scientific method
  criticism of Redfield, 34
  early meteorological interests, 20–21
  and Espy (*See* espy, James P., supporters of)
  European tours, 46–48, 163–64
  on Ferrel's theory, 138
  and Hare, 100, 102–3
  lectures at Princeton, 64
  "Meteorology in Its Connection with Agriculture", 115, 127
  and Myers, 160–62
  as Smithsonian secretary, 75–76, 78
  on teleology, 8
Hensley, J. M., 82
Herrick, E. C., 56
Herschel, John, 23, 36–37, 39, 62, 64 comments at BAAS, 39
Hippocrates of Cos, 5
Historiography, xix-xxi, 170–73
Holland, 4, 154, 165
Holton, C. B., 153
Holyoke, Edward, 6
Hooper, William, 153
Hopkins, Albert, 63
Horr, Asa, 92, 183
Hough, Franklin B., 20
Houzeau, J. C., 165–66
How, Henry, 82
Hudson's Bay, 123, 135
Humboldt, Alexander von, 48
  and Army *Meteorological Register,* 16
Humboldtianism, 48
Humphreys, A. A., 157
Hunter, George, 10
Hurricanes, 22, 24–25, 139
Hutton, James, 31
  theory of rain, 37, 104
Hygrometers, 6, 48, 120. *See also* Daniell, John Frederic; Instruments
Hypsometry. *See* Barometer

Illinois, 17, 122, 129, 134, 149, 176, 179, 181–82. *See also* Systems
India, 64
Indiana, 17, 124, 176–77, 182
  New Harmony, 88–89
Indian Territory, 132
Institute of Mining Engineers (Russia), 168
Instruments, xxi, 19, 76
  Army Medical Department, 70
  calibration of, 118–21
  destroyed in fire, 147
  Espy's nephelescope, 98–99
  Hare's cycloidographs, 102–3
  for Maine observers, 122
  for New York observers, 118, 120
  for Pennsylvania observers, 60
  for physical observatories, 48–49
  lack of, in South, 149
  of Royal Society, 46
  self-registering, 12, 47, 60, 63, 122, 147
  for signal stations, 157
  for Smithsonian observers, 81–87, 175
  supplied by Patent Office, 125
International conference of directors, 169
International developments,
  Belgium, 165–66
  England, 165–68
  France, 165–66, 168
  Germany, 165–67
  Russia, 165–66, 168–69
  in seventeenth and eighteenth centuries, 1
  United States, 165–66, 169–70
Iowa, 92, 109, 129, 132, 182–83
Ireland, 163
Isotherms, 19, 69, 153
Italy, 153–54, 165

Jamestown colonists, 2
Japan, 103
Jefferson, Thomas, xvii, 1
  and climatic change, 3–5
  climatological observations, 9–10
  and James Madison, 9
  meteorological correspondents, 9

and national system, 10
Notes on Virginia, 4
and Panama Canal, 4
and Rev. James Madison, 1, 9
Jefferson College, 27
Jewett, Charles Coffin, 111
Johnson, Walter, 32, 40
Joint committee, xxi, 23, 37, 44, 46, 55–61, 83, 166
and Army Medical Department, 69
demise of, 61
as model, 78–79
and Navy, 62
storm studies, 57–58
weather map, 59
Jones, Alexander, 142
telegraphic weather reports, 142–43
Joule, James P., 98
Journal de Physique, 168
Journals, private. *See* Weather diaries
Jurin, James, 1

Kane, Elisha Kent, 136
Kanold, Johann, 1
Kansas, 132, 179, 184
Kämtz, Ludwig, 169
Kelley, Oliver Hudson, 92, 180
Kendall, Amos, 142
Kennicott, Robert, 135
Kentucky, 12, 177, 180, 182–83
lack of observers, 64
Kepler, Johannes, 8
Kew Observatory, 164
Khrgian, A. Kh., 169
King's College, 82
Kirtland, Jared P., 135
Kupffer, Adolph, 165–66, 168–69
Kutzbach, Gisela, 165

Lafayette College, 126
Lake Survey. *See* Systems
Lalande, J. J., 106
Lansing, Gerrit Y., 19
Lapham, Increase, 83, 152–54
Laplace, Pierre Simon Marquis de, 138
Lathrop, S. Pearl, 93
Latin America, 75
Latour, Bruno, xxi
Lavoisier, Antoine-Laurent, 1
Law of storms, 38–39, 77
Lawson, Thomas. *See* Surgeons General
Lazzaroni, xxii, 23, 105, 108
LeFroy, J. H., 123
Legaux, Peter, 9–10
Leonard, Levi Washburn, 92, 176
Lewis and Clark expedition, 10
Le Verrier, Urbain, 152, 165–66, 168
Liebig, Justus von, 51
Lightning. *See* Atmospheric electricity
Lining, John, 6
Linnean Society, 163
Literary and Philosophical Repertory, 10
Literary and Philosophical Society of New York, 10, 12
Loomis, Elias, xix, 24, 38, 53, 64, 76, 154
Ohio memorial, 55
on plucked chickens, 35
"Report on the Meteorology of the United States", 77–78
storm study, 63
Treatise on Meteorology, 157
Louisiana, 17, 132, 149, 179
Lovell, Joseph, 13, *See also* Surgeons General

M'Clintock, Francis L., 136
MacSparran, James, 3
Madison, James, and Jefferson, 9
Madison, Rev. James, and Jefferson, 1, 9
Magendie, François, 47
Magnetic crusade, 77
Magnetic observatories, 129
Magnetische Verein, 168
Mahlman, Carl H. W., 165–66
Maine, 12–13, 43, 46, 122, 132, 139, 141, 176, 178–79, 181–83. *See also* Systems
Board of Agriculture, 122
House of Representatives, 92
Mann, James, *Medical Sketches*, 13
Mannheim Meteorological Society. *See* Societas Meteorologica Palatinae
Manning, George, 107
Maps
Abbe's, 156
Blodget's, 111
climate, 69, 129–31
Cooper's, 133–34
England, 142

Maps (*cont.*)
English, 167–68
Espy's, 39, 71–72
forest regions, 133–34
Forry's, 69
Henry's, 127
Joint Committee's, 59
Lapham's, 153–54
Loomis's, 159
physical, 133
"Pilot Charts," 107
rainfall, 127–30
Schott's, 129–31
signal stations, 156
Smithsonian, 127–31, 134, 142–43
storm paths, 159–60
telegraph, 143–44, 149
temperature, 129, 131
tornado, 32–34, 88
weather, xxi, 71–72, 142–43, 153–57, 168
Mariotte, Edme, 1
Martin, F. P. B., 157
Maryland, 176, 178, 181, 183
Mascart, Eleuthère, 166, 168
Mason, Charles, 113, 125
Massachusetts, 3, 5–6, 14, 43, 117, 122, 132, 139, 175–76, 178, 181–83. *See also* Systems
Mathematical Monthly, 138–39
Mather, Cotton, 3
Mather, William Williams, 92, 177
Maury, Matthew F., 95
Maxwell, James Clerk, xx
Meade, George, 124
Medical and Agricultural Register, 10
Medical geography, 5–7, 68
Medical Repository, 15
Meigs, Josiah, 17–19
Melloni, Macedonio, 47
Meteorological Society of London, 166–67
Mexico, 75, 134
Michigan, 17, 124, 179, 181
Middlebury College, 10, 12
Minnesota, 132, 176, 180, 182
Mississippi, 9, 17, 27, 43, 88, 134, 177, 181
Mississippi River, 4, 44, 141
Missouri, 17, 122, 132, 180–82. *See also* Systems
Mitchell, Charles, 120
Montana, 184
Monthly Weather Review, 122
Morse, Samuel F. B., 141
Mulkay, Michael, xxii
Myer, Albert J., 155–58, 160–62, 165, 169. *See also* Signal Office

Nashville Journal of Medicine and Surgery, 138
Nason, Henry Bradford, 93
National Academy of Sciences, 159
National Board of Trade, 153
National Educational Association, 92
National styles, 170–73
Natural history and meteorology, 129–36, 170
Natural theology, 8
Nautical Almanac, 139
Navajo Indian campaigns, 155
Naval Observatory, 53, 110, 164
Navy Department, xviii, 55, 61, 76, 78–81, 101, 117. *See also* Systems
and Espy, 25
Nebraska, 149, 184
Networks, xxii, 158, 189. *See also* Telegraph networks
Nevins, Allan, 171
New England, 1, 5, 7, 10–13
New England professors. *See* Systems
Newfoundland, 4, 70, 153
New Hampshire, 92, 176, 183
New Jersey, 32, 180
New Mexico, 181
New Sweden, 2
New York, 20, 24, 34, 43–44, 55, 65, 66, 72, 86, 117–18, 124, 132, 139, 150, 176–77, 179–81, 183–84
academy system (*see* Systems)
regents of the university, 19–21, 118, 166
and War of 1812, 13
Newman, John Frederick, 48
Newton, Sir Isaac, (commissioner of agriculture), xx
Nicholson, W. L., survey of tornado damage, 129
Nicollet, J. N., 69
Niles' Weekly Register, 19
Norddeutsche Seewarte, 166–67
Norddeutschland, 165

North American Review, 12
North Carolina, 182
Nova Scotia, 25, 70, 82, 153

O'Reilly, Henry, 142
Observations, xxi. *See also* Data collection and reduction; Smithsonian observers; Systems
 American leadership, 46
 Europe, 1
 Jeffersonian, 9–10
 by telegraph, 148–49
 term days, 62, 205
 used by Coffin, 136
 used by Loomis, 63
 used by Schott, 129
Observatoire Royal. *See* Brussels Observatory
Oceanography, 106
Ohio, 17, 35, 63, 102, 124, 132, 134–35, 146, 150, 177, 179–82, 184. *See also* Systems
 River, 44
 University, 92
Olmsted, Denison, 23, 36
 debate with Bache and Espy, 56
 debate with Espy, 45
 Huttonian theory, 37
 lectures on meteorology, 64
 Redfield's eulogy, 103
Oregon, 134, 175
Osler, Edward, 49

Pacific Ocean, 4, 36, 103
Paine, Halbert E., 153–56
Panama, 4
Paris Observatory, 166, 168
Parmenides of Elea, 2
Patent Office, 76, 82, 93, 153
 cooperation with Smithsonian, 125–28
Patrons of Husbandry (the Grange), 92
Peirce, Benjamin, 23, 120, 129
 attack on Espy, 53
Peltier, Jean Charles, 51
Pennsylvania, 43, 59, 72, 111, 132, 142, 147, 176, 178–80, 182–83. *See also* Systems
 legislature, 44
 Lyceum, 43–44, 200
Periodization, xvii-xx
Perry, Matthew C., 103
Phenological phenomena, 12, 17
 observations, 2, 82, 133
Philadelphia Lyceum, 43
Phillips, John, 49
Philosophy of Storms. *See* Espy, James P.
Pickard, Josiah Little, 92, 177
Pickering, Roger, 1
Piddington, Henry, 64, 96–97, 102–3
Pike, Benjamin, 119
Pike, Zebulon Montgomery, 10
"Pilot Charts". *See* Maps
Pitcher, Zina, 92, 181
Plumb, Ovid, 88
Poisson, Siméon, 138
Pouillet, C.-S.-M., 23, 47
Powell, John Wesley, 164
Price, Derek J. de Solla, xxii
Princeton College, 21, 64, 88
Prout, William, 8
Prussia, 163
Prussian Meteorological Institute, 165, 166

Quebec, 9, 53
Queens College, 19
Quetelet, Adolphe, 47–48, 165
 and Espy, 53, 104

Rainmaking, 25, 43–45, 61, 66, 71–72, 99–100
Ramsay, David, 5
Randolph, Thomas Mann, Jr., 1
Ravenel, Henry, 146
Raymond, W. S., 157
Reconstruction, of Smithsonian system, 148–50
Redfield, William, xxii, 23–24, 32
 chided by Henry, 34
 correspondence network, 64
 critics (*see* Espy; Hare)
 defensive Baconianism, 25, 38, 52
 and Navy, 97–98
 responses to Forshey, 28–29
 storm theory, 24–25
 on telegraphy, 141
Redfield, William, supporters of,
 BAAS, 23
 Herschel, 36

Redfield, William, supporters of (*cont.*)
New England circle, 23
Olmsted, 36, 45, 103–4
Piddington, 96
Reid, 36–39
Reid, William, 23–24, 36–39, 64, 97
and BAAS, 38–39
Reingold, Nathan, xix, 21
Rhode Island, 3, 43
Ricksecker, Lucius E., 147
Rittenhouse, David, 9
Rocky Mountains, 134
Rogers, Henry D., 57
Rothenberg, Marc, 8
Royal Charter shipwreck, 167
Royal Meteorological Society, 167
Royal Society of Edinburgh, 48
Royal Society of London, 1, 12, 46, 70, 111, 119, 163
Catalog of Scientific Papers, 92, 175
Ruschenberger, W. W., 62
Ruskin, John, 167
Russia, 16, 172. *See also* International developments
Russian Academy of Sciences, 169
Ryerson, Egerton, 123

Sabine, Edward, 109, 146
Saskatchewan, 134
Saxony, 167
Schmauss, August, xxi
Schott, Charles A., 76, 128. *See also* Climatic change; Maps
monographs on rainfall and temperature, 129, 132
Schumard, Benjamin Franklin, 122
Scientific method, 47–48
Scotland, 6, 43, 163–64
Scottish Meteorological Society, 164
Seaman, A. G., 107
Sergeant, John, 59
Service des Ponts et Chaussées, 168
Signal Office, xxii, 82, 122, 164–66. *See also* International developments
budget in 1874, xix
instruments, 157
telegraph stations, 158
training of officers, 157
transfer of other systems, 160–62
weather reports, 157–58
Silliman, Benjamin, 13
Silliman's Journal. See *American Journal of Science*
Sinclair, Bruce, 105
Smith, Archibald, 49
Smithsonian, data collection and reduction, 76, 83, 128
and Department of Agriculture, 150
and Patent Office, 125–27, 218
Smithsonian Institution, xvii, xviii, 2, 21, 56, 75, 109, 133. *See also* Smithsonian meteorological project; Smithsonian observers
and Board of Regents, 76, 112, 114–15, 142
Contributions, 78, 81, 88–89, 102, 114, 136
controversy with Army Medical Department, 110–14
and Espy, 25
fire, 147–48
international exchanges, 75, 136
programme of organization, 76–78
and weather service bill, 155
Smithsonian meteorological project, xxi, 166. *See also* Cooperation; Systems
advice of Espy, 78
advice of Loomis, 77–78
and Canadian observers, 123
and Coast Survey, 128–32
and Coffin, 140
cooperative nature of, 75–76, 117, 122
and Department of Agriculture, 127, 150
funding, 76–77
impact of Civil War on, 145
and Lake Survey, 124
and Maine observers, 122
and Massachusetts observers, 121
and meteorological theory, 173
and naturalists, 129–36
and Navy, 79–81
and New York academies, 20, 111, 117–20
and Patent Office, 125, 127
reconstruction of, 148–50
and storm controversy, 76–78, 95
telegraph stations, 145

Smithsonian observers, 81–93, 108, 148–49
  collective biography, 89–93, 175–84
  instruments, 175
  membership in AAAS, 92, 175
  telegraphers *see* Observations
  transfer to Signal Office, 161–62
Smithson, James, 77, 81, 147
Societas Meteorologica Palatinae, 1, 46, 165
Société Météorologique de France 168
Société Royale de Médicine, 1
South African Literary and Philosophical Institution, 62
South America, 31
South Carolina, 5–6, 132, 146, 149, 183
Spain, 4
Stevelly, John, 49
Stiles, Ezra, 3
Stoddard, O. N., 102
Stone, Lawrence, 89
Storm controversy, xviii, 23–54, 75
  Espy v. Redfield, 35
  Hare vs. Espy and Redfield, 51
  revisited (1848–60), 95–106
  and Smithsonian meteorological project, 76
Storm of 1821, 37, 39
Struve, Otto, 151
Süddeutsche meteorologische Verein, 165–66
Surgeons General
  Barnes, Joseph K., 147, 155, 160
  Finley, Clement A., 147
  Hammond, William A., 147
  Lawson, Thomas, 69, 95, 110, 113–14
  Lovell, Joseph, 15–16, 57
  Tilton, James, 13–14
Swallow, George Clinton, 122
Switzerland, 164
Symons, George, 167
Systems, xxi-xxii, 72, 171, 174, 189
  Albany Institute, 62–63, 205
  Army Medical Department, 13–16, 56, 68–72, 160
  Canada, 123–24
  Coast Survey, 128–29
  General Land Office, 17–19
  Illinois, 122
  as instruments, xxi, 174
  international (*see* International developments)
  Lake Survey, 124, 160
  Maine, 122
  Massachusetts, 120
  Missouri, 122
  Navy Department, 61–62, 205
  New England professors, 10–13
  New York academies, xxi, 2, 19–20, 46, 56, 63, 117–20, 166
  Ohio, 55–56
  Pennsylvania, 56, 60–61, 204. (*see also* Franklin Institute; Joint Committee)
  Texas, 122

Telegraph companies, 142–44, 147, 149, 158, 172. (*See also* Western Union)
Telegraph networks. *See also* Cincinnati Observatory; Signal Office; Smithsonian meteorological project; Western Union
  impact of Civil War on, 146
  reconstruction of, 148–49
  spread of, 143–44
Tennessee, 46, 134, 146, 178–79, 182
Texas, 122, 134, 146, 153, 155, 178–79. *See also* Systems
Thermometers, 12, 47–48, 86, 120. *See also* Instruments
  Fahrenheit, 6
  Heath, 6
  Kendall, 147
Tilton, James. *See* Surgeons General
Tornadoes, 22, 32
  Bache's study of, 32–34
  Brunswick spout, 31–35, 38, 40, 97
  Camanche, 129
  Chapplesmith's study of, 88–89
  Espy's theory on, 25, 28, 32, 139
  Forshey's account of, 27
  and Forshey's questions, 27
  Hare's theory on, 26, 29, 32
  Herschel's theory on, 39
  Loomis's studies of, 35
  Newark, 40
  Nicholson's study of, 129
  Redfield's theory on, 25, 29, 34
  Smithsonian observations of, 82
  Stoddard's study of, 102

Toronto Observatory, 118–19
Torry, John, 170
Treat, John Breck, 10
Tyler, John, 68

U.S. Agricultural Society, 109
U.S. Army Signal Office. *See* Signal Office
U.S. Census Office, 129
U.S. Exploring Expedition, 64
U.S. House of Representatives, 49, 59, 67, 115. *See also* Paine, Halbert E.
U.S. Senate, 67, 104, 109
University of Georgia, 17
University of Iowa, 92
University of Pennsylvania, 26
Utah, 134, 180

Vaughan, Benjamin, 9
Vermont, 10, 176–78
Virginia, 1, 3, 10, 99, 146, 153, 157, 175, 177–79, 182, 184
Virginia Company, 2
Vital statistics, 5
Volunteer observers. *See* Observations; Smithsonian observers; Systems

Wade, Jeptha Homer, 143
Walker, Sears C., 23, 57
  and Espy, 44
  and Neptune, 106
War of 1812, 13
Washington Evening Star, 145
Washington Territory, 183
Waterhouse, Benjamin, 14
Waterspouts. *See* Tornadoes
Watts, Fredrick, 150
Weather Bureau budget, xix
Weather diaries
  Engleman's, 135
  Espy's use of, 67, 70
  General Land Office, 17
  Jeffersonian, 9–10
  Joint Committee, 58
  Loomis's use of, 63
  Meigs's, 17–19
  New England professors, 10–12
  from sailing ships, 64
  Smithsonian's use of, 79, 111
  Waterhouse's, 14
Weather maps. *See* Maps
Weather modification. *See* Rainmaking
Weather reports. *See also* Cincinnati Observatory; Signal Office; Western Union
  in newspaper, 145
  London, 141
  telegraphic, 141, 142–45, 148–49
Weather services. *See also* International developments; Signal Office
  European and Russian, 164–65, 172–73
Weber, Willhelm, 48
Webster, Noah, 6
Webster, Noah B., 108
Wesselovski, Constantine, 169
Western Meteorological Association, 151
Western Reserve College, 35
Western Union, 143, 149, 151–52, 156, 158
  and Cincinnati weather report, 156
Wetterbüchlein, 9
Wheatstone, Charles, 47
Wheeler, Col., 157
Whewell, William, 38, 51
Whirlwinds. *See* Hurricanes; Tornadoes
White, Andrew Dickson, 8
Wild, Heinrich, 169
Wilkes, Charles, 62
William and Mary College, 9
Williams College, 10, 12, 55, 63–64
Williamson, Hugh, 3, 9
Williamson, R. S., 157
Willis, Henry, 122
Wilson, Henry, 156
Wilson, Job, 5
Wisconsin, 86, 92, 124, 153, 176–77, 179
  Academy of Sciences, 152
Witterungs-Anstalten, 165–66
Woeikof, Alexander, 136
Wood, William, 3
World Meteorological Organization, 173
Wotherspoon, A. S., 110
Württemberg, 165

Yale College, 10, 12, 55, 64
Yukon, 134

Zane, Isaac, Jr., 9

Printed in the United States
96340LV00003B/41-100/A

9 780801 863592